Methods of Moments and Semiparametric Econometrics for Limited Dependent Variable Models

Springer
New York
Berlin
Heidelberg
Barcelona
Budapest
Hong Kong
London
Milan
Paris
Santa Clara
Singapore
Tokyo

Myoung-jae Lee

Methods of Moments and Semiparametric Econometrics for Limited Dependent Variable Models

With 20 Illustrations

 Springer

Myoung-jae Lee
Department of Econometrics
Tilburg University
5000 LE, Tilburg
The Netherlands

Library of Congress Cataloging-in-Publication Data
Lee, Myoung-jae.
    Methods of moments and semiparametric econometrics for limited
dependent variable models / Myoung-jae Lee.
      p. cm.
    Includes bibliographical references and index.
    ISBN 0-387-94626-8 (hardcover)
    1. Econometric models.   2. Moments method (Statistics)
3. Estimation theory.   I. Title.
HB141.L433   1996
330'.01'5195—dc20   95-44882

Printed on acid-free paper.

Production managed by Natalie Johnson; manufacturing supervised by Jeffrey Taub.
Camera-ready copy prepared using Springer macro svsing.sty.
Printed and bound by Braun-Brumfield, Inc., Ann Arbor, MI.
Printed in the United States of America.

9 8 7 6 5 4 3 2 1

ISBN 0-387-94626-8 Springer-Verlag New York Berlin Heidelberg   SPIN 10522321

To my father and mother,

Kang-Lee Lee

Hyun-Sook Lim

Preface

This book has grown out of lecture notes for graduate econometric courses that I taught during 1989–1995. Although I've tried to include most of the important topics for the book title, it seems inevitable that this book reflects my own interests. In covering chosen topics, I tried to give credit to whomever it is due, but I may have missed on a number of occasions; my sincere apologies go to them.

I am thankful to the following people for comments on various parts of this book: Hyung-Taik Ahn, Don Andrews, Marcel Das, Mark Kennet, Choon-geol Moon, Dylan Supina, and Mark Wilhelm. I am especially grateful to Bertrand Melenberg and Peter Robinson for reading the entire book and making suggestions that led to a substantial improvement of the book. It goes without saying that I am solely responsible for any errors in this book. Finally, I am grateful to Arie Kapteyn, Arthur Van Soest, and my other colleagues at the Center for Economic Research and the Department of Econometrics at Tilburg University for the ideal environment where I could complete this book in comfort.

Contents

1
Introduction

1.1 Motivations

For a long time, the standard econometric approach consisted of specifying both the systematic component (e.g., regression functions) and the stochastic component (e.g., the distribution of unobservable "error" terms) fully up to some unknown finite-dimensional parameter vector. Estimation and testing were only concerned with this finite-dimensional parameter vector, for example, by applying least squares methods or maximum likelihood. This approach, however, may be far too restrictive by allowing variability only through the finite-dimensional parameter vector, with the model being not flexible enough to give a good approximation to reality. If the approximation is not sufficiently close, inference based on the model under consideration becomes more or less meaningless.

Since the late 1970s, various non- and semiparametric methods have been suggested to overcome this limitation in the classical econometric approach: unknown parameters are allowed to be (partly) infinite dimensional, permitting much more flexible models that can approximate the reality far more accurately. Despite this development, however, application of the methods is largely lacking due to two reasons. One reason is the difficulty of the literature to people accustomed to least squares estimators (LSE) and maximum likelihood estimators (MLE). The other reason is a lack of commercial computer packages. Hence, anybody who wants to apply semiparametric methods has no alternative but to read the literature and write their own computer programs, with the development of easy-to-use matrix languages such as GAUSS helping. Still the former task of understanding the literature seems to be daunting to most people.

Along with the development in semiparametric econometrics, methods-of-moments estimation (MME) took the center stage of econometrics; most of the estimators and tests are viewed as methods of moments now. The purpose of this book is to convey essential ideas in the recent development of econometric theory on methods of moments and semiparametric econometric methods for limited-dependent-variable (LDV) models. By doing so, we hope to see more applications in the future, which will then provide a healthy feedback to the theory, giving a track record on what works and what does not.

The topics dealt with in this book, in addition to many classical parametric methods, include instrumental variable estimation (IVE), generalized method of moments (GMM), extremum estimators, methods of simulated moments, minimum distance estimation, nonparametric density estimation, nonparametric regression, and semiparametric methods for LDV. Time series topics are not covered; our focus is on iid observational data from a random sampling. Topics requiring functional analysis as a prerequisite are only covered in the last section.

The target audience for this book is researchers and graduate students (on average, second year in the United States and first year in Europe). Choosing topics selectively, the entire book may be covered in a semester. If all topics are covered, then it will take a year to cover the book. In the following section, we show the organization of the book.

1.2   Organization of the Book

This book has two parts and one appendix: six chapters for the first part and four chapters for the second (each chapter has about five to ten sections). In the first part, methods of moments and parametric econometric methods for LDV are studied; also asymptotic theories are introduced in a number of places. In the second part, nonparametric regression and semiparametric econometric methods for LDV are studied. In the appendix, 10 computer programs written in GAUSS are provided. Although our emphasis is on estimation, various specification tests are introduced throughout the book.

Chapter 2 is a review of a course that should precede this book, but the review is from a modern method-of-moments perspective. For a linear model, LSE, IVE, and GMM are examined along with testing linear hypotheses. One digression is in Section 8 on testing for instrument legitimacy. In Chapter 3, asymptotic theories for extremum estimators (with or without a nuisance parameter) are provided, which are then used many times in the other chapters. Also methods-of-moment tests are introduced complementing MME of Chapter 2. In Chapter 4, MLE's are reviewed, which is mainly for single equation LDV models; the reader will see later that, for a MLE for each LDV model, there exist semiparametric methods applicable to the same LDV model as discussed later in this book. Also we briefly discuss numerical optimization for extremum estimators in general. Although a parametric model specification can be tested in various ways, including method-of-moment tests, we introduce χ^2 specification tests, which are relatively new and the most general in their applicability. In Chapter 5, MLEs and other parametric methods for multiple equation LDV models are examined. Recent advances in methods of simulated moments and likelihoods are covered, and minimum distance estimation convenient for multiple equations is studied. One digression is in Section 9 on specification tests based on differences of estimators. In Chapter 6, LSE and GMM for nonlinear models are developed; the wide applicability of GMM will be evident there. Specification tests for GMM as well as nonlinear LSE are introduced.

In Chapters 7 and 8, kernel nonparametric density estimation and nonparametric regression are studied; although other nonparametric methods are introduced, we study only kernel methods in depth, because they are the most convenient in applications and developing theories. In Chapter 9, semiparametric econometric methods for LDV models are covered; here the methods require no nonparametric method to obtain estimates, although nonparametric methods may be needed for asymptotic inference. That is, the estimates are not subject to arbitrariness owing to selecting a "smoothing parameter" in the nonparametric method. In Chapter 10, under the heading "semi-nonparametrics," estimators requiring a smoothing parameter are studied; in most cases, they are two-stage estimators with the first stage being nonparametric.

The programs in the appendix use either a small data set or simulated data. The reader needs some basic knowledge of GAUSS. Although only about 10 estimators are treated, the reader will find that understanding the programs will help in implementing the other estimation methods presented in this book. During 1989–1994, drafts of this book have been taught by having the students write their own GAUSS programs to analyze small data sets. This seems to be a good way of teaching, particularly when the students are not theoretically motivated.

Although the book mostly deals with theoretical issues, most estimation methods have been field-tested, and the reader will find various tips for applied work. Typically in estimations, the following are necessary: an objective function to optimize, the gradient of the optimand, and the variance matrix of the estimator. In tests, test statistic and its variance matrix are needed. So whenever possible, those are provided. As for notations and acronyms, they are explained either in the main body of this book, with the exception of some standard ones.

2

Least Squares and Method of Moments

2.1 Introduction

In a linear model $y_i = x_i'\beta + u_i$ with $E(ux) = 0$, where β is a $k \times 1$ parameter vector of interest, u is the error term, x is a $k \times 1$ regressor vector, and (x_i', y_i) are iid, the least squares estimator (LSE) for β is obtained by minimizing

$$(1/N) \sum_i (y_i - x_i'b)^2 \tag{1.1}$$

with respect to (wrt) b. LSE can also be viewed as the solution of the first-order (moment) condition of the minimization

$$(1/N) \sum_i x_i(y_i - x_i'b) = 0. \tag{1.2}$$

There is no need to insist that (1.2) is the first-order condition for (1.1). Imagine starting with a population moment condition $E(xu) = 0 \Leftrightarrow E\{x(y - x'\beta)\} = 0$. This is a restriction on the joint distribution of (x', y). Since β is not observable, the sample version $(1/N) \sum_i x_i(y_i - x_i'\beta)$ for $E\{x(y - x'\beta)\} = 0$ is not feasible. Instead take (1.2) as the *sample analog* for $E\{x(y - x'b)\} = 0$. Then (1.2) defines LSE as one that establishes $E(xu) = 0$ as closely as possible in the sample analog. An estimator that is defined by a sample moment condition is called a *method-of-moments estimator* (MME).

If a vector z satisfies $E(zu) = 0$, then we can estimate β by using $E\{z(y - x'\beta)\} = 0$. In this case, z is called an *instrumental variable* and the estimator based on it is called an *instrumental variable estimator* (IVE). In this chapter, we study MME and IVE. See Hanson (1982), White (1984), Bowden and Turkington (1984) and Manski (1988) for more. MME and IVE, which include LSE, are easy to use and possess certain optimality properties as will be shown later. They have been at the center stage in econometrics for a long time. Perhaps their only drawback is sensitivity to outliers: if one datum takes an excessively large value, it may have a devastating effect on the estimator.

In Section 2, we formally show MME interpretation of LSE. In Section 3, IVE is studied in detail. When there are more instruments than the number of parameters, choosing the right instrument is a problem. This is addressed in Section 4. In Section 5, the generalized method of moments is introduced. In Section 6, the asymptotic properties of the estimators are studied. In Section 7, linear hypothesis testing is examined. In Section 8, we show how to test the legitimacy of instruments. Finally in Section 9, we study the prediction problem in general, and introduce "extremum estimators" to motivate further study.

2.2   LSE as a Method-of-Moments Estimator (MME)

For the linear model with $E(ux) = 0$, define $Y \equiv (y_1 \ldots y_N)'$, $U \equiv (u_1 \ldots u_N)'$, and $X \equiv (x_1 \ldots x_N)'$, where $x_i \equiv (x_{i1} \ldots x_{ik})'$, so that

$$X \equiv \begin{bmatrix} x_i' \\ \cdots \\ x_N' \end{bmatrix} = \begin{bmatrix} x_{11} \; x_{12} \cdots x_{1k} \\ \cdots\cdots\cdots\cdots \\ x_{N1} \; x_{N2} \cdots x_{Nk} \end{bmatrix};$$

X is a $N \times k$ matrix. In this matrix notation, we have

$$(1/N) \sum_i (y_i - x_i' b)^2 = (1/N)(Y - Xb)'(Y - Xb) \tag{2.1}$$

and the linear model is $Y = X\beta + U$. Premultiply $Y = X\beta + U$ by $(1/N)X'$ to get

$$(1/N)X'Y = (1/N)X'X \cdot \beta + (1/N)X'U. \tag{2.2}$$

According to the law of large numbers (LLN, see Section 6),

$$(1/N)X'U = \left[(1/N) \sum_i u_i x_{i1} \cdots (1/N) \sum_i u_i x_{ik} \right]'$$

$$\equiv (1/N) \sum_i u_i x_i =^P E(ux) = [E(u \cdot x_1) \cdots E(u \cdot x_k)]' = 0 \tag{2.3}$$

owing to the assumption $E(ux) = 0$, where $=^p$ is convergence in probability; see also Section 6. Observe that the dimension of $X'U$ is $(k \times N) \cdot (N \times 1) = k \times 1$. Hence, we get LSE estimator b_{LSE} from (2.2) and (2.3):

$$b_{\text{LSE}} = (X'X/N)^{-1}(X'Y/N) = (X'X)^{-1}(X'Y)$$

$$= \left\{ (1/N) \sum_i x_i x_i' \right\}^{-1} (1/N) \sum_i x_i y_i. \tag{2.4}$$

The asymptotic property of LSE will be examined in Section 6.

2.3   Instrumental Variable Estimator (IVE)

For the model $y = x'\beta + u$, assume a $k \times 1$ moment condition $E(uz) = 0$. With Z being an instrument matrix (Z is a $N \times k$ matrix) corresponding to z (z is a $k \times 1$ vector), we have the following analog to (2.2):

$$(1/N)Z'Y = (1/N)Z'X \cdot \beta + (1/N)Z'U. \tag{3.1}$$

Using $(1/N)Z'U =^p E(uz) = 0$ and assuming that z and x are correlated and that $(Z'X)/N$ is invertible, we have an IVE b_{IV}:

$$b_{\text{IV}} = (Z'X/N)^{-1}(Z'Y/N) = (Z'X)^{-1}(Z'Y)$$

$$= \left\{ (1/N) \sum_i z_i x_i' \right\}^{-1} (1/N) \sum_i z_i y_i. \tag{3.2}$$

This includes LSE as a special case when $Z = X$. While IVE in its broad sense includes any estimator using instuments, here we define IVE in its narrow sense as the b_{IV} in (3.2).

A good instrument should be highly correlated with x while uncorrelated with u. The latter is necessary for b_{IV} to be consistent for β (see Section 6) and the former makes b_{IV} more efficient: the variance of b_{IV} $[V(b_{\text{IV}})]$ is proportional to $(Z'X)^{-1}$, so if $Z'X$ is large due to a high correlation between z and x, then $V(b_{\text{IV}})$ is small. The asymptotic property of IVE will be studied in Section 6.

IVE can also be cast in a minimization problem. The sample analog of $E(zu)$ is $(1/N) \sum_i z_i u_i$. Since u_i is unobservable, replace u_i by $y_i - x_i'b$ to get $(1/N) \sum_i z_i(y_i - x'b)$. We can get IVE by minimizing the deviation of $(1/N) \sum_i z_i(y_i - x'b)$ from 0, but the problem is that $(1/N) \sum_i z_i(y_i - x'b)$ is a $k \times 1$ vector. So convert it to a scalar, using the squared Euclidean norm. Ignoring $(1/N)$, we get

$$\left\{ \sum_i z_i(y_i - x'b) \right\}' \cdot \sum_i z_i(y_i - x'b) = \{Z'(Y - X'b)\}' \cdot Z'(Y - X'b) \tag{3.3}$$

$$= (Y - Xb)'ZZ'(Y - Xb) = Y'ZZ'Y - 2b'X'ZZ'Y + b'X'ZZ'Xb. \quad (3.4)$$

The first-order condition of minimization is

$$-2X'ZZ'Y + 2X'ZZ'Xb = 0.$$

Using the invertibility of $Z'X$, solve this to get

$$b_{\text{IV}} = (Z'X)^{-1}Z'Y. \qquad (3.5)$$

One may wonder where we get the instruments in real life. First of all, any variables in x which are uncorrelated with u can be used as a (self) instrument, including the intercept. In the errors-in-variable model, suppose $y = \beta'x^* + v$ but only $x = x^* + \varepsilon$ is observed, where y is consumption, x^* is the "permanent" income, and x is an error-ridden measure of x^*. Then substituting $x^* = x - \varepsilon$ into y equation, we get

$$y = x'\beta + (v - \beta'\varepsilon) \equiv x'\beta + u, \qquad x = x^* + \varepsilon.$$

Hence, x is correlated with the error term u. If "schooling years" is correlated with x but uncorrelated with u, then it qualifies as an instrument. In time series data with $E(x_t u_t) \neq 0$, x_{t-j} with $j > 0$ can serve as an instrument if $E(x_{t-j} u_t) = 0$.

One example of IVE is the *generalized least squares estimator* (GLS). Recalling GLS briefly, consider the linear model in matrix notation:

$$Y = X\beta + U, \qquad E(UU') = \Omega, \qquad (3.6)$$

where Ω is a $N \times N$ matrix. The term "generalized" refers to the variance matrix Ω, which is not the scalar matrix $\sigma^2 I_N$ as is often assumed in the linear model. Let $\Omega = QDQ'$, the eigenvalue decomposition where D is the diagonal matrix of the eigenvalues and Q is the matrix of the orthonormal eigenvectors such that $Q' = Q^{-1}$. Assuming that Ω is known, GLS transforms the data by multiplying (3.6) by $\Omega^{-1/2} = QD^{-1/2}Q'$:

$$\Omega^{-1/2}Y = \Omega^{-1/2}X\beta + \Omega^{-1/2}U \Leftrightarrow Y^* = X^*\beta + U^*. \qquad (3.7)$$

Then $E(U^*) = 0_N$ and $E(U^*U^{*\prime}) = I_N$ for $\Omega^{-1/2}\Omega\Omega^{-1/2} = I_N$. Apply LSE to (3.7):

$$b_{\text{GLS}} = (X'\Omega^{-1}X)^{-1}(X'\Omega^{-1}Y) \equiv (Z'X)^{-1}(Z'Y), \qquad (3.8)$$

where $Z \equiv \Omega^{-1}X$. Z may be viewed as an instrument. In practice, Ω is unknown, but it can be estimated using LSE residuals so long as the form of Ω is specified. For instance, suppose there is no correlation among u_i's, then $\Omega = \text{diag}\{V(u_1)\ldots V(u_N)\}$. If $V(u_i \mid x_i) = \sigma^2 x_i^2$, then Ω can be estimated by replacing $V(u_i \mid x_i)$ with $\hat{\sigma}^2 x_i^2$ where $\hat{\sigma}^2 = (1/N)\sum_i(y_i - x_i'b_{\text{LSE}})^2$.

Another example of IVE is the *seemingly unrelated regression* (SUR). Suppose we have two equations under homoskedasticity:

$$y_1 = x_1'\beta_1 + u_1, \qquad y_2 = x_2'\beta_2 + u_2;$$

$$E(u_m x_n) = 0, \qquad m, n = 1, 2, \tag{3.9}$$

$$E(u_1^2 \mid x_1, x_2) = \sigma_1^2, \quad E(u_2^2 \mid x_1, x_2) = \sigma_2^2, \quad E(u_1 u_2 \mid x_1, x_2) = \sigma_{12},$$

where x_1 and x_2 are, respectively, $k_1 \times 1$ and $k_2 \times 1$ vectors. In the single equation $y = x'\beta + u$, we use the word "heteroskedasticity" when the conditional variance $V(u|x)$ is not a constant but a function of x (and instruments); in this multiequation setup, homoskedasticity means σ_{12} as well as σ_1^2 and σ_2^2 are constants. The name SUR is used because the two equations can be related through their error term correlation. Define $y \equiv (y_1, y_2)'$, $\beta \equiv (\beta_1', \beta_2')'$, and $u \equiv (u_1, u_2)'$ to rewrite (3.9) as

$$y = x'\beta + u, \quad E(xu) = 0, \quad E(uu') = C,$$

$$x' = \begin{bmatrix} x_1' & 0 \\ 0 & x_2' \end{bmatrix}, \quad \beta = \begin{bmatrix} \beta_1 \\ \beta_2 \end{bmatrix}, \quad C = \begin{bmatrix} \sigma_1^2 & \sigma_{12} \\ \sigma_{12} & \sigma_2^2 \end{bmatrix}. \tag{3.10}$$

In matrix notation, this becomes

$$Y = X\beta + U, \qquad E(UU') = I_N \otimes C, \tag{3.11}$$

where $\otimes$ is the Kronecker product. Note that the dimension of X is $2N \times (k_1 + k_2)$. Apply GLS to (3.11); $(I_N \otimes C)^{-1} X = (I_N \otimes C^{-1}) X$ can be regarded as an instrument. The (mn)th component, $m, n = 1, 2$, of C can be estimated by $(1/N) \sum_i (y_{mi} - x_{mi}' \hat{b}_m)(y_{ni} - x_{ni}' \hat{b}_n)$, where $\hat{b}_m$ is the LSE of the mth equation.

2.4   Best Instrumental Variable Estimator (BIV)

Consider a random variable z following a Poisson distribution with its parameter λ. Then $E(z) = \lambda$ and $V(z) = \lambda$. One way to estimate λ is the maximum likelihood estimation (MLE) by maximizing the log-likelihood

$$\ln \left(\prod_i e^{-\lambda} \cdot \lambda^{z_i} / z_i! \right) = \sum_i \{ -\lambda + z_i \ln \lambda - \ln(z_i!) \}$$

$$= -\lambda N + (\ln \lambda) \cdot \sum_i z_i - \sum_i \ln(z_i!)$$

wrt λ. The first-order condition is $-N + (1/\lambda) \sum_i z_i = 0$, which yields the MLE $\lambda_{\text{MLE}} = (1/N) \sum_i z_i$, the sample mean. Another way to estimate λ

is to use the moment information, which then yields two estimators for λ: the sample mean $(1/N) \sum_i z_i$ and the sample variance $(1/N) \sum_i (z_i - \bar{z}_i)^2$. In the second approach, there the question arises as to which one to use, or better yet, how to combine the two estimators for one parameter. In general, suppose we have s $(> k)$ moment conditions $E(zu) = 0$ for a $k \times 1$ parameter vector. Then there arises the question of selecting or combining the more than enough moment conditions to get k equations.

Consider $(1/N) \sum_i z_i (y_i - x_i'b) = (1/N)Z'(Y - Xb)$. We want to make this as close to zero as possible in view of $E(zu) = 0$. Since the sample moment is a $s \times 1$ vector, we would like to convert it to a scalar using a norm. Suppose that we use a quadratic norm with a $s \times s$ positive definite (p.d.) matrix W^{-1}: ignoring $(1/N)$,

$$\{Z'(Y - X'b)\}'W^{-1}\{Z'(Y - X'b)\} = (Y - Xb)'ZW^{-1}Z'(Y - Xb); \quad (4.1)$$

this generalizes (3.3) $\{Z'(Y - Xb)\}'I_s\{Z'(Y - Xb)\}$. The choice of W determines how to combine the moment conditions. In this section, we discuss the choice of W under homoskedasticity. In the next section, we deal with the case of heteroskedasticity of unknown form.

If z is a $s \times 1$ possible instrument vector, then "x fitted by z,"

$$\hat{X} \equiv Z(Z'Z)^{-1}Z'X,$$

is the best instrument matrix under homoskedasticity, "best" in the sense that the resulting IVE has the smallest variance under $E(uz) = 0$ and homoskedasticity as to be shown in Section 5. Note that the dimension of $\hat{X}$ is $N \times k$. Then the *best instrumental variable estimator* b_{BIV} is

$$b_{\text{BIV}} = (\hat{X}'X)^{-1}\hat{X}'Y = [\{Z(Z'Z)^{-1}Z'X\}'X]^{-1}\{Z(Z'Z)^{-1}Z'X\}'Y$$
$$\quad (4.2)$$
$$= \{X'Z(Z'Z)^{-1}Z'X\}^{-1}X'Z(Z'Z)^{-1}Z'Y,$$

which is also obtained by setting $W = (Z'Z)$ in (4.1) and minimizing

$$\{Z'(Y - Xb)\}'(Z'Z)^{-1}\{Z'(Y - Xb)\} = (Y - X'b)'Z(Z'Z)^{-1}Z'(Y - X'b).$$
$$\quad (4.3)$$

The name BIV for (4.2) is not necessarily agreed upon in the literature, although we will use the name from now on. Often we call (4.2) IVE (IVE in a wide sense).

Compare ZZ' in (3.4) with $Z(Z'Z)^{-1}Z'$ in (4.3). The later is called the *(linear) projection matrix* formed by Z and is often denoted as $\mathbf{P}_Z$. Then b_{BIV} can be written as

$$b_{\text{BIV}} = \{(P_Z X)'P_Z X\}^{-1}(P_Z X)'Y, \quad (4.4)$$

since P_Z is symmetric and idempotent: $P_Z'P_Z = P_Z P_Z = P_Z$. If X is included in Z, then we can show that $\hat{X} = P_z X = X$, so BIV = LSE.

One example of b_{BIV} is the *two-stage least squares estimator* (2SLSE) in a structural simultaneous equation system. Suppose that the first structural equation is $y_1 = w_1'\gamma + x_1'\beta + u_1$, where w_1 and x_1 are, respectively, the endogenous and exogenous variables in the first equation. Let z denote all the exogenous variables of the system; z includes x_1. Then the first stage of 2SLSE is regressing w_1 on z to get its LSE fitted value $\hat{w}_1$. The second stage is regressing y_1 on $\hat{w}_1$ and x_1 to estimate γ and β. Defining $x \equiv (w_1', x_1')'$, 2SLSE is equivalent to (4.4), for $P_Z W_1 = \hat{W}_1$ and $P_Z X_1 = X_1$.

2.5 Generalized Method-of-Moments Estimator (GMM)

In the previous section, we gave an answer to the question of how to combine more than enough moment conditions using a quadratic norm under homoskedasticity. In this section, allowing heteroskedasticity of unknown form and generalizing $E\{z(y - x'\beta)\} = 0$ into $E\psi(y, x, z, \beta) = 0$, which may be nonlinear in β, we give the answer that includes b_{BIV} as a special case for homoskedasticity and the linear moment condition. Under these conditions, BIV inherits its efficiency from the efficiency of the "GMM estimator" introduced in this section; the meaning of "efficiency" will soon become clear. We will abbreviate $E\psi(y, x, z, \beta)$ as $E\psi(\beta)$.

Suppose there is a $k \times 1$ parameter vector β and $s \ (\geq k)$ population moment conditions $E\psi(\beta) = 0$, which may be nonlinear in β. Then the *generalized method-of-moments estimator* (GMM) is a class of estimators indexed by W minimizing

$$(1/N) \sum_i \psi(b)' \cdot W^{-1} \cdot (1/N) \sum_i \psi(b), \qquad (5.1)$$

where W is a p.d. $s \times s$ matrix. The question in GMM is which W to use in (5.1). Hansen (1982) shows that the W that yields the smallest variance for the resulting estimator is

$$V\left\{(1/\sqrt{N}) \sum_i \psi(\beta)\right\},$$

which becomes $E\{\psi(\beta)\psi(\beta)'\}$ for iid samples.

In order to understand this, regard $(1/N) \sum_i \psi(b)$ as an "error" vector, for it deviates from the mean $E\psi(b)$, which is 0 if $b = \beta$. The intuition for $W = V\left\{(1/\sqrt{N}) \sum_i \psi(b)\right\}$ is that, in minimizing (5.1), it is better to standardize $(1/N) \sum_i \psi(b)$, otherwise one component with a high variance can unduly dominate the minimand. Standardizing the vector will give equal weight to each component. This intuition will also be useful in relation to testing. Since we always use (5.1) with $W = V\left\{(1/\sqrt{N})\sum_i \psi(\beta)\right\}$, often the optimal GMM is simply called (the) GMM.

Looking at (5.1), it seems that we may be able to do better than GMM by using a criterion function other than the quadratic norm. But Chamberlain (1987) shows that the GMM is the efficient estimator under the given moment condition $E\psi(\beta) = 0$ with iid observations.

In order to get a concrete idea of GMM, consider

$$E\psi(\beta) = E(zu) = E\{z(y - x'\beta)\} = 0. \qquad (5.2)$$

With (5.2), (5.1) becomes

$$\begin{aligned}
\{Z'(Y - Xb)\}'W^{-1}\{Z'(Y - Xb)\} &= (Z'Y - Z'Xb)'W^{-1}(Z'Y - Z'Xb) \\
&= (Y'ZW^{-1} - b'X'ZW^{-1}) \cdot (Z'Y - Z'Xb) \\
&= Y'ZW^{-1}Z'Y - 2b'X'ZW^{-1}Z'Y + b'X'ZW^{-1}Z'Xb.
\end{aligned}$$

$$\qquad (5.3)$$

From the first-order condition of minimization, we get

$$X'ZW^{-1}Z'Y = X'ZW^{-1}Z'Xb$$

$$\Leftrightarrow b_{\text{GMM}} = (X'ZW^{-1}Z'X)^{-1} \cdot (X'ZW^{-1}Z'Y). \qquad (5.4)$$

As noted already, $W = V(zu)$ is the optimal choice:

$$W = E(zz'u^2) = E_z\{zz'(E_{u|z}u^2)\}. \qquad (5.5)$$

Under homoskedasticity $(E_{u|z}(u^2) = \sigma^2)$,

$$W = \sigma^2 \cdot E(zz') =^p \sigma^2 \cdot (Z'Z/N).$$

But any scalar in W will be canceled in (5.4), for both W and W^{-1} appear in (5.4). Hence, setting $W = Z'Z$ is enough. Thus, b_{GMM} becomes b_{BIV} of (4.2) under homoskedasticity. Under heteroskedasticity, (5.5) does not yield such a simple form as $Z'Z$, but $W = E(zz'u^2)$ can be estimated by

$$(1/N)\sum_i z_i z_i' r_i^2 = (1/N)Z'DZ, \qquad (5.6)$$

where $r_i = y_i - x_i'b_{\text{BIV}}$ and $D = \text{diag}(r_1^2 \ldots r_N^2)$.

Note that under homoskedasticity, we do not need an initial estimate. But under heteroskedasticity, the GMM is obtained in two stages: First apply BIV to get the residuals r_i's, then substitute $\sum_i z_i z_i' r_i^2$ into W in (5.4). For this reason, the GMM is sometimes called a "two-stage IVE." Also notable in (5.5) is that the heteroskedasticity of u is wrt z, where z may include variables other than x.

As an example of GMM for (5.2), consider the following rational expectation model (Nijman, 1990):

$$y_t = \rho \cdot E(y_{t+1} \mid I_t) + x_t'\beta + \varepsilon_t, \qquad t = 1, \ldots, T,$$

where I_t is the information available up to period t including x_t, y_{t-1}, $x_{t-1}, \ldots$, and $E(\varepsilon_t x_{t-j}) = 0$ and $E(\varepsilon_t y_{t-j}) = 0$ for all $j = 1 \ldots t$. One way to estimate ρ, called "errors-in-variable approach," is to replace $E(y_{t+1} \mid I_t)$ by y_{t+1}:

$$\begin{aligned} y_t &= \rho y_{t+1} + x_t'\beta + \varepsilon_t + \rho\{E(y_{t+1} \mid I_t) - y_{t+1}\} \\ &\equiv \rho y_{t+1} + x_t'\beta + u_t. \end{aligned}$$

Then $y_{t-1}, x_{t-1}, y_{t-2}, x_{t-2}, \ldots$ are all valid instruments by the rational expectation hypothesis: the prediction error $E(y_{t+1} \mid I_t) - y_{t+1}$ should be orthogonal to all available information up to t. If $E(x_t \varepsilon_t) = 0$, then x_t is also a good instrument. Redefining $\beta \equiv (\rho, \beta')'$ and $x_t \equiv (y_{t+1}, x_t')'$ we have

$$y_t = x_t'\beta + u_t, \qquad t = 1, \ldots, T - 1.$$

For this, b_{GMM} was proposed by Cumby, Huizinga, and Obstfeld (1983).

While more on GMM with a nonlinear ψ will be discussed in the chapter on nonlinear regression, we can summarize our analysis for the linear model under $E(zu) = 0$ and heteroskedasticity of unknown form as follows. First, the efficient estimator when $s \geq k$ is

$$b_{\mathrm{GMM}} = \{X'Z(Z'DZ)^{-1}Z'X\}^{-1}X'Z(Z'DZ)^{-1}Z'Y, \qquad (5.7)$$

where D is in (5.6). If homoskedasticity holds,

$$b_{\mathrm{BIV}} = \{X'Z(Z'Z)^{-1}Z'X\}^{-1}X'Z(Z'Z)^{-1}Z'Y \qquad (5.8)$$

is the efficient estimator to which b_{GMM} becomes asymptotically equivalent. If $s = k$, and $\{(1/N)\sum_i z_i x_i'\}^{-1}$ exists, then b_{GMM} becomes exactly equal to b_{IV}:

$$b_{\mathrm{GMM}} = \{X'Z(Z'DZ)^{-1}Z'X\}^{-1}X'Z(Z'D'Z)^{-1}Z'Y$$

$$= (Z'X)^{-1}(Z'DZ)(X'Z)^{-1}X'Z(Z'DZ)^{-1}Z'Y = (Z'X)^{-1}(Z'Y). \qquad (5.9)$$

Furthermore, if $Z = X$, then $b_{\mathrm{GMM}} = b_{\mathrm{IV}} = b_{\mathrm{LSE}}$.

Since GMM is efficient under the condition $E(zu) = 0$, BIV is also efficient under homoskedasticity. IVE inherits the efficiency from GMM when the dimension of z is the same as that of x, and LSE is efficient when $z = x$.

2.6   Asymptotic Properties

If x is a known constant and the distribution of u, F_u, is known, then deriving the asymptotics of MME is easy. But in economics, almost always,

x is random and F_u is unknown. Assumption of a normally distributed error term is not as easily justifiable as in the natural sciences, where data are generated through well-controlled experiments. Hence, we need to derive asymptotic distributions using LLN's and central limit theorems (CLT).

There are many versions of LLN and CLT. In this section, we introduce only a few. Unless otherwise noted, random variables (rv) below are defined on a probability space $(\Omega, \mathbf{A}, \mathbf{P})$, where Ω is a sample space, $\mathbf{A}$ is a σ-algebra (or event-space), and $\mathbf{P}$ is a probability defined on $\mathbf{A}$. A σ-algebra is a class of subsets of Ω satisfying certain conditions, and $\mathbf{P}$ assigns a probability to each member of $\mathbf{A}$. Often a generic element of Ω will be denoted by ω.

(6.1) *Law of Large Numbers.* Let $\{x_i\}$ be a sequence of iid rv. Then a necessary and sufficient condition that $\bar{x}_N \equiv (1/N) \sum_i x_i$ converges to $E(x)$ a.e. (almost everywhere) is that $E(x)$ exists. If $\{x_i\}$ is an inid (independent but non-identically distributed) sequence, then $\sum_{i=1}^{\infty} E|x_i - Ex_i|^{p_n}/i^{p_n} < \infty$ for $1 \le p_n \le 2$ is sufficient for $(1/N) \sum_i (x_i - Ex_i)$ to converge to 0 a.e.

The first LLN is called the "Kolmogorov LLN" and the second is from Chow and Teicher (1988, p. 124).

In the probability space $(\Omega, \mathbf{A}, \mathbf{P})$, denote the almost everywhere convergence as $=^{ae}$ wrt $\mathbf{P}$, or $=^{ae}$ $[\mathbf{P}]$ (often $[\mathbf{P}]$ is omitted); it means that there exists a set $A \in \mathbf{A}$ such that the convergence holds on each ω in A and $\mathbf{P}(A) = 1$. The set A can be strictly smaller than Ω, but we can ignore the difference A^c for $\mathbf{P}(A^c) = 0$. Any set B is called a *support* of $\mathbf{P}$ if $\mathbf{P}(B) = 1$ and $B \in \mathbf{A}$; the smallest such set is "the" support. Almost everywhere convergence is also called "almost sure convergence" (denoted as $=^{as}$) or strong consistency.

A weaker concept of convergence, called "convergence in probability," is available; for the sample mean, it is

$$P(|\bar{x}_N - E(x)| < \varepsilon) \to 1 \ \text{ as } N \to \infty, \text{ for any constant } \varepsilon > 0.$$

It is often called "(weak) consistency" and is denoted as $=^p$. More explanation on the convergence concepts and probability theory in general can be found in Serfling (1980), Billingsley (1986), Chow and Teicher (1988), and Dudley (1989), among many others.

In the inid case, LLN was stated as

$$(1/N) \sum_i \{x_i - E(x_i)\} = (1/N) \sum_i x_i - (1/N) \sum_i E(x_i) =^{as} 0.$$

To be precise, this expression is not the same as

$$(1/N) \sum_i x_i =^{as} (1/N) \sum_i Ex_i,$$

for the latter requires $(1/N) \sum_i E(x_i)$ to be convergent. But we will not be fastidious about this, and simply say that $(1/N) \sum_i x_i$ converges to

$(1/N) \sum_i E(x_i)$ a.s. in inid cases. An inid case can occur when there is heteroskedasticity and "sampling is exogenous" [x_i fixed and then y_i drawn from the conditional distribution $F(y \mid x_i)$], for $V(y \mid x_i)$ varies across i; if the sampling is random [$(x_i', y_i)'$ drawn together from the joint distribution], heteroskedasticity does not imply inid.

To better understand the condition $\sum_{i=1}^{\infty} E|x_i - E(x_i)|^{p_n}/i^{p_n} < \infty$ for the inid LLN, suppose $p_n = 2$ for all n to have $\sum_{i=1}^{\infty} V(x_i)/i^2 < \infty$. Now imagine $V(x_i) = i$, growing with i. Then the condition becomes $\sum_{i=1}^{\infty} 1/i$, which is divergent. However, since $\sum_{i=1}^{\infty} 1/(i^{1+\varepsilon}) < \infty$ for any $\varepsilon > 0$, $\sum_{i=1}^{\infty} V(x_i)/i^2 < \infty$ so long as $V(x_i)$ increases at a rate $i^{1-\varepsilon}$. Hence we can allow different (and growing) $V(x_i)$ across i. For iid cases, the condition is trivially satisfied.

(**6.2**) *Lindeberg CLT.* Let $\{x_i\}$ be independent zero-mean rv. Define $s_N^2 \equiv \sum_{i=1}^{N} \sigma_i^2$, where $\sigma_i^2 \equiv V(x_i)$. If for any $\varepsilon > 0$

$$\sum_{i=1}^{N} E(x_i^2 1[|x_i| > \varepsilon s_N]) \to 0 \qquad \text{as } N \to \infty,$$

then

$$(1/\sqrt{N}) \sum_{i=1}^{N} x_i/s_N \Rightarrow N(0,1).$$

If $E(x_i) \neq 0$, redefine x_i as $x_i - E(x_i)$ to apply (6.2). The condition in (6.2) is called the "Lindeberg condition." One sufficient condition for the Lindeberg condition is

$$\sum_{i=1}^{N} E|x_i|^{2+\delta}/s_N^{2+\delta} \to 0 \qquad \text{as } N \to \infty$$

for some $\delta > 0$, which yields the "Liapunov CLT." In an iid case with $E|x|^{2+\delta} < \infty$, the Liapunov condition is easily satisfied:

$$N \cdot E|x|^{2+\delta}/\{N \cdot \sigma^2\}^{(2+\delta)/2} \to 0, \qquad \text{as } N \to \infty.$$

In the CLT, the normalized sum follows $N(0,1)$ as $N \to \infty$, where x_i can be just about anything. For instance, x_i can be an indicator function for the head of a coin in its ith toss or the income of a person i. Tossing a coin may bear no relation to income (they may not be defined on the same probability space), but still the normalized sums follow $N(0,1)$. This concept of convergence in distribution is often denoted by "$=^d$" as well as "$\Rightarrow$" in (6.2).

Usually we do not have inid data in econometrics. But if there is heteroskedasticity and sampling is exogenous, then data otherwise iid under random sampling become inid as mentioned already. This is the reason why we provide LLN and CLT for inid cases as well as iid cases. Once those conditions are met, estimation and statistical inference in inid cases are almost the same as those in iid cases. From now on, we will deal with only iid cases unless otherwise noted.

If $x_i = c$, a constant, then $(1/N)\sum_i x_i = c$, but $(1/\sqrt{N})\sum_i x_i = \sqrt{N}\cdot c$, which goes to ∞ as $N \to \infty$. This shows that LLN is rather intuitive, while CLT is special in that the randomness in x_i compensates the $\sqrt{N}$ rate of growth to result in a normal distribution. Often the CLT is stated as

$$(1/\sqrt{N}) \sum_i x_i \Rightarrow N(E(x), V(x)),$$

which may be easier to grasp than (6.2).

Using LLN and CLT, we check the asymptotics of b_{IV}:

$$b_{\mathrm{IV}} = (Z'X)^{-1}(Z'Y) = (Z'X)^{-1}\{Z'(X\beta + U)\}$$
$$= \beta + (Z'X/N)^{-1}(Z'U/N) =^{\mathrm{ae}} \beta + E^{-1}(zx')\cdot E(zu) = \beta.$$

Hence, b_{IV} converges to β a.e. as $N \to \infty$. Asymptotic distribution of $b_{\mathrm{IV}} - \beta$ is derived by multiplying $b_{\mathrm{IV}} - \beta$ by $\sqrt{N}$ with $E^{-1}(\ ) \equiv \{E(\ )\}^{-1}$,

$$\sqrt{N}(b_{\mathrm{IV}} - \beta) = (Z'X/N)^{-1}(Z'U/\sqrt{N}) =^d E^{-1}(zx')\cdot N(0, E(zz'u^2))$$
$$= N(0, E^{-1}(zx')\cdot E(zz'u^2)\cdot E^{-1}(xz'))$$
$$= N(0, \{E(xz')\cdot E^{-1}(zz'u^2)\cdot E(zx')\}^{-1}).$$

(6.3)

The distribution of b_{LSE} can be obtained from (6.3) with $z = x$:

$$\sqrt{N}(b_{\mathrm{LSE}} - \beta) =^d N(0, E^{-1}(xx')\cdot E(xx'u^2)\cdot E^{-1}(xx')).\qquad (6.4)$$

If we assume homoskedasticity $E(u^2 \mid z) = \sigma^2$, we get

$$E(zz'u^2) = E_z\{zz'E_{u|z}(u^2)\} = E_z(zz'\sigma^2) = \sigma^2 E(zz').$$

Thus, under homoskedasticity,

$$\sqrt{N}(b_{\mathrm{IV}} - \beta) =^d N(0, \sigma^2 \cdot E^{-1}(zx')\cdot E(zz')\cdot E^{-1}(xz'))$$

(6.5)

$$= N(0, \sigma^2 \cdot \{E(xz')\cdot E^{-1}(zz')\cdot E(zx')\}^{-1}),$$

$$\sqrt{N}(b_{\mathrm{LSE}} - \beta) =^d N(0, \sigma^2 \cdot E^{-1}(xx')).\qquad (6.6)$$

Note that for any $k \times 1$ vector γ,

$$\gamma' E(xx')\gamma = E(\gamma'xx'\gamma) = E\{(x'\gamma)'(x'\gamma)\} = E(x'\gamma)^2 \geq 0.$$

Hence, $E(xx')$ is p.s.d., and by assuming that the rank of $E(xx')$ is k, $E(xx')$ becomes p.d. Often we just assume that $E(xx')$ is p.d. [$\Leftrightarrow E^{-1}(xx')$ is p.d.], which is called the *full rank condition*.

To compare the variance matrices in (6.5) and (6.6), note that

$$E(xx') - E(xz') \cdot E^{-1}(zz') \cdot E(zx') = E(x - \gamma'z)(x - \gamma'z)' \geq 0, \quad (6.7)$$

where

$$\gamma \equiv E^{-1}(zz') \cdot E(zx'), \qquad (6.8)$$

and "≥ 0" in (6.7) means that the matrix difference is p.s.d. From the first equality in (6.7), we get

$$E(xx') = E(xz') \cdot E^{-1}(zz') \cdot E(zx') + E(x - \gamma'z)(x - \gamma'z)'. \qquad (6.9)$$

This is a decomposition of $E(xx')$ into two parts, one explained by z and the other unexplained by z; regard $x - \gamma'z$ as the "residual." The coefficient γ is called the *linear projection coefficient* of x on z and $\gamma'z$ is the linear projection of x on z. Compared with this, we may call $E(x \mid z)$ the *projection of x on z*; however, sometimes the linear projection is simply called the projection.

The inequality

$$E(xx') \geq E(xz') \cdot E^{-1}(zz') \cdot E(zx') \qquad (6.10)$$

is often called *generalized Cauchy–Schwarz inequality*, which shows that the "explained variation" [x explained by z, the right-hand side of (6.10)] is not greate than the "total variation." From the inequality, if $E(xz')$ is invertible, we get

$$E^{-1}(xx') \leq E^{-1}(zx') \cdot E(zz') \cdot E^{-1}(xz'), \qquad (6.11)$$

which shows that, under homoskedasticity, LSE is more efficient than IVE; recall (6.5) and (6.6). Hence, under homoskedasticity, there is no reason to use IVE unless $E(xu) \neq 0$. With heteroskedasticity, we do not have such a result, since the comparison depends on $V(u \mid x, z)$.

One way to deal with heteroskedasticity is to specify the form of heteroskedasticity such as $E(u^2 \mid z) = h(z)$ and estimate $h(z)$ to apply GLS. Another (better) way is to estimate *heteroskedasticity consistent covariance* [see Eicker (1963) and White (1980)]: the covariance in (6.3) is estimated by

$$\{(1/N)Z'X\}^{-1}(1/N)Z'DZ\{(1/N)X'Z\}^{-1}, \qquad (6.12)$$

where

$$D \equiv \operatorname{diag}(r_1^2, \ldots, r_N^2),$$

$r_i \equiv y_i - x_i'b_N$ and b_N is an initial consistent estimator. If homoskedasticity prevails, then the above matrix converges to the one in (6.5). So it is safe to use the above matrix as the covariance of $\sqrt{N}(b_{\text{IV}} - \beta)$, which provides the correct standard errors under both homoskedasticity and heteroskedasticity.

Similarly to $b_{\text{IV}} =^{\text{as}} \beta$, it is easy to show that b_{BIV} in (4.2) also converges to β a.s. As for the distribution,

$$\sqrt{N}(b_{\text{BIV}} - \beta) = \{(X'Z/N) \cdot (Z'Z/N)^{-1} \cdot (Z'X/N)\}^{-1}$$

$$\cdot(X'Z/N) \cdot (Z'Z/N)^{-1} \cdot (Z'U/\sqrt{N})$$

$$=^d \{E(xz') \cdot E^{-1}(zz') \cdot E(zx')\}^{-1} E(xz') \cdot E^{-1}(zz') \cdot N(0, E(zz'u^2))$$

$$=^d N(0, M \cdot E(zz'u^2) \cdot M'), \qquad (6.13)$$

where M is the long matrix in front of $N(\cdot, \cdot)$. Under homoskedasticity, we have $E(zz'u^2) = \sigma^2 E(zz')$, which simplifies the covariance matrix drastically to yield

$$\sqrt{N}(b_{\text{BIV}} - \beta) =^d N(0, \sigma^2 \{E(xz') \cdot E^{-1}(zz') \cdot E(zx')\}^{-1}). \qquad (6.14)$$

$E(xz')$ can be estimated by $(1/N)\sum_i x_i z_i' = (1/N)X'Z$. Likewise $E(zz'u^2)$ can be estimated by $(1/N)\sum_i z_i z_i' r_i^2 = (1/N)Z'DZ$, where $r_i = y_i - x_i' b_{\text{BIV}}$. Substituting $Y = X\beta + U$ into (5.4), GMM in (5.4) can be written as

$$b_{\text{GMM}} = \beta + (X'ZW^{-1}Z'X)^{-1}(X'ZW^{-1}Z'U); \qquad (6.15)$$

$$\sqrt{N}(b_{\text{GMM}} - \beta) = \{(X'Z/N) \cdot W^{-1} \cdot (Z'X/N)\}^{-1}(X'Z/N) \cdot W^{-1} \cdot (Z'U/\sqrt{N})$$

$$=^d \{E(xz') \cdot W^{-1} \cdot E(zx')\}^{-1} E(xz') \cdot W^{-1} \cdot N(0, E(zz'u^2)) \equiv N(0, C_W), \qquad (6.16)$$

where

$$C_W \equiv \{E(xz') \cdot W^{-1} \cdot E(zx')\}^{-1} \cdot E(xz') \cdot W^{-1}$$

$$\cdot E(zz'u^2) \cdot W^{-1} \cdot E(zx') \cdot \{E(xz') \cdot W^{-1} \cdot E(zx')\}^{-1}. \qquad (6.17)$$

With the optimal $W = E(zz'u^2)$, this variance matrix becomes

$$C_{\text{GMM}} = \{E(xz') \cdot E^{-1}(zz'u^2) \cdot E(zx')\}^{-1}. \qquad (6.18)$$

Under homoskedasticity, C_{GMM} becomes the variance matrix of b_{BIV}. But in a finite sample, the estimate for C_{GMM} may be slightly different from that of the variance matrix of b_{BIV}. Note that an estimator for C_{GMM} is easily obtained in the optimal GMM: the first part of b_{GMM} in (5.7) times N is an estimate for C_{GMM}.

2.7   Testing Linear Hypotheses

Suppose $\sqrt{N}(b_N - \beta) =^d N(0, C)$, where b_N is any of the previously described estimators based on a sample of size N. Often we want to test linear hypotheses in the form $R\beta = r$, where R is a $g \times k$ constant matrix with its rank equal to or less than k and r is a $g \times 1$ constant vector. Since b_N converges to β a.e., Rb_N converges to $R\beta$, a.e. [more generally, $f(b_N)$ converges to $f(\beta)$ a.e. for any continuous function $f(\cdot)$]. If $R\beta = r$ is true, Rb_N should be close to r. Hence, testing $R\beta = r$ is based on the difference $Rb_N - R\beta = Rb_N - r$ under $H_0 \colon R\beta = r$. More precisely, the test statistic will be derived from $\sqrt{N}(Rb_N - r)$.

Let $RCR' = H\Lambda H'$, where H is a matrix whose g columns are the orthonormal vectors of RCR' and Λ is the diagonal matrix of the eigenvalues of RCR'. The eigenvalue decomposition removes the dependence among the linear restrictions in R (the rows in R). Define $S \equiv H\Lambda^{-0.5}H'$. Then $S'S = (RCR')^{-1}$. Using the distribution of b_N,

$$\sqrt{N} \cdot R(b_N - \beta) =^d N(0, RCR'), \qquad (7.1)$$

$$\sqrt{N} \cdot SR(b_N - \beta) =^d N(0, I_g), \qquad \text{since } SRCR'S' = I_g, \qquad (7.2)$$

$$N(Rb_N - R\beta)'S'S(Rb_N - R\beta) = N(Rb_N - R\beta)'(RCR')^{-1}(Rb_N - R\beta) =^d \chi_g^2, \qquad (7.3)$$

for (7.3) is a sum of squared uncorrelated $N(0, 1)$ rv. Then, under $H_0 \colon R\beta = r$, the test statistic (7.3) becomes

$$N(Rb_N - r)'(RCR')^{-1}(Rb_N - r) =^d \chi_g^2. \qquad (7.4)$$

The test with (7.4) is called a *Wald test*. Note that, under H_0, $\sqrt{N}(Rb_N - r)$ has the variance RCR', and the matrix $(RCR')^{-1}$ in the middle of (7.4) standardizes the vector $\sqrt{N}(Rb_N - r)$. The idea is similar to choosing the optimal weighting matrix in GMM.

For LSE under homoskedasticity with $E(u^2 \mid x) = \sigma^2$, C is $\sigma^2 E^{-1}(xx') =^{ae} \sigma^2(X'X/N)^{-1}$. Substitute this into (7.4) to get

$$(Rb_N - r)'(R(X'X)^{-1}R')^{-1}(Rb_N - r)/\sigma^2 =^d \chi_g^2. \qquad (7.5)$$

In practice, σ^2 is to be replaced by a consistent estimator such as $(1/N) \sum(y_i - x_i'b_{\text{LSE}})^2$.

If we want to calculate the power of the test, we need to specify H_a. One choice is $H_a \colon R\beta = r + (\delta/\sqrt{N})$, where δ is a $g \times 1$ vector. Since H_a converges to H_0 as $N \to \infty$, H_a is called a *local alternative* or *Pitman drift*. Substituting $H_a \colon R\beta = r + \delta/\sqrt{N}$ into (7.2),

$$\sqrt{N} \cdot S(Rb_N - r) - S\delta =^d N(0, I_g) \Rightarrow \sqrt{N} \cdot S(Rb_N - r) =^d N(S\delta, I_g). \quad (7.6)$$

Hence, the test statistic (7.4) has a noncentral χ_g^2 with the noncentrality parameter (NCP) $\delta'S'S\delta = \delta'(RCR')^{-1}\delta$, which is non-negative because

$(RCR')^{-1}$ is p.s.d. The larger the NCP, the higher the power is, for it becomes easier to tell H_0 from H_a; imagine H_0 and H_a as being, respectively, "centered" at 0 and the NCP.

Consider two estimators with the variance matrices C_0 and C_1, where $C_0 \le C_1$. Then $RC_1R' - RC_0R' = R(C_1 - C_0)R'$ is p.s.d., which implies that $(RC_1R')^{-1} \le (RC_0R')^{-1}$; that is, $(RC_0R')^{-1} - (RC_1R')^{-1}$ is p.s.d. Hence, for any δ,

$$\delta'\{(RC_0R')^{-1} - (RC_1R')^{-1}\}\delta = \delta'(RC_0R')^{-1}\delta - \delta'(RC_1R')^{-1}\delta \ge 0. \quad (7.7)$$

So the more efficient estimator will have the higher power.

If we had used H_a, which is not convergent to H_0, as $N \to \infty$, the NCP would have N in it, which means that the NCP will go to ∞. Then the power of the test (7.4) will be 1 no matter what H_a looks like. So the local alternative is good for a meaningful comparison of power as illustrated in (7.7).

2.8   Testing Instrument Legitimacy

Suppose that we have a $k_1 \times 1$ legitimate instrument vector z_1 for the model $y = x'\beta + u$. If a new instrument vector w is available but we are not sure of its legitimacy [i.e., $\text{COV}(w, x) \ne 0$ and $\text{COV}(w, u) = 0$], then it is possible to test if w is a good instrument vector. Using w in addition to x as instruments does not make IVE inconsistent even if $\text{COV}(w, x) = 0$, while using w and x causes a bias if $\text{COV}(w, u) \ne 0$. So the more important question is whether $\text{COV}(w, u)$ is zero or not, which this section addresses.

Let Z_1 denote the matrix for z_1, and let Z_2 denote the matrix for z_1 and w combined. Also let $b_{1,\text{ive}}$ and $b_{2,\text{ive}}$ denote the IVE using Z_1 and Z_2, respectively. Then, for $t = 1, 2$,

$$b_{t,\text{ive}} \equiv \{X'Z_t(Z_t'Z_t)^{-1}Z_t'X\}^{-1} \cdot X'Z_t(Z_t'Z_t)^{-1}Z_t'Y. \quad (8.1)$$

Also GMM b_t, $t = 1, 2$, are

$$b_t \equiv \{X'Z_t(Z_t'D_tZ_t)^{-1}Z_t'X\}^{-1} \cdot X'Z_t(Z_t'D_tZ_t)^{-1}Z_t'Y \quad (8.2)$$

where D_t is $\text{diag}(\hat{U}_t)$ with $\hat{U}_t \equiv Y - X \cdot b_{t,\text{ive}}$.

Using the vector notation, instead of the matrix, we get

$$b_{t,\text{ive}} - \beta =^p \left\{ \sum_i x_i z_{it}'/N \cdot \left(\sum_i z_{it} z_{it}'/N \right)^{-1} \cdot \sum_i z_{it} x_i'/N \right\}^{-1}$$

$$\cdot \sum_i x_i z_{it}'/N \cdot \left(\sum_i z_{it} z_{it}'/N \right)^{-1} \cdot \sum_i z_{it} u_i/N; \quad (8.3)$$

$$b_t - \beta =^p \left\{ \sum_i x_i z'_{it}/N \cdot \left(\sum_i z_{it} z'_{it} u_i^2/N \right)^{-1} \cdot \sum_i z_{it} x'_i/N \right\}^{-1}$$

$$\cdot \sum_i x_i z'_{it}/N \cdot \left(\sum_i z_{it} z'_{it} u_i^2/N \right)^{-1} \cdot \sum_i z_{it} u_i/N. \tag{8.4}$$

Under $H_0 : E(wu) = 0$, both b_1 and b_2 are consistent and $b_1 =^p b_2$. Thus, analogously to the Wald test, which compares the distance between Rb_N and r, we can consider testing the H_0 by looking at the distance between b_1 and b_2. Following Hausman's (1978) test idea, the variance of $b_1 - b_2$ can be shown to be $V[b_1 - \beta] - V[b_2 - \beta]$ which is p.d. under H_0, since b_2 is more efficient than b_1 under H_0. In a later chapter, we will explain Hausman's (1978) test in more detail. The test statistic for $H_0 : E(wu) = 0$ is

$$(b_1 - b_2)' \cdot \{V[b_1 - \beta] - V[b_2 - \beta]\}^{-1} \cdot (b_1 - b_2) \Rightarrow \chi_k^2. \tag{8.5}$$

Sometimes in practice, the inverted matrix in the middle of (8.5) fails to be p.d. Then the following somewhat lengthy step is necessary to rectify the problem. Define A_t and ψ_{it}, $t = 1, 2$, such that

$$b_t - \beta \equiv (1/N) \sum_i A_t z_{it} u_i \equiv (1/N) \sum_i \psi_{it}; \tag{8.6}$$

that is, A_t is the $k \times k_t$ matrix in front of $\sum_i z_{it} u_i/N$ in (8.4). Then

$$V[b_1 - b_2] =^p (1/N) \sum_i (\psi_{i1} - \psi_{i2})(\psi_{i1} - \psi_{i2})'/N. \tag{8.7}$$

2.9   Prediction and Extremum Estimator (OPT)

Consider predicting y with a function $r(x)$ of x. Under the quadratic loss function of misprediction, we seek to find $r(x)$ minimizing

$$E\{y - r(x)\}^2 = E_x E_{y|x}\{y - r(x)\}^2. \tag{9.1}$$

Differentiating $E_{y|x}\{y - r(x)\}^2$ wrt $r(x)$ and setting the derivative to zero,

$$(-2) \cdot E_{y|x}\{y - r(x)\} = 0 \Leftrightarrow E_{y|x}(y) = E(y \mid x) = r(x). \tag{9.2}$$

Since this holds for any x, we minimize (9.1) by setting $r(x) = E(y \mid x)$.

In (9.1), we used a quadratic loss function where misprediction is penalized by the squared distance. In general, we can think of minimizing $E|y - r(x)|^p$, $p > 0$. There is no a priori reason to set $p = 2$ except for analytic convenience. Perhaps it is more intuitive to set $p = 1$ and minimize

$$E|y - r(x)| = E_x E_{y|x}|y - r(x)| = E_x \left[\int_{-\infty}^{r(x)} \{r(x) - y\} f(y \mid x) dy \right.$$

$$+ \int_{r(x)}^{\infty} \{y - r(x)\} f(y \mid x) dy \Big].$$

(9.3)

Differentiate the term in $[\cdot]$ wrt $r(x)$ and use the Leibniz's rule to get

$$\int_{-\infty}^{r(x)} f(y \mid x) dy - \int_{r(x)}^{\infty} f(y \mid x) dy = 0;$$

$r(x) = \text{Med}(y \mid x)$ satisfies this to minimize (9.3) if y is continuous.

With $r(x) = x'\beta$, the estimator obtained by minimizing the sample version of (9.3)

$$(1/N) \sum_{i} |y_i - x_i'b|$$

(9.4)

is called the *least absolute deviation estimator* (LAD). LAD has as long a history as LSE has; however, because the estimator cannot be gotten in a closed form, it has not been popular. In fact, as will be clear later, LSE and other GMM estimators in the linear model are rare cases where we have an explicit form for the estimators. With advances in computer technology, LAD can be easily calculated these days. In LAD, $x'\beta$ is the conditional median. In this regard, we use the word "regression" for a location measure in the conditional distribution of $y \mid x$, not just for $E(y \mid x)$. See Koenker and Bassett (1978) and Bloomfield and Steiger (1983) for more on the *median regression*.

Generalizing the median regression, suppose we use an asymmetric loss function that penalizes the positive and negative errors differently: $\alpha|y - r(x)|$ if $y - r(x) > 0$ and $(1 - \alpha)|y - r(x)|$ if $y - r(x) < 0$ where $0 < \alpha < 1$. In this case, the expected loss becomes

$$E\{\alpha \cdot (y - r(x)) \cdot 1[y > r(x)] + (1 - \alpha) \cdot (r(x) - y) \cdot 1[y < r(x)]\}$$

$$= E_x \left[\alpha \int_{r(x)}^{\infty} \{y - r(x)\} f(y \mid x) dy + (1 - \alpha) \int_{-\infty}^{r(x)} \{r(x) - y\} f(y \mid x) dy \right],$$

(9.5)

where $1[A]$ is the *indicator function*, taking 1 if A holds and 0 otherwise.

Differentiating the term in $[\cdot]$ in (9.5) wrt $r(x)$, we get

$$-\alpha \int_{r(x)}^{\infty} f(y \mid x) dy + (1 - \alpha) \int_{-\infty}^{r(x)} f(y \mid x) dy = 0.$$

(9.6)

Now choose

$$r(x) = \alpha\text{th quantile} \equiv \inf\{y^* : F(y^* \mid x) \geq \alpha\};$$

if $f(y \mid x)$ is continuous, then the αth quantile is defined by $F(y^* \mid x) = \alpha$. Assuming the continuity of $f(y \mid x)$, $\int_{-\infty}^{r(x)} f(y \mid x) = \alpha$. Then (9.6) becomes

$-\alpha(1 - \alpha) + (1 - \alpha)\alpha = 0$. Therefore, under the asymmetric loss function in (9.5), the αth quantile minimizes the expected loss (9.5). Note that if $\alpha = 1/2$, then we have the median regression. The sample version for (9.5) with $r(x) = x'b$ is

$$(1/N) \sum_i \{\alpha(y_i - x_i'b) \cdot 1[y_i > x_i'b] + (1 - \alpha)(x_i'b - y_i) \cdot 1[y_i < x_i'b]\} \quad (9.7)$$

$$= (1/N) \sum_i (y_i - x_i'b) \cdot (\alpha - 1[y_i - x_i'b < 0]). \quad (9.8)$$

The function $a \cdot (\alpha - 1[a < 0])$ in (9.8) is called the "check function." An application of this αth *quantile regression* appears in Buchinsky (1994).

Suppose the loss function is $c \cdot 1[|y - r(x)| > \delta]$ where δ is a positive constant. That is, if the prediction falls within $\pm\delta$ of y, then there is no loss. Otherwise, the loss is a constant c. Then the expected loss is

$$E\{1[|y - r(x)| > \delta]\} = 1 - E\{1[|y - r(x)| \leq \delta]\}$$

$$= 1 - E_x[F_{y|x}\{r(x) + \delta\} - F_{y|x}\{r(x) - \delta\}]. \quad (9.9)$$

This is minimized by choosing $r(x)$ such that the interval $[r(x) - \delta, r(x) + \delta]$ captures the most probability mass under $f_{y|x}$. Manski (1991) calls it "δ-mode." If $f_{y|x}$ is unimodal and δ is small, then $r(x)$ is approximately equal to the mode of $f_{y|x}$. Lee (1989)'s *mode regression* estimator is obtained by maximizing the following sample version:

$$(1/N) \sum_i 1[|y_i - x_i b| \leq \delta]. \quad (9.10)$$

If the reader thinks that the distinction between the various measures of central tendency in a distribution is trivial, consider the following litigation (Freedman, 1985). In the early 1980s, a number of lawsuits were filed by railroad companies against several state taxing authorities in the United States. The companies argued that their property tax rate should be equalized to the median of the other property tax rates, while the state authorities argued that the mean is more appropriate. The problem was that the probability distribution of the property tax rate has a long right tail. As a result, the median was smaller than the mean, and the difference had an implication of millions of dollars. Eventually, the states won the case, not because mean is a better measure than median, but because the Courts concluded that the word "average" in the law (so-called "4-R Act") meant the mean, not the median.

In general, an estimator can be defined implicitly by

$$b_N \equiv \text{argmin } Q_N(z, b) = (1/N) \sum_i q(z_i, b), \quad (9.11)$$

where argmin stands for "argument minimizing," $Q_N(z, b)$ is the sample analog of $Q(z, b) \equiv E\{q(z, b)\}$, and z is $(x', y)'$. An estimator defined as in (9.11) is called an *extremum estimator*. Maximum likelihood estimator with the linear model is a prime example:

$$b_{\text{MLE}} \equiv \operatorname{argmax}(1/N) \sum_i \ln\{f_u(y_i - x_i'b)\}, \tag{9.12}$$

where f_u is the density function of u in $y = x'\beta + u$.

There are many other estimators defined in this fashion, e.g., (9.4) and (9.10). Since they do not have a closed-form expression as LSE and IVE do, studying their asymptotic properties is more demanding. Also, while the maximand for MLE in (9.12) is differentiable, (9.4) and (9.10) are not. The necessary tools for extremum estimators with differentiable maximands, which are the ones we will face most of the time, are to be studied in the next chapter.

3

Extremum Estimators and Method-of-Moments Estimators

3.1 Introduction

LSE and IVE are rare cases where the estimators are written in closed forms. Often estimators are defined implicitly by

$$b_N \equiv \text{argmax}_{b \in B} Q_N(b), \qquad (1.1)$$

where B is a parameter space and

$$Q_N(b) \equiv (1/N) \sum_i q(x_i, y_i, b) = (1/N) \sum_i q(z_i, b) = (1/N) \sum_i q(b); \quad (1.2)$$

$z_i \equiv (x_i', y_i)'$ and we often omit z in $q(z, b)$. In LSE, $q(z, b) = -(y - x'b)^2$. If the regression function is nonlinear in β, say $\rho(x, \beta)$, then $q(z, b) = -\{y - \rho(x, b)\}^2$, which renders a nonlinear LSE. An estimator defined by (1.2) is called an *extremum estimator* (OPT for "optimization"). MLE is also an extremum estimator, where $q(z_i, b)$ is the log-likelihood function for z_i evaluated at b; "likelihood function" is either a density or a distribution function depending on whether the random variables involved are continuous or discrete [both are a "derivative of a probability measure" with respect to (wrt) a measure].

For any estimator b_N, usually four questions arise. The first is identification (ID): for which population parameter is b_N designed? The second is consistency: does b_N converge to the parameter say β? The third is the asymptotic distribution of $\sqrt{N}(b_N - \beta)$; if this is normally distributed (as it almost always is), what is the variance? The fourth is estimation of the asymptotic variance, which typically involves β and other unknown components: how do we estimate the variance then? We discuss these one by one.

First, for the OPT b_N of (1.1), its parameter β is identified if β satisfies uniquely

$$\beta \equiv \operatorname{argmax}_{b \in B} Q(b) = \operatorname{argmax} Eq(z, b), \tag{1.3}$$

where the uniqueness holds by restricting q, z or the parameter space B; more on these restrictions will be discussed later in Section 5.

Second, with ID holding, the following three conditions together imply the consistency ($b_N =^p \beta$ or $b_N =^{as} \beta$): the compactness of B, the continuity of $Eq(b)$ in b, and a uniform law of large numbers

$$\sup_{b \in B} \left| (1/N) \sum_i q(z_i, b) - Eq(z_i, b) \right| =^p 0. \tag{1.4}$$

To see this, first observe that

$$Eq(b_N) =^P (1/N) \sum_i q(b_N) \geq (1/N) \sum_i q(\beta) =^p Eq(\beta)$$

$$\Rightarrow Eq(b_N) \geq Eq(\beta), \tag{1.5}$$

where the first equality is due to (1.4), and the second inequality is due to the definition of b_N. But (1.3) implies $Eq(\beta) \geq Eq(b_N)$. Combining this with (1.5) leads to $Eq(b_N) =^p Eq(\beta)$. For any open neighborhood U of β, $Eq(b)$ attains a maximum on U^c at a point β^* due to the continuity of $Eq(b)$ and the compactness of U^c. But since ID implies $Eq(\beta^*) < Eq(\beta)$, b_N cannot stay out of U while satisfying $Eq(b_N) =^p Eq(\beta)$ however small U may be. This means $b_N =^p \beta$; if (1.4) holds a.s., then $b_N =^{as} \beta$. Note that, for the first equality of (1.5), a usual (pointwise) LLN under iid is not sufficient, for $q(z_i, b_N)$, $i = 1 \ldots N$, are dependent on one another through b_N. But when a pointwise LLN holds, usually the uniform LLN holds as well; see Andrews (1987a) and Pötcher and Prucha (1989).

Third, the asymptotic distribution $\sqrt{N}(b_N - \beta)$ is almost always $N(0, C)$ for some matrix C. Showing this and estimating C (which is the fourth question) will be discussed in Sections 2–4; the major portion of this chapter is devoted to deriving the asymptotic variance.

One important generalization of (1.1) is a two-stage OPT

$$b_N \equiv \operatorname{argmax}_{b \in B} Q_N(b, a_N), \tag{1.6}$$

where a_N is a first-stage estimator for a *nuisance parameter* α, which is not of interest but should be estimated before β. For instance, in the generalized LSE (GLS), the variance matrix of the error terms is not interesting per se. But if we want to get GLS, then we need to estimate the variance matrix first. For OPT in (1.6), its asymptotic distribution may be affected by $a_N - \alpha$. Finding when the first stage affects the second stage, and if it does, then in which way, are interesting questions.

Usually $q(b)$ in (1.2) is differentiable wrt b. Then b_N satisfies

$$(1/N) \sum_i q_b(z_i, b_N) = 0, \tag{1.7}$$

where $q_b(z_i, b_N) \equiv \partial q(z_i, b)/\partial b\big|_{b=b_N}$. Here, b_N can be regarded as a MME with the population moment condition $E\{q_b(\beta)\} = 0$. If there is a nuisance parameter α with an estimator a_N, then instead of (1.7), we get

$$(1/N) \sum_i q_b(z_i, b_N, a_N) = 0, \tag{1.8}$$

which yields a two-stage MME with a nuisance parameter. Although MME seems to include OPT as a subclass, there is a difference: more than one b_N may satisfy (1.7) even when b_N in (1.1) is unique. Still, the MME interpretation is helpful, particularly when we derive the asymptotic distribution of the OPT in (1.1) and (1.6).

The rest of this chapter is organized as follows. In Section 2, asymptotic distribution of OPT is examined. In Section 3, two-stage estimators with a nuisance parameter are studied. In Section 4, method-of-moments tests are introduced that are related to MME with a nuisance parameter. In Section 5, ID is examined in detail. See Newey and McFadden (1994) for a more rigorous account of the topics in this chapter.

3.2   Asymptotic Distribution of Extremum Estimators

Both $1/n$ and $1/n^2$ converge to 0, but the speed of convergence is different. Weak or strong consistency tells us that $b_N \to \beta$ as $N \to \infty$. Then, a natural question to arise is "at which rate?" Usually b_N is "$\sqrt{N}$-consistent," which means that $\sqrt{N}(b_N - \beta)$ converges to a Op(1) random variable with a nondegenerate distribution.

Consider the following OPT with the linear regression function:

$$b_N \equiv \text{argmax}_{b \in B}(1/N) \sum_i q(y_i - x_i'b), \tag{2.1}$$

where $q(u)$ is twice continuously differentiable with the derivatives q_u and q_{uu}. The first-order condition of maximization is

$$(1/N) \sum_i q_u(y_i - x_i'b_N)(-x_i) = 0. \tag{2.2}$$

Applying the mean value theorem to this around β, we get

$$0 = -(1/N) \sum_i q_u(y_i - x_i'\beta)x_i + (1/N) \sum_i q_{uu}(y_i - x_i'b_N^*) \cdot x_i x_i'(b_N - \beta).$$

(2.3)

where $b_N^* \in (b_N, \beta)$; note that the mean value theorem applies to each component of (2.3) separately, which means that each component of (2.3) may need a different b_N^* although this is not explicit in (2.3). Multiply both sides by $\sqrt{N}$ to get

$$(1/\sqrt{N}) \sum_i q_u(y_i - x_i'\beta)x_i = (1/N) \sum_i q_{uu}(y_i - x_i'b_N^*) \cdot x_i x_i' \cdot \sqrt{N}(b_N - \beta).$$

(2.4)

Invert the second-order matrix to solve this for $\sqrt{N}(b_N - \beta)$:

$$\sqrt{N}(b_N - \beta) = \left\{ (1/N) \sum_i q_{uu}(y_i - x_i'b_N^*)x_i x_i' \right\}^{-1}$$

$$\cdot (1/\sqrt{N}) \sum_i q_u(y_i - x_i'\beta)x_i.$$

(2.5)

This equation is the key for the asymptotic distribution of the OPT (2.1). It can be shown that, using (1.4) and $b_N^* =^P \beta$ ($\Leftarrow b_N = \beta$),

$$(1/N) \sum_i q_{uu}(y_i - x_i'b_N^*) \cdot x_i x_i' =^P E\{q_{uu}(y - x'\beta)xx'\}.$$

(2.6)

Then (2.5) becomes

$$\sqrt{N}(b_N - \beta) =^p E^{-1}\{q_{uu}(y - x'\beta)xx'\} \cdot (1/\sqrt{N}) \sum_i q_u(y_i - x_i'\beta)x_i.$$

(2.7)

Apply a central limit theorem (CLT) to $(1/\sqrt{N}) \sum_i (\cdot)$ to obtain

$$\sqrt{N}(b_N - \beta) =^d N(0, E^{-1}\{q_{uu}(u)xx'\} \cdot E\{q_u(u)^2 xx'\}$$

$$\cdot E^{-1}\{q_{uu}(u)xx'\}).$$

(2.8)

It is straightforward to derive the distribution of b_{LSE} as a special case of (2.8) with $q(y - x'b) = -(y - x'b)^2$.

In (2.7) and (2.8), the presence of x and xx' is due to the linear regression function. For a generic OPT for $(1/N) \sum_i q(z_i, b)$, we have

$$\sqrt{N}(b_N - \beta) =^d N(0, E^{-1}\{q_{bb}(\beta)\} \cdot E\{q_b(\beta)q_b(\beta)'\} \cdot E^{-1}\{q_{bb}(\beta)'\}) \quad (2.9)$$

$$= N(0, [E\{q_{bb}(\beta)'\} \cdot E^{-1}\{q_b(\beta)q_b(\beta)'\} \cdot E\{q_{bb}(\beta)\}]^{-1}). \quad (2.10)$$

Note that the form of the variance matrix is the "outer-product of the first-order vector surrounded by the inverted second-order matrix." This result will be repeatedly used later.

In most cases, analogously to (2.6), we have

$$(1/N) \sum_i g(z_i, b_N) =^p (1/N) \sum_i g(z_i, \beta) =^p E\{g(z, \beta)\}, \qquad (2.11)$$

where $g(z, b)$ is a "smooth" function of z and b. This makes it possible to estimate a variance matrix [say, $E\{g(z, \beta)\}$ by its sample version with β replaced by b_N $[(1/N) \sum_i g(z_i, b_N)]$. One caution is that, although (2.11) holds almost always,

$$(1/\sqrt{N}) \sum_i g(z_i, b_N) \neq^p (1/\sqrt{N}) \sum_i g(z_i, \beta). \qquad (2.12)$$

The difference between (2.11) and (2.12) is the norming factors N versus $\sqrt{N}$. It is important to be aware of this difference.

There exists an interesting interpretation of the variance matrix in (2.10). Let $f(z, \beta)$ denote the likelihood function of z. Observe that, from the first-order condition $E\{q_b(\beta)\} = 0$ for the OPT for (2.9), we get

$$\int [\partial\{q_b(z, \beta) \cdot f(z, \beta)\}/\partial\beta'] \cdot dz = \int [\{\partial q_b(z, \beta)/\partial\beta'\} \cdot f(z, \beta)] \cdot dz$$

$$+ \int [q_b(z, \beta) \cdot \{\partial f(z, \beta)/\partial\beta'\}] \cdot dz$$

$$= \int q_{bb}(z, \beta) \cdot f(z, \beta) dz + \int q_b(z, \beta)$$

$$\cdot [\{\partial f(z, \beta)/\partial\beta'\}/f(z, \beta)] \cdot f(z, \beta) \cdot dz. \qquad (2.13)$$

Take $\partial(\cdot)/\partial\beta$ out of the integral in $\int [\partial\{q_b(z, \beta) \cdot f(z, \beta)\}/\partial\beta'] \cdot dz$ to get

$$0 = \partial E(q_b)/\partial\beta = E(q_{bb}) + E(q_b s_b'), \qquad (2.14)$$

where $s_b \equiv f_b(z, \beta)/f(z, \beta)$, the so-called *score function*. Note that we need a regularity condition to interchange the order of $\partial(\cdot)/\partial\beta$ and $\int$ in (2.13); for instance, $|q_b(z, \beta) f(z, \beta)| < g(z)$ with $Eg(z) < \infty$. From (2.14),

$$E(q_{bb}) = -E(q_b s_b'). \qquad (2.15)$$

Using this, the inverse of the variance matrix in (2.10) becomes

$$E(s_b q_b') \cdot E^{-1}\{q_b q_b'\} \cdot E\{q_b s_b'\} \qquad (2.16)$$

which is the "square" of the projection of s_b on q_b.

With $q(z, \beta) = \ln\{f(z, \beta)\}$ for MLE, (2.15) becomes

$$E[\partial \ln\{f(z, \beta)\}/\partial bb'] = -E(s_b s_b'). \tag{2.17}$$

This equality is often called the *information equality*, while (2.15) is called the *generalized information equality*. With (2.17), (2.16) becomes

$$E(s_b q_b') \cdot E^{-1}\{q_b q_b'\} \cdot E\{q_b s_b'\} = E(s_b s_b'); \tag{2.18}$$

that is,

$$V[\sqrt{N}(b_{\text{MLE}} - \beta)] = E^{-1}\{s_b(z, \beta)s_b(z, \beta)'\}, \tag{2.19}$$

which is the inverse of the expected outer-product of the score function. $E(s_b s_b')$ [or $N \cdot E(s_b s_b')$] is called the *information matrix*. Since

$$E(s_b q_b') \cdot E^{-1}\{q_b q_b'\} \cdot E\{q_b s_b'\} \le E(s_b s_b'), \tag{2.20}$$

OPT is less efficient than MLE in general. But the obvious advantage of an OPT is that there is no need to specify the likelihood function.

3.3 Extremum Estimators with Nuisance Parameters

Imagine a two-stage OPT b_N defined by

$$b_N = \text{argmax} \ (1/N) \sum_i q(b, a_N), \tag{3.1}$$

where a_N is a first-stage estimator consistent for a nuisance parameter α. The two-stage OPT b_N satisfies the first-order condition

$$(1/N) \sum_i q_b(b_N, a_N) = 0. \tag{3.2}$$

Applying the mean value theorem to this around β, analogously to (2.3),

$$(1/\sqrt{N}) \sum_i q_b(\beta, a_N) + (1/N) \sum_i q_{bb}(b_N^*, a_N) \cdot \sqrt{N}(b_N - \beta) = 0 \tag{3.3}$$

$$\Rightarrow \sqrt{N}(b_N - \beta) = \left[-(1/N) \sum_i q_{bb}(b_N^*, a_N) \right]^{-1} (1/\sqrt{N}) \sum_i q_b(\beta, a_N). \tag{3.4}$$

Now expand $(1/\sqrt{N}) \sum_i q_b(\beta, a_N)$ around α to get

$$(1/\sqrt{N}) \sum_i q_b(\beta, a_N) = (1/\sqrt{N}) \sum_i q_b(\beta, \alpha)$$

$$+ (1/N) \sum_i q_{ba}(\beta, a_N^*) \cdot \sqrt{N}(a_N - \alpha). \tag{3.5}$$

Substitute this into (3.4). Using (2.11), replace $(1/N) \sum_i q_{bb}(b_N^*, a_N)$ and $(1/N) \sum_i q_{ba}(\beta, a_N^*)$, respectively, by $E\{q_{bb}(\beta, \alpha)\}$ and $E\{q_{ba}(\beta, \alpha)\}$ to get

$$\sqrt{N}(b_N - \beta) =^p -E^{-1}(q_{bb}) \cdot \left[(1/\sqrt{N}) \sum_i q_b + E(q_{ba}) \cdot \sqrt{N}(a_N - \alpha) \right]. \tag{3.6}$$

The distribution of $\sqrt{N}(b_N - \beta)$ depends on the two terms on the right-hand side (rhs).

Suppose

$$\sqrt{N}(a_N - \alpha) =^p (1/\sqrt{N}) \sum_i \eta_i, \tag{3.7}$$

which implies that $\sqrt{N}(a_N - \alpha) =^d N(0, E(\eta\eta'))$. For instance, if a_N is the LSE for $z_i = w_i'\alpha + \varepsilon_i$, then η_i is $E^{-1}(ww') \cdot w_i \varepsilon_i$ so that the variance matrix becomes $E(\eta\eta')$. The idea here is that if $\sqrt{N}(a_N - \alpha)$ has an asymptotic variance Ω, we can think of a random vector η such that $E(\eta\eta') = \Omega$, which then leads to an expression like (3.7). With η_i, (3.6) becomes

$$\sqrt{N}(b_N - \beta) =^p -E^{-1}(q_{bb}) \cdot (1/\sqrt{N}) \sum_i \{q_b(z_i) + E(q_{ba}) \cdot \eta_i\}. \tag{3.8}$$

In $\{\cdots\}$ on the rhs of (3.8), η_i is the first-stage error, and $E(q_{ba})$ may be called the "link" for the first and second stages. If $E(q_{ba}) = 0$, then there is no first-stage effect on the second.

The term $(1/\sqrt{N}) \sum_i \{q_b(z_i) + (Eq_{ba}) \cdot \eta_i\}$ follows $N(0, C)$ where

$$C = E(q_b q_b') + E(q_b \eta')E(q_{ba}') + E(q_{ba})E(\eta q_b') + E(q_{ba})E(\eta\eta')E(q_{ba}').$$

Hence

$$\sqrt{N}(b_N - \beta) =^d N(0, E^{-1}(q_{bb}) \cdot C \cdot E^{-1}(q_{bb})). \tag{3.9}$$

Although C looks complicated, it can be estimated simply by $C_N \equiv (1/N) \sum_i \delta_i \delta_i'$, where

$$\delta_i \equiv q_b(b_N, a_N) + \left\{ (1/N) \sum_i q_{ba}(b_N, a_N) \right\} \cdot \eta_i(a_N),$$

where $\eta_i(a_N)$ is an estimate for $\eta_i = \eta_i(\alpha)$. For instance, if a_N is the LSE for $z_i = w_i'\alpha + \varepsilon_i$, then $\eta_i(\alpha) = E^{-1}(ww') \cdot w_i \varepsilon_i = E^{-1}(ww') \cdot w_i(z_i - w_i'\alpha)$, and so

$$\eta_i(a_N) = \left\{ (1/N) \sum_i w_i w_i' \right\}^{-1} w_i(z_i - w_i'a_N).$$

In the following, we examine a few examples.

Suppose $\text{COV}(\sqrt{N}(a_N - \alpha), (1/\sqrt{N}) \sum_i q_b) = 0$, which is equivalent to $E(q_b \eta') = 0$. Then we get

$$\sqrt{N}(b_N - \beta) =^d N(0, E^{-1}(q_{bb})\{E(q_b q_b') + E(q_{ba})E(\eta \eta')E(q_{ba}')\}E^{-1}(q_{bb})). \tag{3.10}$$

This case can occur if a_N is estimated by another sample not used for b_N, or a_N is a function of a term that is "orthogonal" to $(1/\sqrt{N}) \sum_i q_b$. If $E(\eta \eta') = 0$, then (3.9) becomes (2.9).

Another special case of (3.9) is $E(q_{ba}) = 0$. Then (3.9) becomes again (2.9). One example is the weighted LSE or the feasible generalized LSE minimizing

$$(1/N) \sum_i \{(y_i - x_i'b)/s_i\}^2, \tag{3.11}$$

where $E(u_i \mid x_i) = 0$ and s_i^2 is an estimator for $\sigma_i^2 \equiv E(u_i^2 \mid x_i)$. Suppose

$$\sigma_i^2 = (\alpha_1 + \alpha_2 x_{ki})^2, \quad s_i^2 = (a_1 + a_2 x_{ki})^2, \quad a_N \equiv (a_1, a_2)'. \tag{3.12}$$

Differentiating (3.11) wrt b, we get

$$(1/N) \sum_i (-2) \cdot x_i(y_i - x_i'b) \cdot s_i^{-2}. \tag{3.13}$$

Differentiate this wrt a_N to get

$$(1/N) \sum_i 4 \cdot x_i(y_i - x_i'b) \cdot s_i^{-3} \cdot \partial s_i/\partial a_N. \tag{3.14}$$

Evaluating this at β and α, (3.13) becomes $(1/N) \sum_i u_i g(x_i)$ for a function $g(x_i)$, for s_i and $\partial s_i/\partial a_N$ are functions of x_i. But

$$(1/N) \sum_i u_i g(x_i) =^p E\{u g(x)\} = E\{g(x) \cdot E(u \mid x)\} = 0. \tag{3.15}$$

Therefore $E(q_{ba}) = 0$. This explains why the feasible generalized LSE has the same asymptotic distribution as the (infeasible) generalized LSE.

As another example of $E(q_{ba}) = 0$, consider the following two-stage LSE (2SLSE) model:

$$y = x_1'\beta_1 + x_2'\beta_2 + u \equiv x'\beta + u, \qquad E(u \mid x_2) = 0, \tag{3.16}$$

where x_j is a $k_j \times 1$ vector, $j = 1, 2$, and $E(x_1 u) \neq 0$. Here (3.16) can be regarded as the first equation of a simultaneous equation system, and x_1 is the endogenous regressors (variables) included in the first equation. Let a $s \times 1$ ($s \geq k_1 + k_2$) vector z denote the exogenous variables in the system $[E(zu) = 0]$; z is an instrument for the endogenous regressor x_1. Then a_N is the LSE of x_1 on z, and α is $E^{-1}(zz')E(zx_1')$, a $s \times k_1$ matrix of the projection coefficient of x_1 on z. (3.2) for 2SLSE is

$$(1/N) \sum_i (y_i - x_i'b_N) \cdot ((z_i'a_N), x_{i2})' = 0. \tag{3.17}$$

Note that the dimension of $(z_i'a_N)'$ is the same as that of x_1 ($k_1 \times 1$), and the dimension of $((z_i'a_N)', x_{i2}')'$ is $(k_1 + k_2) \times 1$. In the moment condition (3.17), the instrument $z_i'\alpha$ for x_{i1} is estimated by $z_i'a_N$.

Stack up the $s \times k_1$ matrix α as a $(s \times k_1) \times 1$ vector α^*; see the next paragraph for an example. Differentiate (3.17) wrt a_N^*, the version of a_N stacked analogously to α^*. Then we get a $(k_1 + k_2) \times (s \cdot k_1)$ matrix $(1/N) \sum_i q_{ba}$ whose typical element is either 0 or

$$(1/N) \sum_i (y_i - x_i'b_N)z_{ij} = 0, \qquad j = 1, \ldots, s. \tag{3.18}$$

Since (3.18) $=^p E\{(y - x'\beta)z_j\} = E(uz_j) = 0$, $E(q_{ba})$ is zero in 2SLSE, meaning no first-stage estimation effect on the second stage. This result shows that estimating instruments does not affect the second stage.

To be specific about (3.18), let $s = 3$ and $k_1 = 2$. Then α is a 3×2 matrix and α^* can be set as

$$\alpha^* = (\alpha_{11}, \alpha_{21}, \alpha_{31}, \alpha_{12}, \alpha_{22}, \alpha_{32})'. \tag{3.19}$$

Differentiating $((z_i'a_N)', x_{i2}')'$ wrt a_N^*, $\partial(z_i'a_N)/\partial a_N^*$ is nonzero, while $\partial x_2/\partial a_N^* = 0$. Observe that (omitting N in a_N)

$$z_i'a = (z_{i1}a_{11} + z_{i2}a_{21} + z_{i3}a_{31}, \; z_{i1}a_{12} + z_{i2}a_{22} + z_{i3}a_{32}).$$

Differentiate this wrt $a^* = (a_{11}, a_{21}, a_{31}, a_{12}, a_{22}, a_{32})'$ to get

$$\begin{bmatrix} z_{i1} & z_{i2} & z_{i3} & 0 & 0 & 0 \\ 0 & 0 & 0 & z_{i1} & z_{i2} & z_{i3} \end{bmatrix}.$$

Attaching a $k_2 \times 6$ zero matrix at the bottom for $\partial x_2/\partial a_N^* = 0$, we get the desired $(k_1 + k_2) \times (s \cdot k_1) = (2 + k_2) \times 6$ matrix.

As an example of $E(q_b\eta') \neq 0$ and $E(q_{ba}) \neq 0$, consider

$$y = x'\beta + u, \tag{3.20}$$

where the kth variable x_k is not observable. Suppose $x_k = E(w \mid z) = z'\alpha$, where z is a $g \times 1$ vector with $E(zu) = 0$. Then x_{ik} is consistently estimated by

$$\hat{x}_{ik} = z_i'a_N, \tag{3.21}$$

where a_N is the LSE of w on z. Let

$$\hat{x} \equiv (x_1, \ldots, x_{k-1}, \hat{x}_k)'.$$

The issue here is the effect of using $\hat{x}$ instead of x in the LSE b_N of y on $\hat{x}$ to estimate β. It is easy to prove $b_N =^p \beta$; the first-stage error $a_N - \alpha$ matters only for the variance of b_N.

The first-order condition of the LSE, (3.2) in this example, is

$$(1/N) \sum_i \hat{x}_i(y_i - \hat{x}_i' b_N) = 0. \tag{3.22}$$

Differentiating this wrt a_N to get $(1/N) \sum_i q_{ba}$,

$$(1/N) \sum_i q_{ba} = \left[\begin{array}{c} 0_{(k-1) \times g} \\ (1/N) \sum_i (y_i - \hat{x}_i' b_N) z_i' \end{array} \right] - b_k \cdot (1/N) \sum_i \hat{x}_i z_i' \tag{3.23}$$

$$=^p -\beta_k \cdot E(xz') \neq 0.$$

Hence the first-stage error is felt in the second stage. This shows that *estimating explanatory variables affects the second-stage variance, while estimating instruments does not as in 2SLSE.*

One caution is that estimating explanatory variables is not the same as the "errors-in-variable" problem where the parameters cannot even be consistently estimated. In the errors-in-variable problem, x_k is observed as $x_k + \varepsilon$, where ε does not converge to 0, while $\hat{x}_k$ of (3.21) can be written as $x_k + v$, where v is an $o_p(1)$ error. See Pagan and Ullah (1988) for the same point made for the typical erroneous practice of using a risk term as a regressor in macrofinance literature.

3.4 Method-of-Moments Tests

Suppose that we consider a model $y = x'\alpha + u$ with a suspicion that w may be omitted in the model. One way to test the possible omission is to see if $E\{(y - x'\alpha)w\} = 0$. If w is indeed omitted, $y - x'\alpha$ should be correlated with w, resulting in $E\{(y - x'\alpha)w\} \neq 0$. More generally, suppose that α is supposed to satisfy a moment condition $E\{m(z, \alpha)\} = 0$, which is implied by the model specification but not used in getting a_N. Then we can test the validity of the model specification by checking if

$$(1/\sqrt{N}) \sum_{i=1}^{N} m(z_i, a_N) \quad \text{is centered at zero,} \tag{4.1}$$

because (4.1) will have $E\{m(z, \alpha)\}$ as its mean. Testing model specifications using moment conditions is called a *method-of-moments test* [MMT, Newey (1985a), Tauchen (1985), and Pagan and Vella (1989)].

As method-of-moments estimators include many known estimators as special cases, MMT includes many known tests as special cases. In this section, we examine MMT where deriving the asymptotic distribution of the test statistic (4.1) is a main task. Since the form of (4.1) is almost the same as (1.8) except that there is no second-stage estimator b_N in (4.1) and

(4.1) has $(1/\sqrt{N})$ instead of $(1/N)$, the technique of the previous section can be applied with a simple modification. Namely, if $\sqrt{N}(a_N - \alpha) =^p (1/\sqrt{N}) \sum_i \eta_i$ holds, then

$$(1/\sqrt{N}) \sum_i m(z_i, a_N) =^p (1/\sqrt{N}) \sum_i \{m(z_i, \alpha) + E(m_a) \cdot \eta_i\} =^d N(0, C),$$
(4.2)

where

$$C = E(mm') + E(m\eta')E(m_a') + E(m_a)E(\eta m') + E(m_a)E(\eta\eta')E(m_a');$$
(4.3)

m and m_a are evaluated at α. This is essentially the same as (3.8) and (3.9) if we replace q_b, q_{bb}, and q_{ba} there, respectively, by m, $-I$ and m_a. C can be estimated by

$$C_N \equiv (1/N) \sum_i \delta_i \delta_i', \quad \delta_i \equiv m(z_i, a_N) + \left\{ (1/N) \sum_i m_a(z_i, a_N) \right\} \cdot \eta_i(a_N);$$
(4.4)

recall (3.9) and the following discussion there. In the rest of this section, we examine a couple of examples for MMT; we may omit either z or a in $m(z, a)$ and $m_a(z, a)$.

Recall the MMT for H_0: w is not omitted in $y = x'\alpha + u$. Let x be a $p \times 1$ vector and w be a $k \times 1$ vector. Assume that a_N is the LSE. Then, a test statistic is

$$(1/\sqrt{N}) \sum_i w_i(y_i - x_i' a_N) = (1/\sqrt{N}) \sum_i m(z_i, a_N), \qquad (4.5)$$

where $z \equiv (y, x', w')'$. Then $m_a = -wx'$ which is a $k \times p$ matrix. Since a_N is the LSE of y on x, we have

$$\sqrt{N}(a_N - \alpha) =^p (1/\sqrt{N}) \sum_i E^{-1}(xx') \cdot x_i u_i = (1/\sqrt{N}) \sum_i \eta_i. \qquad (4.6)$$

Thus (4.2) becomes

$$(1/\sqrt{N}) \sum_i \{w_i u_i - E(wx') \cdot E^{-1}(xx') \cdot x_i u_i\}. \qquad (4.7)$$

In estimating the variance matrix, u_i can be replaced by $y_i - x_i' a_N$.

In addition to the preceding MMT, another (easier) way to test omission of variables is a so-called *artificial regression*; see MacKinnon (1992) for a survey. With $H_0: y = x'\alpha + u$, suppose there is a reason to believe that w may be a relevant variable for y. Then we may consider an alternative $y = x'\alpha + w'\gamma + u$. More generally, we may set up

$$H_a: y = x'\alpha + \eta \cdot g(w) + u, \qquad (4.8)$$

which nests H_0 with $\eta = 0$, where $g(w)$ is a known function of w. Here, α and η can be easily estimated and tested with the LSE of y on x and $g(w)$. By employing a sufficiently general $g(w)$, we can detect departures from H_0 into various directions. If η is significantly different from 0, then the model $y = x'\alpha + u$ must be wrong. As a matter of fact, we can try almost anything in the place of $g(w)$. In this sense, $g(w)$ is artificial: we do not necessarily think that the model in H_a is true, but so long as $g(w)$ can detect a misspecification, using $g(w)$ is justified. This explains the name "artificial regression."

Consider the linear model $y = x'\alpha + u$, where u has the density function f_u. Suppose that we assumed the symmetry of f_u but estimated α by LSE a_N, which does not use the symmetry assumption. A test of symmetry can be done for $H_0: E(u^3) = 0$, with

$$(1/\sqrt{N}) \sum_i (y_i - x_i' a_N)^3 \equiv (1/\sqrt{N}) \sum_i r_i^3, \tag{4.9}$$

since the symmetry implies $E(u^3) = 0$. Note that we cannot test $E(u) = 0$, which is also implied by the symmetry, because we used this condition to obtain the LSE. Rejecting $E(u^3) = 0$ negates symmetry, but accepting $E(u^3) = 0$ does not necessarily imply symmetry. Observe

$$m_a(z_i, a_N) = -3r_i^2 x_i \Rightarrow m_a(z_i, \alpha) = -3u_i^2 x_i. \tag{4.10}$$

Thus, using (4.6), (4.2) becomes

$$(1/\sqrt{N}) \sum_i \{u_i^3 - 3 \cdot E(u^2 x') \cdot E^{-1}(xx') \cdot x_i u_i\}. \tag{4.11}$$

Note that we cannot estimate the variance matrix by $(1/N) \sum_i r_i^6$, for

$$(1/N) \sum_i r_i^6 =^p (1/N) \sum_i u_i^6 =^p E(u^6), \tag{4.12}$$

which ignores the second term in (4.11).

Besides the above omitted variable and symmetry tests, there are other examples that can be thought of easily. For instance, if we suspect that the error terms may be correlated (in time series data), we may test if $E(u_i u_{i-1}) = 0$; the appropriate test statistic is $(1/\sqrt{N}) \sum_{i=2}^N r_i r_{i-1}$. If we want to test for homoskedasticity, then we may examine if $E\{x(u^2 - \sigma^2)\} = 0$, which holds if if $E(u \mid x) = \sigma^2$, a constant. The test statistic is $(1/\sqrt{N}) \sum_i x_i (r_i^2 - s^2)$ where $s^2 = (1/N) \sum_i r_i^2$.

More generally, a conditional moment condition $E(v \mid z) = 0$ implies $E\{v \cdot g(z)\} = 0$ for any function $g(z)$. This can be tested by $(1/\sqrt{N}) \sum_i \hat{v}_i g(z_i)$, where $\hat{v}_i$ is an estimate for v_i. The test includes the above homoskedasticity test as a special case with $v_i = u_i^2 - \sigma^2$ and $g(z_i) = x_i$. When the moment condition in a MMT is derived from a conditional moment condition, the MMT may be called a *conditional moment test* (Newey, 1985).

For a conditional moment test with $E(v \mid z) = 0$, one can use many different functions for $g(z)$ and test $E\{v \cdot g(z)\} = 0$. In principle, if we use sufficiently many functions, say $g_1(z), \ldots, g_\nu(z)$, for $g(z)$ such that any function of z can be well approximated by $g_j(z)$, $j = 1 \ldots \nu$, then a test testing all of $E\{v \cdot g_j(z)\} = 0$, $j = 1 \ldots \nu$, may be as good as an (infeasible) test testing $E(v \mid z) = 0$; see Bierens (1990) and De Jong and Bierens (1994) for such tests. In practice however, since we will be using only a finite ν, there will be a set D for z such that $E(\nu \mid z) \neq 0$ when $z \in D$, which are however not detected by the test with the finite ν.

3.5 Identification

For a parameter space B and a random vector z in a model,

> identification (ID) is selecting a subset in B that characterizes some aspects of the probability distribution of z.

In $y = x'\beta + u$, if $E(y \mid x) = x'\beta$, then β characterizes the conditional distribution of $y \mid x$ where $z = (x', y)'$. The identified subset is mostly a unique point in B, but it can also be a set with finite or infinite elements. For instance, if we observe only $1[y \geq 0]$, where y is generated by $y = x'\beta + u$, then β is not fully identified since the scale of y is not observed. In this case, the set

$$\{b: b = \beta \cdot \gamma, \ \gamma \text{ is any positive scalar}\}$$

with infinite elements can be identified. Even when ID is "set-valued" as in this example however, typically we impose restrictions on the identified set to isolate a unique parameter. So from now on, our main focus will be the case where the identified set is a point.

In principle, ID is a separate issue from estimation; one can prove that certain parameters are identified without showing how to estimate them [for instance, Elbers and Ridder (1982)]. But if the ID of a parameter is done by a "discriminating function" or "a separating function," then ID naturally leads to estimation. For example, suppose that β is a unique argmax of $Q(b)$. This is equivalent to

$$\beta = \{\gamma: Q(\gamma) > Q(b) \text{ for all } b \neq \beta \text{ and } b \in B, \gamma \in B\}. \tag{5.1}$$

Here $Q(\cdot)$ is a discriminating function that separates β from the rest in B. In MLE, $Q(b) = E[\ln\{f(z, b)\}]$, where $f(z, \beta)$ is the log-likelihood function for z. In LSE, $Q(b) = -E\{(y - x'b)^2\}$. In both examples, the natural estimators are extremum estimators, the maximizers of the corresponding sample moments. The fact that a parameter may be estimated by multiple estimators suggests that there can be multiple discriminating functions. If "$b \in B$" in (5.1) is replaced by "$b \in N_\beta$ where N_β is a neighborhood of β," then the ID in (5.1) is local, not global. Unless otherwise mentioned, we will use the term ID only for the global ID. If the discriminating function is a vector m of zero moments, then

$$\beta = \{\gamma \colon E\{m(z,\gamma)\} = 0, \ \gamma \in B\}. \tag{5.2}$$

This naturally leads to the method-of-moments estimators. If β in (5.2) is unique, β is identified; otherwise, β is only locally identified at best.

Let $F(z,b)$ denote the probability distribution of z when the parameter is $b \in B$. In the literature of ID, the set $\{F(z,b); \ b \in B\}$ is called a "model" while the single $F(z,b)$ is called a "structure" (Rothenberg, 1971). If $F(z,\gamma) = F(z,\beta)$ for all z, then we cannot separate β from γ, since observations on z cannot tell beyond $F(z,\cdot)$. In this case, γ and β are said to be "observationally equivalent." Thus β is identifiable if there are no other observationally equivalent elements in B. This shows that the "maximal" discriminating function is $F(z,\cdot)$, or equivalently, the likelihood function $f(z,\cdot)$. Suppose we specify the form of $f(z,\cdot)$. Then, for the true parameter β to be identified, it is necessary to have

$$P_\beta(Z_\beta) > 0 \text{ where } Z_\beta \equiv \{z \colon f(z,\beta) \neq f(z,b),$$

$$\text{for any } b \neq \beta \text{ and } b \in B\} \tag{5.3}$$

where P_β is the probability when z follows $F(z,\beta)$.

To link (5.3) to MLE, define the "Kullback–Leibler information number"

$$H(\beta,b) = E_\beta[\ln\{f(z,\beta)/f(z,b)\}], \tag{5.4}$$

where E_β means that the integration is taken under $f(z,\beta)$. Using (5.3), Jensen's inequality, and the fact that $\ln(\cdot)$ is a strictly concave function, $H(\beta,b)$ can be shown to be nonnegative for any $b \in B$ and zero iff $b = \beta$. This means that the ID of β can be viewed as a minimization problem of $H(\beta,b)$ wrt $b \in B$ where $H(\beta,\beta) = 0$ is the minimum value. The sample version of (5.4) is

$$(1/N) \sum_i \ln\{f(z,\beta)\} - (1/N) \sum_i \ln\{f(z,b)\}. \tag{5.5}$$

Minimizing this wrt b is equivalent to maximizing the second term wrt b, which then renders MLE. Since $H(\beta,b) = 0$ iff $b = \beta$, β is identified in MLE. This way of viewing ID in parametric models with the Kullback–Leibler information number appears in Bowden (1973).

Although (5.3) is the minimal requirement for ID, invoking it for MLE requires specifying the form of the likelihood function. Instead, we can use

other discriminating functions and moment conditions. The potential drawback is that some parameters in the model that do not change the discriminating function (or the moment condition) at hand cannot be identified.
However, if we either use sufficiently many discriminating functions or the
discriminating functions are close to the likelihood function in some sense,
the shortcoming can be overcome while avoiding the potential danger of
misspecifying $f(z, \cdot)$. In the following, we show this line of approach, examining ID with method-of-moments estimators and extremum estimators;
the latter will be discussed only under regression frameworks.

As a simple example, consider y_i, $i = 1 \ldots N$, where y_i follows $N(0, \sigma^2)$.
The first moment is useless to identify σ. We can however use the second
moment to identify σ: $E(y^2) = \sigma^2 \Leftrightarrow E(y^2) - \sigma^2 = 0$, a moment condition.
Suppose that y_i follows a more general distribution $G(y, \beta)$ whose moment
generating function $M(t) = E(e^{ty})$ exists for all t, $|t| \le t_0$ for some t_0. Then
the sequence of moments uniquely determines $G(y, \beta)$. In this case, the
method of moments can identify everything that MLE can identify, because
for any component of β there will be some moment that can separate it
from the rest in B. Recall also (2.20): with $E(q_b) = 0$ viewed as a moment
condition, if q_b can approximate the score function s_b sufficiently well, then
the method of moments can be as good as MLE.

We defined ID as choosing a subset in B that reflects certain aspects of
$F(z)$, where $F(z)$ is the distribution function of z. For extremum estimator,
we can specialize the definition as

$$\beta \text{ maximizes } Q(b) \equiv Eq(z, b) \text{ and } \beta \text{ characterizes } F(z), \qquad (5.6)$$

In regression analysis with $z = (x', y)'$ where y and x are, respectively,
the dependent and independent variables, however, we usually want β to
characterize the conditional distribution $F_{y|x}$ of $y \mid x$, going beyond simply
maximizing $Eq(z, b)$.

Consider a scalar $\nu(x, \beta)$ that characterizes $F_{y|x}$; for instance, $E(y \mid x) =
\nu(x, \beta)$. The following definition of ID for extremum estimator with $\nu(x, \beta)$
seems general enough for our purpose: for all $b \ne \beta$ and $b \in B$,

$$\text{Conditional maximum: } E_{y|x}q\{y, \nu(x, \beta)\} > E_{y|x}q\{y, \nu(x, b)\}; \qquad (5.7)$$

$$\text{Separation: } P(x \mid \nu(x, \beta) \ne \nu(x, b)) > 0. \qquad (5.8)$$

The condition (5.7) identifies $B^* \equiv \{b: \nu(x, b) = \nu(x, \beta), b \in B\}$, and (5.8)
separates β from the rest of B^*. Then for any $b \ne \beta$ and $b \in B$,

$$Eq\{y, \nu(x, \beta)\} = E[q\{y, \nu(x, \beta)\} \cdot 1[X_{\text{id}}]]$$

$$+ E[q\{y, \nu(x, \beta)\} \cdot 1[X_{\text{id}}^c]] > Eq\{y, \nu(x, b)\}. \qquad (5.9)$$

Usually the conditional maximum (5.7) is proved easily by differentiation along with some model assumptions. Often (5.8) is more difficult to

establish than (5.7). Consider the LSE for the linear model $y = x'\beta + u$ with $E(u \mid x) = 0$. Condition (5.7) follows from $E(y \mid x) = x'\beta$ and the squared loss function. (5.8) can be proved by the p.d. of $E(xx') < \infty$: for any $b \neq \beta$ and $b \in B$,

$$(\beta - b)'E(xx')(\beta - b) = E\{(\beta - b)'xx'(\beta - b)\} = E\{(\beta - b)'x\}^2 > 0$$

$$\Rightarrow P(x'\beta \neq x'b) > 0. \tag{5.10}$$

In LSE, the discriminating function is $E(y - x'b)^2$. If we apply the least absolute deviation estimator to $y = x'\beta + u$ with $\mathrm{Med}(u \mid x) = 0$, then the discriminating function is $E|y - x'b|$. Both functions have nothing to do with the likelihood function $f(z, \cdot)$, but both can be equivalent to a likelihood function under an appropriate condition on u. As is well known, if u is independent of x and follows $N(0, \sigma^2)$, then the LSE becomes MLE. If u is independent of x and follows a double exponential distribution, then the least absolute deviation estimator becomes MLE.

Consider a nonlinear model $y = \nu(x, \beta) + u$ with $E(u \mid x) = 0$ where the dimension of β can be larger than that of x. (5.7) holds as in the linear model. Assuming that $\nu(b)$ is continuously differentiable, (5.8) can be written as

$$P\{x \mid \nu_b(x, b^*)'(b - \beta) \neq 0\} > 0, \tag{5.11}$$

where ν_b is the first derivative and $b^* \in (b, \beta)$. Then analogously to the linear model, $E[\nu_b(x, b^*)\nu_b(x, b^*)']$ being p.d. is sufficient for (5.8). But since b and b^* are arbitrary, this should hold for any $b \in B$; that is, the following condition is sufficient for (5.8):

$$E[\nu_b(x, b) \cdot \nu_b(x, b)'] \text{ is p.d. for any } b \text{ in } B. \tag{5.12}$$

This is rather restrictive. So ID in the nonlinear model demands more than in the linear model, which is understandable particularly well for the case where the dimension of β is larger than that of x.

4

Maximum Likelihood Estimation

4.1   Introduction

Let $\{(x_i', y_i)\}_{i=1}^N$ be an iid sample drawn from a known distribution $F(x_i, y_i, \beta)$, where β is a $k \times 1$ vector of unknown parameters. Let $f_{y|x}(y, \beta)$ denote the *likelihood function* of $y \mid x$, which is the density function of $y \mid x$ if $y \mid x$ is continuous or the probability of $y \mid x$ if $y \mid x$ is discrete. Define $f_x(x)$ analogously, which is not a function of β. The *maximum likelihood estimator* (MLE) maximizes the likelihood of the sample:

$$P\{(x_1, y_1) \ldots (x_N, y_N)\} = \prod_{i=1}^N P(x_i, y_i, b) = \prod_{i=1}^N f_{y|x_i}(y_i, b) \cdot f_x(x_i)$$

with respect to (wrt) b. Equivalently, MLE maximizes

$$(1/N) \sum_i \ln\{f_{y|x_i}(y_i, b) \cdot f_x(x_i)\}$$

$$= (1/N) \sum_i [\ln\{f_{y|x_i}(y_i, b)\} + \ln\{f_x(x_i)\}]. \tag{1.1}$$

Dropping $f_x(x)$, which is not a function of b, MLE maximizes the log of the conditional likelihood

$$(1/N) \sum_i \ln\{f_{y|x_i}(y_i, b)\}, \tag{1.2}$$

which still depends on x_i as well as on y_i and b. Written in this way, MLE is an extreme estimator. If we observe only $\{y_i\}$, then (1.2) cannot be obtained. But if the marginal likelihood $f_y(y_i)$ of y is a function of β, β may be estimable by maximizing $(1/N) \sum_i \ln\{f_y(y_i)\}$. This shows that there are different likelihood functions: joint, conditional, marginal and so on. Which likelihood to use in practice will depend on data availability and the parameter we want to know. Unless otherwise mentioned, we will always refer to the joint likelihood function for $z = (x', y)'$ as in (1.1).

As an example, consider a nonlinear regression model

$$y = \rho(x, \beta) + u, \tag{1.3}$$

where $\rho(x, \cdot)$ is a known nonlinear function of β. In (1.3), $y \mid x$ is equal to $u \mid x$ up to the "constant" $\rho(x, \beta)$ so that $f_{y|x}(y) = f_{u|x}\{y - \rho(x, \beta)\}$. Thus (1.2) becomes

$$(1/N) \sum_i \ln\{f_{u|x_i}(y_i - \rho(x_i, b))\}. \tag{1.4}$$

If u has a known form of heteroskedasticity, say $u \mid x \cong N(0, e^{2x'\gamma})$, then (1.4) becomes

$$(1/N) \sum_i \ln[(2\pi \cdot e^{2x'g})^{-1/2} \cdot \exp[-(1/2)\{(y - \rho(x_i, b))/e^{x'g}\}^2]], \tag{1.5}$$

which is to be maximized wrt b and g. This includes the homoskedastic case when $x = (1, x_2 \ldots x_k)'$ and $\gamma = (\gamma_1, 0 \ldots 0)'$.

The disadvantage of MLE is clear: we need to specify the distribution of $y \mid x$; if heteroskedasticity is present, its form should be spelled out as in (1.5), differently from some method-of-moments estimators discussed in preceding chapters. The advantage of MLE is twofold: one is its applicability to a variety of problems that are hard to approach without taking advantage of the specified likelihood functions, and the other is its asymptotic efficiency among a wide class of estimators. Owing to these advantages, MLE is popular in applied works and MLE can serve as a benchmark when we compare different estimators.

Since identification and asymptotics for MLE have been discussed in the preceding chapter, we will discuss specific topics for MLE in a single equation framework; see Amemiya (1985) for more. In Section 2, three classical tests for linear and nonlinear hypotheses are introduced. In Section 3, we show how to implement MLE (and extremum estimators in general) numerically. The remaining sections of this chapter show examples of MLE except the last section; more MLEs involving multiple equations will appear in the following chapter. In Sections 4 and 5, binary response and ordered response models are studied, respectively. In Section 6, truncated and censored models are examined. In Section 7, basic duration models are introduced along with Weibull MLE. Finally in Section 8, we present χ^2 specification tests that are applicable to most parametric models.

4.2   Testing Linear and Nonlinear Hypotheses

Suppose we want to test $H_0: R\beta = r$, where R is a $g \times k$ constant matrix with its rank $g \leq k$, and r is a $g \times 1$ constant vector; both R and r do not involve β. Denoting MLE by b_N and $V[\sqrt{N}(b_N - \beta)]$ by C, we get a *Wald test* statistic

$$N(Rb_N - r)'(RCR')^{-1}(Rb_N - r) =^d \chi_g^2 \qquad (2.1)$$

as in Chapter 2. The basic idea of the Wald test is comparing two estimators: one good under H_a only and the other good under both H_0 and H_a. To see the idea of the Wald test, imagine that we try to estimate $R\beta$. Under H_0, r is a valid estimator for $R\beta$. Under both H_0 and H_a, Rb_N is a valid estimator for $R\beta$. If H_0 holds, then $Rb_N - r =^p 0$; otherwise, $Rb_N - r$ will converge to a nonzero constant. The Wald test is based on this (normalized) distance $Rb_N - r$.

Before we discuss Wald tests for nonlinear hypotheses, we will introduce the δ-*method*, which is convenient in deriving the asymptotic distribution of $h(b_N)$ where h is a $g \times 1$ continuously differentiable function of b. Actually this method has been used already in deriving the asymtotic distribution of extremum estimators. Suppose $\sqrt{N}(b_N - \beta) =^d N(0, C)$. Expand $\sqrt{N} \cdot h(b_N)$ as

$$\sqrt{N} \cdot h(b_N) = \sqrt{N} \cdot h(\beta) + \sqrt{N} \cdot h_b(b_N^*)(b_N - \beta), \qquad (2.2)$$

where $b_N^* \in (b_N, \beta)$ and $h_b(\cdot)$ is the first derivative of dimension $g \times k$. Since $b_N =^p \beta$, we have $b_N^* =^p \beta$. Owing to the continuity of h_b, $h_b(b_N^*) =^p h_b(\beta)$. Therefore,

$$\sqrt{N}\{h(b_N) - h(\beta)\} =^p h_b(\beta) \cdot \sqrt{N}(b_N - \beta) =^d N(0, h_b(\beta) \cdot C \cdot h_b(\beta)'). \quad (2.3)$$

As usual, the variance can be estimated by $h_b(b_N) \cdot C \cdot h_b(b_N)'$.

Using the δ-method, we can test a $g \times 1$ nonlinear hypothesis $H_0: h(\beta) = r$, which includes $R\beta = r$ as a special case. Rewrite $h(\beta) = r$ as

$$h(b_N) - r = h(b_N) - h(\beta) \Rightarrow \sqrt{N}\{h(b_N) - r\} = \sqrt{N}\{h(b_N) - h(\beta)\}$$

$$=^d N(0, h_b(b_N) \cdot C \cdot h_b(b_N)'). \qquad (2.4)$$

Hence the Wald test statistic is

$$N \cdot \{h(b_N) - r\}' \cdot \{h_b(b_N) \cdot C \cdot h_b(b_N)'\}^{-1} \cdot \{h(b_N) - r\} \Rightarrow \chi_g^2. \qquad (2.5)$$

As an example of a nonlinear hypothesis, suppose we want to test H_0: $\beta_2\beta_3 = 1$ and $\beta_4/\beta_5 = \beta_6/\beta_7$ where $g = 2$ and $k = 7$. Rewrite H_0 as

$$\begin{bmatrix} \beta_2\beta_3 \\ \beta_4\beta_7 - \beta_5\beta_6 \end{bmatrix} = \begin{bmatrix} 1 \\ 0 \end{bmatrix} \qquad (2.6)$$

$$\Rightarrow \frac{h_b(\beta)}{2 \times 7} = \begin{bmatrix} 0, & \beta_3, & \beta_2, & 0, & 0, & 0, & 0 \\ 0, & 0, & 0, & \beta_7, & -\beta_6, & -\beta_5 & \beta_4 \end{bmatrix}, \qquad (2.7)$$

where $r = (1,0)'$ and $h(b_N) = (b_2 b_3, b_4 b_7 - b_5 b_6)'$ with $b_N \equiv (b_1 \ldots b_k)'$. Substitute r, (2,6), $h_b(b_N)$, and $h(b_N)$ into (2.5) to implement the test.

There is a (small-sample) problem in Wald tests with nonlinear hypotheses. The second part of H_0, $\beta_4/\beta_5 = \beta_6/\beta_7$, can be reformulated in many algebraically equivalent ways; one example is already shown in the second row of (2.6). If we had used $\beta_4/\beta_5 - \beta_6/\beta_7 = 0$ in the second row of (2.6), then the second row of $h_b(\beta)$ would be

$$0, \quad 0, \quad 0, \quad \beta_5^{-1}, \quad -\beta_4 \beta_5^{-2}, \quad -\beta_7, \quad \beta_6 \beta_7^{-2}.$$

Using this will render a different value for the Wald test (2.5), although this problem will disappear as $N \to \infty$. A more drastic example is testing $H_0 \colon \beta_2 = 1$, which can be rewritten as nonlinear hypotheses $\beta_2^2 = 1$, $\beta_2^3 = 1, \ldots, \beta_2^{1000} = 1$ (Lafontaine and White, 1986).

The following two answers may be useful in choosing a nonlinear hypothesis in practice (Gregory and Veall, 1985, Phillips and Park, 1988). First, there may be a hypothesis more natural than others; in the preceding example, $\beta_2 = 1$ is a more natural choice than $\beta_2^{1000} = 1$, for we do not actually think that β_2 is exactly one (β_2^{1000} will be either 0 or ∞ depending on whether $\beta_2 < 1$ or $\beta_2 > 1$). Second, a nonlinear hypothesis in a multiplicative form seems better than that in a ratio form; thus, we used $\beta_4 \beta_7 - \beta_5 \beta_6 = 0$ rather than $\beta_4/\beta_5 - \beta_6/\beta_7 = 0$ in (2.6).

Suppose that b_N maximizes $(1/N) \sum_i q(b)$ and that b_R maximizes $(1/N) \sum_i q(b)$ subject to $R\beta = r$, where $q(b) = \ln\{f(b)\}$, a likelihood function; namely, b_R maximizes

$$(1/N) \sum_i \ln\{f(b)\} + \lambda'(Rb - r), \qquad (2.8)$$

where λ is a $g \times 1$ Lagrangian multiplier. Apply a Taylor's expansion to $\sum_i q(b_R)$ twice around b_N to get

$$\sum_i q(b_R) = \sum_i q(b_N) + \sum_i q_b(b_N)' \cdot (b_R - b_N)$$
$$+ (1/2) \cdot (b_R - b_N)' \cdot \sum_i q_{bb}(b_N^*) \cdot (b_R - b_N).$$

Since $\sum_i q_b(b_N) = 0$ by the definition of b_N, this can be rewritten as

$$2 \cdot \left\{ \sum_i q(b_N) - \sum_i q(b_R) \right\} = (b_R - b_N)' \cdot \left\{ -\sum_i q_{bb}(b_N^*) \right\} \cdot (b_R - b_N)$$
$$= \sqrt{N}(b_R - b_N)' \cdot \left\{ -(1/N) \sum_i q_{bb}(b_N^*) \right\} \cdot \sqrt{N}(b_R - b_N)$$

$$=^p \ \sqrt{N}(b_R - b_N)' \cdot [-E\{q_{bb}(\beta)\}] \cdot \sqrt{N}(b_R - b_N) \qquad (2.9)$$

for $b_N =^p \beta$, $b_R =^p \beta$ and $b_N^* =^p \beta$ under H_0. Denoting the information matrix as I_f, we can show that (see the chapter for nonlinear LSE)

$$\sqrt{N}(b_R - b_N) = -I_f^{-1} R' (R I_f^{-1} R')^{-1} \cdot \sqrt{N}(R b_N - r). \qquad (2.10)$$

Substitute (2.10), $q(b) = \ln\{f(b)\}$ and $-E\{q_{bb}(\beta)\} = I_f$ into (2.9) to get

$$2 \cdot \left[\sum_i \ln\{f(b_N)\} - \sum_i \ln\{f(b_R)\} \right]$$

$$=^p \ N(R b_N - r)' \cdot (R I_f^{-1} R')^{-1} \cdot (R b_N - r) \Rightarrow \chi_g^2 \qquad (2.11)$$

due to (2.1). The difference between the two maximands

$$2 \cdot \left[\sum_i \ln\{f(b_N)\} - \sum_i \ln\{f(b_R)\} \right] =^d \chi_g^2 \qquad (2.12)$$

is called the *likelihood ratio* (LR) test statistic in MLE. LR is also good for nonlinear hypotheses so long as b_R satisfies the hypothesis. The LR test has a disadvantage of requiring both b_N and b_R, although it has good "invariance properties" such as invariance to reparametrization of the model; see Dagenais and Dufour (1991) and the references therein for more on invariance.

When we discussed method of moments tests (MMT), the idea was to test (zero) moment conditions implied by the model that are not used in obtaining the estimates. Applying the idea to MLE, we can devise *score test* or *Lagrangian multiplier test* (LM). When we get b_R under H_0, we may not use all the first-order conditions used in getting b_N. Thus, if H_0 is correct, then b_R should satisfy the first-order conditions for b_N not used in getting b_R.

Denote the score vector evaluated at b_R as

$$s_i(b_R) \equiv \partial \ln\{f(b_R)\}/\partial b. \qquad (2.13)$$

Then the LM test statistic is

$$\left\{ \sum_i s_i(b_R)' \right\} \cdot \left\{ \sum_i s_i(b_R) s_i(b_R)' \right\}^{-1} \cdot \left\{ \sum_i s_i(b_R) \right\} =^d \chi_g^2, \qquad (2.14)$$

which requires only b_R. Note that the dimension of $s_i(b_R)$ is $k \times 1$, while the degree of freedom in χ^2 is still g. To give justice to the name LM, consider the first-order condition satisfied by b_R in (2.8):

$$(1/N) \sum_i s_i(b_R) = -R' \cdot \lambda. \qquad (2.15)$$

Substitute this into (2.14) to get

$$\lambda' \cdot \left[R \cdot \left\{ \sum_i s_i(b_R) s_i(b_R)' \right\}^{-1} \cdot R' \right] \cdot \lambda; \qquad (2.16)$$

note that the dimension of λ is $g \times 1$. Under $H_0: R\beta = r$, the matrix in $[\cdot]$ can be shown to be the variance matrix of λ.

As an example, consider H_0: the first k_1 components of β are zero. Define

$$\beta \equiv (\beta_1', \beta_2')', \quad s_i(b) \equiv (s_{1i}(b)', s_{2i}(b)')', \qquad (2.17)$$

where the dimension of β_j and s_j is $k_j \times 1$, $j = 1, 2$. The condition $(1/N) \sum_i s_{2i}(b_R) = 0$ is used to get b_R, and the LM test examines if b_R satisfies $(1/N) \sum_i s_{1i}(b_R) = 0$. Since estimating b_R is easier than b_N in this example, the LM test has a practical advantage over the Wald and LR tests. This advantage, however, will not hold for a complicated H_0.

Rewrite (2.14) as

$$\left\{ \sum_i 1 \cdot s_i(b_R)' \right\} \cdot \left\{ \sum_i s_i(b_R) s_i(b_R)' \right\}^{-1} \cdot \left\{ \sum_i s_i(b_R) \cdot 1 \right\}. \qquad (2.18)$$

Now imagine regressing 1_N, the unit vector of N many one's, on the $N \times k$ vector composed of $s_i(b_R)$ with no intercept. The total variation in the dependent variable is N, and the explained variation is (2.18). Thus,

$$R^2 = (2.14)/N \Leftrightarrow N \cdot R^2 = (2.14) =^d \chi_g^2. \qquad (2.19)$$

Hence, the LM statistic can be obtained by $N \cdot R^2$, regressing 1 on s_i with no intercept. This way of doing a test with $N \cdot R^2$ from an (artificial) regression is sometimes called a "regression-based test."

In the example (2.17), we showed that only a part of the score vector $s_i(b)$ is effectively used. If we follow the idea of MMT, we should test, not $(1/N) \sum_i s_i(b_R) =^p 0$, but only

$$(1/N) \sum_i s_{1i}(b_R) =^p 0. \qquad (2.20)$$

The asymptotic distribution of the test statistic $(1/\sqrt{N}) \sum_i s_{1i}(b_R)$ can be found analogously to those of the MMT's in Chapter 3. Although we will use the names score test and LM test interchangeably, in view of (2.16) and (2.20), it seems fitting to use the term LM only for (2.15), while labeling MMT and (2.20) as score tests.

Suppose we want to test (2.20). Although the asymptotic distribution for $(1/\sqrt{N}) \sum_i s_{1i}(b_R)$ can be found, it may be complicated due to having

b_R, not β: the variance of $\sqrt{N}(b_R - \beta)$ will appear in the asymptotic distribution. However, there is a way to avoid this problem. In the following we will show that

$$(1/\sqrt{N}) \sum_i [s_{1i}(b_2) - I_{12}I_{22}^{-1}s_{2i}(b_2)] =^d N(0, (I_{11} - I_{12}I_{22}^{-1}I_{21})^{-1}), \quad (2.21)$$

where b_2 is any $\sqrt{N}$-consistent estimator for β_2, $I_{12} \equiv E\{s_1(\beta)s_2(\beta)'\}$, and I_{11}, I_{22}, and I_{21} are analogously defined.

Apply the mean value theorem to the lhs of (2.21) around β_2 to get

$$(1/\sqrt{N}) \sum_i \{s_{1i}(\beta_2) - I_{12}I_{22}^{-1} \cdot s_{2i}(\beta_2)\} + \sqrt{N}(b_2 - \beta_2)'$$

$$\cdot \left[(1/N) \sum_i \partial s_{1i}(b_2^*)/\partial b_2 - I_{12}I_{22}^{-1} \cdot (1/N) \sum_i \partial s_{2i}(b_2^*)/\partial b_2 \right]. \quad (2.22)$$

Using

$$(1/N) \sum_i \partial s_{1i}(b_2^*)/\partial b_2 =^p E\{\partial s_1(\beta_2)/\partial b_2\},$$

$$(1/N) \sum_i \partial s_{2i}(b_2^*)/\partial b_2 =^p E\{\partial s_2(\beta_2)/\partial b_2\}, \quad (2.23)$$

and $E(ss') = -E(\partial s/\partial b)$, we get

$$E\{\partial s_1(\beta_2)/\partial b_2\} = -I_{12}, \quad E\{\partial s_2(\beta_2)/\partial b_2\} = -I_{22}. \quad (2.24)$$

Substituting (2.23) and (2.24) into (2.22) makes the term in $[\cdots]$ of (2.22) equal to 0 [up to an $o_p(1)$ term]. Hence, only the first term in (2.22) remains, establishing (2.21). (2.21) is convenient, for the covariance matrix does not depend on the variance of $\sqrt{N}(b_2 - \beta_2)$. In (2.21), instead of using s_1, we use the part of s_1 not explained by s_2. The test using the effective score (s_1 not explained by s_2) in (2.21) is called the *Neyman's $C(\alpha)$ test*.

The Wald, LR, and score tests are called the three classical tests. All three follow χ_g^2 asymptotically under H_0, but they differ in terms of computational ease, performance in small samples, and invariance properties. The three tests are based on different kinds of distances that are small under H_0 and large otherwise (Engle, 1984):

(i) Wald: distance between the estimators $|b_N - b_R|$;

(ii) LR: distance between the maximands $|Q_N(b_N) - Q_N(b_R)|$;

(iii) Score: distance between the slopes

$$|\partial Q_N(b_R)/\partial b - \partial Q_N(b_N)/\partial b| = |\partial Q_N(b_R)/\partial b - 0|. \quad (2.25)$$

Asymptotically the three tests for MLE are equivalent in having the same size under $H_0: R\beta = r$ and the same power against a local alternative $H_a: R\beta = r + \delta/\sqrt{N}$. Under this H_a, they follow the same noncentral χ^2 with the noncentrality parameter (NCP) $\delta'(RI_f^{-1}R')^{-1}\delta$.

4.3 Numerical Optimization

For LSE, estimators are written in closed forms. However, MLE is defined implicitly by the b that maximizes (1.1). So, obtaining MLE in practice requires some types of numerical searching process. That is, we substitute a number for b in the sample maximand and evaluate the maximand. Then we decide whether the number is the maximizer or other numbers for b could increase the maximand. If the latter is the case, we choose another number for b and repeat the process. Searching for an estimator in this trial and error fashion is called *numerical maximization*. The discussion in this section is good for extremum estimators, in general, not just for MLE.

Intuitively, numerical searching is like being deserted in a foggy mountainous area and wanting to get to the highest point of the area. If the sky were clear, it would be easier to locate the peak visually. But with the foggy sky, our vision is impaired, and we need to decide two things: which direction to move and how far to move in that direction. If we go too far in the wrong direction, it is difficult to come back. But if we don't go far enough, we will never know what lies in the area.

Let $Q(b)$ be a quadratic maximand where b is a scalar. Then b attains the maximum only if $Q'(b) = 0$. If $Q'(b)$ is positive (negative), we should increase (decrease) b, for we are to the left (right) of the peak. The direction to move is decided by the first derivative, and the magnitude of our move depends on $Q''(b)$. In the following, we formalize this idea to obtain the "Newton–Rhapson" algorithm. See Press et al. (1986) for the numerical maximization methods appearing in this section.

Suppose that $Q(b)$ is (approximately) quadratic with a n.d. second derivative matrix. Let b_0 and b_1, respectively, denote the initial and the next estimates. Choose b_1 by maximizing the following wrt b:

$$Q(b) \cong Q(b_0) + G(b_0)'(b - b_0) + (1/2)(b - b_0)' \cdot H(b_0) \cdot (b - b_0), \quad (3.1)$$

where G is the first derivative vector (gradient) and H is the second derivative matrix (Hessian). Then, differentiating the right-hand side (rhs) of (3.1) wrt b, we get

$$b_1 = b_0 - H(b_0)^{-1}G(b_0). \quad (3.2)$$

Repeating (3.2) while updating b_0 with b_1 each time is the *Newton–Rhapson algorithm*.

The end of iteration (3.2) can be determined in various ways. Three popular stopping criteria are:

(i) stop when $|b_1 - b_0| < \varepsilon$ (such as $\varepsilon = 0.0001$),

(ii) stop when $|Q(b_1) - Q(b_0)| < \varepsilon$,   (3.3)

(iii) stop when $|\{(1/N) \sum_i s_i\}'\{(1/N) \sum_i s_i\}| < \varepsilon$, where the choice of ε is arbitrary. The advantage of (i) over (ii) and (iii) is that using (i) provides a uniform criterion to compare different numerical optimization algorithms maximizing different maximands. The disadvantage of (i) is that, depending on the scales of the explanatory variables, the scales of the elements in b are different. One element in b can dominate the others, and we may end up minimizing the one largest element in b.

For LSE, $H(b_0) = -(1/N) \sum_i x_i x_i'$. Although $-E(xx')$ is negative definite by assumption, its sample version may not be so. If $H(b_0)$ is not n.d., Newton–Rhapson can fail. One way to avoid this problem is to use $(1 - \lambda)H(b_0) + \lambda M$ instead of $H(b_0)$, where M is a chosen n.d. matrix; the scalar weight λ should be chosen too. Depending on λ and M, many variations of Newton–Rhapson are possible.

Sometimes $H(b)$ does not give a good "magnitude of the movement" (or *step size*) along the direction given by $G(b)$. With the step size too small, it will take a long time to reach the peak. With the step size too large, we may overstep, going from one side of the peak to the other side resulting in an oscillation around the peak. The latter is the more severe problem. One way to avoid this is to modify (3.2) as

$$b_1 = b_0 - \eta \cdot H(b_0)^{-1} G(b_0), \tag{3.4}$$

where η is a positive constant. The smaller η is, the smaller is the step size. The choice of η is arbitrary.

Often $H(b)$ is complicated. One way to simplify $H(b)$ for iteration is to use only the terms in $H(b)$ that do not disappear in $E\{H(\beta)\}$. This is called the *method of scoring*. Owing to this approximation of $H(b)$, the method of scoring may be slower in areas away from β. But near β, the ignored term is almost zero, so that the method of scoring should perform comparably to the Newton–Rhapson. In MLE, $-H(\beta)$ is the same as the expected outer product of the score function. So we often use

$$-(1/N) \sum_i \{\partial \ln f_z(z; b)/\partial b\} \cdot \{\partial \ln f_z(z; b)/\partial b'\}, \tag{3.5}$$

for $H(b)$ which saves the burden of deriving $H(b)$ analytically. Thus, we only need to get the first derivative to get MLE. One advantage of (3.5) is that it is always n.s.d.

Suppose that $Q(b)$ is differentiable only once or that $Q''(b)$ is too complicated to obtain analytically. In this case, we have only the gradient available. For MLE, we can use (3.5) for $Q''(b)$, but for other extremum estimators in general, no such approximation is possible. In this case Newton–Rhapson-type iteration is impossible. In the following, we present an algorithm only using the gradient.

Let b_0 denote the current estimate and $b_1 = b_0 + \eta \delta$ denote the next candidate with its vector direction δ and scalar step size η. Then

$$Q(b_0 + \eta \delta) - Q(b_0) \cong \eta \cdot G(b_0)' \delta, \tag{3.6}$$

where G denotes the gradient. For the rhs to be positive, δ should be chosen such that $G(b_0)'\delta$ is always positive. One obvious choice is $\delta = G(b_0)$. Hence, the direction of improvement is determined. Since η is a positive scalar, it is not difficult to find the optimal step size for the direction δ. "Grid search" is a possibility. Better yet, "line search by bracketing" [see Press et al. (1986)] provides an efficient way to find the optimal η.

There is no proof that the Newton–Rhapson-type algorithms are superior to algorithms combining the gradient and a line search method. Some algorithms, such as "downhill simplex," do not use even the gradient. Perhaps due to historical reasons, the Newton–Rhapson-type algorithms have been more popular. But we need to bear in mind that if $Q(b)$ is not approximately quadratic, the Newton–Rhapson can be misleading; if $Q(b)$ is shaped like a normal density, the Newton–Rhapson may continue to search forever without finding the optimal b. Such failure depends on the starting point of the algorithm. Hence employing an algorithm using only the gradient or no gradient at all can be more robust.

Although we showed that β attains a unique global maximum in MLE, there may be multiple global maxima or local maxima in a sample likelihood function. Since numerical search procedures may stop at any maxima, local or global, they may stop the iteration prematurely. The only way to avoid this pitfall is to try a number of different starting values and obtain the local maximum for each starting value. If the starting values are scattered enough to be "dense" in the entire parameter space, then one of the local maxima is likely to be a global maximum. Thus by choosing the estimate that yields the maximum among the local maxima, we will get a better chance of finding a global maximizer. Except for a few cases (Sections 4–6 in this chapter), usually the log-likelihood function of a MLE is not globally concave with multiple local minima.

4.4 Binary Response

As an example of MLE, take the *binary response* model:

$$y = 1[y^* \geq 0] = 1[x'\beta + u \geq 0], \quad E(u) = 0 \text{ and } V(u) = \sigma^2$$

$$\Leftrightarrow P(y = 1 \mid x) = 1 - G(-x'\beta) = G(x'\beta) \tag{4.1}$$

where G is a twice differentiable symmetric distribution function of u that is independent of x. In (4.1), y^* is the unobserved underlying (or latent) continuous variable, and only $(x', y)'$ is observed. Define v as

$$v \equiv y - E(y \mid x) = y - G(x'\beta). \tag{4.2}$$

Then v satisfies $E(v \mid x) = 0$. (4.2) is a nonlinear model $y = G(x'\beta) + v$.

The following are examples of the binary response model:

(i) Loan approval: $y = 1$ if a loan application is approved (and 0 otherwise), and x is a list of the characteristics of the applicant and the loan. Here y^* is "loan-worthiness."

(ii) Accepting an offer: $y = 1$ if an offer is accepted, and x is a list of the characteristics of the offer and the decision maker.

(iii) Surviving a situation: $y = 1$ if survival, and x is a list of the characteristics of the subject and the situation. Here y^* is the difference between the hardship and the durability of the subject.

To apply MLE to (4.1), we need to specify G. Although $E(u) = 0$ and $V(u) = \sigma^2$ are assumed, σ is still unknown. Divide $x'\beta + u$ by σ to get $x'(\beta/\sigma) + (u/\sigma)$, and $V(u/\sigma) = 1$. If G is indexed by the mean and variance (as in normal distributions), then u/σ follows the standardized distribution Φ of G. Thus, we have

$$P(y = 1 \mid x) = \Phi(x'\beta/\sigma) = \Phi(x'\alpha), \qquad \alpha \equiv \beta/\sigma.$$

With this, we can specify the likelihood and do MLE, but what is estimated by the MLE is $\alpha \equiv \beta/\sigma$ not β. Still the sign of β_i and ratios β_i/β_j can be estimated from the sign of α_i and ratios α_j/α_j, respectively.

If u has heteroskedasticity, say $V(u \mid x) = \sigma(x)^2$, where $\sigma(x)$ is a function of x, then we need to divide $x'\beta + u$ by $\sigma(x)$ to get $\{x'\beta/\sigma(x)\} + \{u/\sigma(x)\}$ to have the standardized error term. Suppose $x'\beta/\sigma(x) \cong x'\delta$. Then we will be estimating δ with the MLE; δ is a mixture of the mean and variance function parameters and it will be impossible to find β with the estimates for δ unless the functional form of $\sigma(x)$ is known. This limitation in the identification of the mean function in MLE for binary models should be borne in mind.

One choice of the distribution function $G(\cdot)$ of u is the logistic distribution:

$$G(u, \delta) = e^{u/\delta}/(1 + e^{u/\delta}) = 1/(1 + e^{-u/\delta}),$$

which has mean 0 and variance $\delta^2\pi^2/3$. Since $G(u) = 1 - G(-u)$, it is symmetric around 0. Choosing $\delta = \sqrt{3}/\pi$ renders the standardized logistic distribution. With this logistic distribution, the MLE for (4.1) is called *logit*. From now on, however, we will use normal distributions where Φ is the $N(0, 1)$ distribution function. With Φ, the MLE for (4.1) is called *probit*. Denote the $N(0, 1)$ density function as ϕ.

Usually, $y = 1[x'\beta + u \geq c]$, where c is an unknown constant and $x_1 = 1$ for all i. We can absorb c into the intercept β_1 to have $y = 1[x'\beta + u \geq 0]$, where β_1 is redefined as $\beta_1 - c$. The cost of this is that the intercept estimate for $(\beta_1 - c)/\sigma$ cannot be interpreted unless c is known. Combining the problems of σ and c, we often say that the parameters in the binary model are identified up to an additive constant (c) and a positive multiplicative constant ($1/\sigma$).

The sample likelihood function is

$$L = \prod_{i=1}^{N} \Phi(x_i'a)^{y_i} \{1 - \Phi(x_i'a)\}^{1-y_i} \tag{4.3}$$

and $(1/N)$ times $\ln L$ is

$$Q_N(a) = (1/N) \sum_i \{y_i \ln \Phi(x_i'a) + (1 - y_i) \ln(1 - \Phi(x_i'a))\}. \tag{4.4}$$

Denote $\Phi(x_i'a)$ and $\phi(x_i'a)$ as Φ_i and ϕ_i, respectively, to get

$$\partial Q_N(a)/\partial a = (1/N) \sum_i (y_i - \Phi_i)\phi_i x_i / \{\Phi_i(1 - \Phi_i)\}. \tag{4.5}$$

$\sqrt{N}(a_{\text{MLE}} - \alpha)$ follows $N(0, I_f^{-1})$, where

$$I_f \equiv E[\phi(x'\alpha)^2 \cdot xx'[\{y - \Phi(x'\alpha)\}/\{\Phi(x'\alpha) \cdot (1 - \Phi(x'\alpha))\}]^2]$$

$$= E[\phi(x'\alpha)^2 \cdot xx'/\{\Phi(x'\alpha) \cdot (1 - \Phi(x'\alpha))\}]. \tag{4.6}$$

The equation $(4.5) = 0$ can be viewed as a moment condition $(1/N) \sum_i v_i w_i = 0$, where $v_i = y_i - \Phi_i$ and $w_i \equiv x_i \phi_i / \{\Phi_i(1 - \Phi_i)\}$ is an instrument. Alternatively, regard $x_i \phi_i / \{\Phi_i(1-\Phi_i)\}^{1/2}$ as the instrument and $v_i / \{\Phi_i(1-\Phi_i)\}^{1/2}$ as the (weighted, or standardized) residual, for $V(v \mid x_i) = \Phi_i(1 - \Phi_i)$.

4.5   Ordered Discrete Response

Suppose that y^* $(= x'\beta + u)$ is continuous with $u \cong N(0, \sigma^2)$, but the observed response y is ordered and discrete with R categories:

$$y = r - 1 \text{ if } \gamma_{r-1} \leq y^* < \gamma_r, \quad r = 1, \dots, R, \ \gamma_0 = -\infty, \ \gamma_R = \infty; \tag{5.1}$$

that is,

$$
\begin{aligned}
y = 0 \quad & \text{if } x'\beta + u < \gamma_1 \quad &\Leftrightarrow u/\sigma < -x'\beta/\sigma + \gamma_1/\sigma \\
= 1 \quad & \text{if } \gamma_1 \leq x'\beta + u < \gamma_2 \Leftrightarrow \gamma_1/\sigma - x'\beta/\sigma \leq u/\sigma < \gamma_2/\sigma - x'\beta/\sigma \\
& \qquad\qquad \cdots\cdots \\
= R - 1 \quad & \text{if } \gamma_{R-1} \leq x'\beta + u \quad \Leftrightarrow \gamma_{R-1}/\sigma - x'\beta/\sigma \leq u/\sigma.
\end{aligned}
$$

One example is income data in which individual income is not recorded, but the bracket to which the income belongs is known. Another example is the number of durable goods purchased (car or TV). Depending on restrictions placed on the γ_r's, various specifications are possible: γ_r's may be known, or unknown but limited in its range, or completely unknown except for their ordering. See Lee (1992a) for a partial survey.

Suppose the γ_r's are unknown. Subtract γ_1 from $\gamma_{r-1} \leq x'\beta + u < \gamma_r$ and divide the inequality by σ to get

$$(\gamma_{r-1} - \gamma_1)/\sigma \leq x'\beta/\sigma - \gamma_1/\sigma + u/\sigma < (\gamma_r - \gamma_1)/\sigma, \quad r = 1, \ldots, R.$$

Here γ_1 is absorbed into β_1 and what can be estimated is

$$(\beta_1 - \gamma_1)/\sigma, \quad \beta_j/\sigma \text{ with } j = 2, \ldots k, \quad (\gamma_{r-1} - \gamma_1)/\sigma \text{ with } r = 3, \ldots, R.$$

Denote these simply as

$$\alpha \equiv ((\beta_1 - \gamma_1)/\sigma, \ \beta_2/\sigma, \ldots, \beta_k/\sigma)',$$

$$\tau_{r-1} \equiv (\gamma_{r-1} - \gamma_1)/\sigma, \quad r = 3, \ldots, R. \tag{5.2}$$

Denote the $N(0,1)$ distribution function and the density as Φ and ϕ, respectively. Under the independence between u and x, we get

$$P(y = r-1 \mid x) = P(\gamma_{r-1} \leq x'\beta + u < \gamma_r \mid x) = \Phi(\tau_r - x'\alpha) - \Phi(\tau_{r-1} - x'\alpha). \tag{5.3}$$

Also define

$$y_{ir} = 1 \quad \text{if the } i\text{th person has } y = r$$
$$= 0 \quad \text{otherwise.} \tag{5.4}$$

Assuming the γ_r's are unknown, *ordered probit* maximizes

$$Q_N(a, t) \equiv (1/N) \sum_{i=1}^{N} \sum_{r=1}^{R} y_{i(r-1)} \ln\{\Phi(t_r - x_i'a) - \Phi(t_{r-1} - x_i'a)\} \tag{5.5}$$

wrt a and t_r. The first derivatives are

$$\partial Q_N / \partial b = (1/N) \sum_{i=1}^{N} \sum_{r=1}^{R} y_{i(r-1)}(-x_i)$$

$$\cdot [\{\phi(t_r - x_i'a) - \phi(t_{r-1} - x_i'a)\} / \{\Phi(t_r - x_i'a) - \Phi(t_{r-1} - x_i'a)\}];$$

$$\partial Q_N / \partial t_r = (1/N) \sum_{i=1}^{N} \phi(t_r - x_i'a) \cdot [y_{i(r-1)} / \{\Phi(t_r - x_i'a) - \Phi(t_{r-1} - x_i'a)\}$$

$$- y_{ir} / \{\Phi(t_{r+1} - x_i'a) - \Phi(t_r - x_i'a)\}] \quad \text{for } r = 2, \ldots, R - 1. \tag{5.6}$$

Using this, ordered probit can be easily implemented.

In the example of the number of durable goods purchased, y depends on the period and timing over which y is measured. If the period and the timing are the same for all i, then they can be ignored; otherwise they should be accounted for. For instance, suppose that y is for 6 months from

1/1/1990 for person A and 12 months from 1/1/1992 for person B, then a duration variable w_i (6 for A and 12 for B) should be used as an additional regressor. Also if 1990 is a year of economic downturn while 1992 is a year of economic upturn, y can be lower in 1990. This can be accommodated by using yearly dummies or a dummy variable for downturn and upturn. The point worth repeating here is that care should be taken whenever the observations have to do with time.

Another line of MLE approach to ordered discrete response is the *Poisson MLE* where we assume that $y \mid x$ follows a Poisson distribution with parameter $\lambda(x) > 0$:

$$P(y = r \mid x) = \{\lambda(x)^r / r!\} \cdot e^{-\lambda(x)}, \quad r = 0, 1, 2, \ldots . \tag{5.7}$$

For the Poisson distribution, $E(y \mid x) = \lambda(x)$ and $V(y \mid x) = \lambda(x)$. Focus on the first term $\lambda(x)^r / r!$. Since $r!$ increases much faster than $\lambda(x)^r$ as $r \to \infty$, the probability of y taking a large integer decreases rapidly. The second term $e^{-\lambda(x)}$ is a normalizing factor for the sum of the first term over r: $\sum_{r=0}^{\infty} \lambda(x)^r / r! = e^{\lambda(x)}$.

To ensure $\lambda(x) > 0$ while keeping the linear model, a popular specification for $\lambda(x)$ is

$$E(y \mid x) = \lambda(x) = \exp(x'\beta), \tag{5.8}$$

which yields the following log-likelihood:

$$Q_N(b) = (1/N) \sum_i \{y_i(x_i'\beta) - \exp(x_i'b) - \ln(y_i!)\}. \tag{5.9}$$

Differentiate this wrt b to get

$$\partial Q_N(b) / \partial b = (1/N) \sum_i \{y_i - \exp(x_i'b)\} x_i. \tag{5.10}$$

With this, the Poisson MLE can be implemented. Differentiating the gradient wrt b again, we get

$$\partial^2 Q_N(b) / \partial b \partial b' = (1/N) \sum_i \{-\exp(x_i'b)\} \cdot x_i x_i', \tag{5.11}$$

which will be almost always n.d. for all b. Hence, for the Poisson MLE, a Newton–Rhapson-type algorithm should converge straightforwardly. At β, $y - \exp(x'\beta) = y - E(y \mid x)$. Defining $v = y - E(y \mid x)$, the first-order condition (5.10) can be looked upon as a moment condition $(1/N) \sum_i v_i x_i = 0$. Since $E(v \mid x) = 0$ implies $E\{v \cdot g(x)\} = 0$ for any function of x, the Poisson MLE specification can be tested by a method of moment test in the preceding chapter. Observe that $V(v \mid x) = V(y \mid x) = \lambda(x)$ and so v is heteroskedastic by definition.

One problem of the Poisson MLE is the restriction $E(y \mid x) = V(y \mid x)$, which is unlikely to be satisfied in practice; frequently, we have $V(y \mid x) >$

$E(y \mid x)$, a problem of "over-dispersion." To see why this occurs, consider an omitted variable v independent of x to observe

$$E(y \mid x, v) = e^{x'\beta + v} = e^{x'\beta} \cdot e^v$$
$$= e^{x'\beta} E(e^v) \cdot \{e^v / E(e^v)\} = e^{x'\beta + \ln\{E(e^v)\}} \cdot w, \quad (5.12)$$

where $w \equiv e^v / E(e^v)$; note that $E(w) = 1$. With $x'\beta = \beta_1 + \beta_2 x_2 + \cdots + \beta_k x_k$, redefine β_1 as $\beta_1 + \ln\{E(e^v)\}$ to rewrite (5.12) as $E(y \mid x, w) = e^{x'\beta} \cdot w$. From this,

$$E(y \mid x) = e^{x'\beta} \cdot E(w) = e^{x'\beta}. \quad (5.13)$$

Now observe that

$$E(y^2 \mid x) = E_{w|x}\{E(y^2 \mid x, w)\} = E_w\{E(y^2 \mid x, w)\}$$
$$= E_w\{V(y \mid x, w) + E^2(y \mid x, w)\} = E_w(e^{x'\beta} w + e^{2x'\beta} w^2)$$
$$= e^{x'\beta} + e^{2x'\beta} E(w^2). \quad (5.14)$$

With $E^2(y \mid x) = e^{2x'\beta}$,

$$V(y \mid x) = e^{x'\beta} + e^{2x'\beta} E(w^2) - e^{2x'\beta} = e^{x'\beta} + e^{2x'\beta}\{E(w^2) - 1\}$$
$$= e^{x'\beta} + e^{2x'\beta}\{E(w^2) - E^2(w)\} > e^{x'\beta} = E(y \mid x). \quad (5.15)$$

This is analogous to the following from the usual linear model: if $y = x_1'\beta_1 + x_2'\beta_2 + u$ holds where x_1 and x_2 are independent, then ignoring x_2 does not cause a bias in estimating β_1 with x_1 alone, but the error term variance increases since now $x_2'\beta_2 + u$ becomes the error term.

Although we showed that an unobservable term causes $V(y \mid x) > E(y \mid x)$ in the Poisson model, this should not be taken as $V(y \mid x) > V(y \mid x, w)$. To see this, observe that, for any random variables y and x,

$$V(y) = V\{E(y \mid x)\} + E_x\{V(y \mid x)\}; \quad (5.16)$$

this can be proven by taking $E_x(\cdot)$ on $V(y \mid x) = E(y^2 \mid x) - E^2(y \mid x)$ and using $V\{E(y \mid x)\} = E_x\{E^2(y \mid x)\} - E_x^2\{E(y \mid x)\}$. If $E(y \mid x) = 0$, then $V(y)$ is a "weighted average" of $V(y \mid x)$, which can be larger or smaller than $V(y)$.

Cameron and Trivedi (1986) and Winkelmann and Zimmermann (1995) show various parametric approaches generalizing the Poisson MLE while relaxing the restriction $E(y \mid x) = V(y \mid x)$; they also list many applied works for ordered response. Another (better) way to avoid the restriction is to view (5.8) $E(y \mid x) = e^{x'\beta}$ just as a nonlinear regression function without $V(y \mid x)$ specified. Then following the distribution theory of extremum estimators with the minimand $(1/N) \sum_i \{y_i - \exp(x_i'b)\}^2$, we get the asymptotic normality with $V[\sqrt{N}(b_N - \beta)]$ being

$$E^{-1}\{xx' \exp(x'\beta)\} \cdot E[\{y - \exp(x'\beta)\}^2 xx'] \cdot E^{-1}\{xx' \exp(x'\beta)\}. \quad (5.17)$$

While ordered probit is based on the latent continuous variable y^*, Poisson is not. The main objective in ordered discrete response is to link discrete y to possibly continuous $x'\beta$. In ordered probit, $x'\beta$ affects y through $\Phi(\cdot)$, while in the Poisson MLE $x'\beta$ affects y through $\lambda(\cdot)$. Both $\Phi(\cdot)$ and λ can take continuous values.

4.6   Censored and Truncated Models

Suppose $y^* = x'\beta + u$ is censored at 0 from below where u is independent of x with $u \cong N(0, \sigma^2)$. Then

$$y = \max(y^*, 0) = \max(x'\beta + u, 0). \tag{6.1}$$

This has been called a *censored regression model*, or *Tobit* in econometrics. A nontrivial portion of the sample has $y_i = 0$. For those observations with $y_i^* \leq 0$ ($\Leftrightarrow u \leq -x'\beta$), we only observe $y_i = 0$. Hence, the likelihood for $y_i = 0$ is $\Phi(-x'\beta/\sigma) = 1-\Phi(x'\beta/\sigma)$ as in the binary choice model. For those with $y_i > 0$, y_i^* is fully observed. Hence, the likelihood is $\sigma^{-1}\phi\{(y-x'\beta)/\sigma\}$. The log-likelihood for the full sample is

$$Q_N(b, s) = (1/N) \sum_i [(1 - 1_i) \cdot \ln \Phi(-x_i'b/s) + 1_i \ln\{s^{-1}\phi((y_i - x_i'b)/s)\}]$$

$$= (1/N) \sum_i [(1 - 1_i) \cdot \ln \Phi(-x_i'b/s) \tag{6.2}$$

$$+ 1_i\{(-1/2)\ln(2\pi) - (1/2)\ln(s^2) - (2s^2)^{-1}(y_i - x_i'b)^2\}],$$

which is to be maximized over b and s where $1_i \equiv 1[y_i > 0]$. Defining $\phi_i \equiv \phi(x_i'b/s)$ and $\Phi_i \equiv \Phi(x_i'b/s)$, the gradient is

$$\partial Q_N/\partial b = (1/N) \sum_i [-(1 - 1_i)\phi_i x_i/\{s(1 - \Phi_i)\} + 1_i(1/s^2)(y_i - x_i'b)x_i],$$

$$\partial Q_N/\partial s = (1/N) \sum_i [(1-1_i)x_i'b \cdot \phi_i/\{s^2(1 - \Phi_i)\} - 1_i/s + 1_i(y_i - x_i'b)^2/s^3].$$

$$\tag{6.3}$$

There are many examples of the censored model:

(i) Female labor supply: A large percentage of females do not work, resulting in many zero observations in y.

(ii) Expenditure on durable goods: Many households report zero expenditure on durable goods in a given period.

(iii) Duration of unemployment: Suppose we follow up N unemployed people to observe the durations, but the study ends while still a certain percentage remains unemployed. Then the only thing known for them is that each duration of those is greater than a number.

The last example is $y_i^* \geq c_i$, where c_i is known and varies across i. This case can be converted to (6.1) in the following way:

$$y_i = \max(x_i'\beta + u_i, c_i) \Leftrightarrow y_i - c_i = \max(x_i'\beta - c_i + u_i, 0);$$

c_i can be treated as a regressor with the known coefficient -1.

Define the score functions s_{bi}, s_{si} and $s_i \equiv (s_{bi}', s_{si}')'$ such that

$$\partial Q_N/\partial b = (1/N)\sum_i s_{bi} \quad \text{and} \quad \partial Q_N/\partial s = (1/N)\sum_i s_{si}. \tag{6.4}$$

Note that $V\{\sqrt{N}(b_N - \beta)\}$ is not obtained by $E^{-1}(s_b s_b') =^p \{(1/N)\sum_i s_{bi} s_{bi}'\}^{-1}$. To see this, define $\gamma \equiv (\beta', s')'$ and $g_N = (b_N', s_N')'$ to observe

$$V[\sqrt{N}(g_N - \gamma)] = I_f^{-1} = \begin{bmatrix} E(s_b s_b') & E(s_b s_s') \\ E(s_s s_b') & E(s_s s_s') \end{bmatrix}^{-1}; \tag{6.5}$$

we use s and s' to allow a vector s for more generality. Using the partitioned matrix inverse, the upper left $k \times k$ submatrix of I_f^{-1} is

$$[E(s_b s_b') - E(s_b s_s') \cdot E^{-1}(s_s s_s') \cdot E(s_s s_b')]^{-1}$$

$$= [E(s_b - \gamma' s_s)(s_b - \gamma' s_s)']^{-1} \equiv E^{-1}(s_b^* s_b^{*'}) \tag{6.6}$$

where $\gamma \equiv E^{-1}(s_s s_s') \cdot E(s_s s_b')$, and s_b^* is called the *effective score function* for b. Hence,

$$\sqrt{N}(b_N - \beta) =^d N(0, E^{-1}(s_b^* s_b^{*'})); \tag{6.7}$$

the variance becomes $E^{-1}(s_b s_b')$ only when $E(s_b s_s') = 0$.

In (6.6), γ is the regression coefficient of s_b on s_s, and s_b^* is the "residual," or the part of s_b not explained by s_s. Since $\gamma' s_s$ makes (6.6) larger, the estimation of β is hampered by the correlation between s_b and s_s. For a future reference, note that

$$\sqrt{N}(g_N - \gamma) =^p (1/\sqrt{N})\sum_i E^{-1}(ss') \cdot s_i = (1/\sqrt{N})\sum_i I_f^{-1} s_i, \tag{6.8}$$

$$\sqrt{N}(b_N - \beta) =^p (1/\sqrt{N})\sum_i E^{-1}(s_b^* s_b^{*'}) \cdot s_{bi}^*, \tag{6.9}$$

$$\sqrt{N}(s_N - \sigma) =^p (1/\sqrt{N})\sum_i E^{-1}(s_s^* s_s^{*'}) \cdot s_{si}^*. \tag{6.10}$$

Using (6.9), $V[\sqrt{N}(b_N - \beta)]$ is

$$E^{-1}(s_b^* s_b^{*'}) \cdot E(s_b^* s_b^{*'}) \cdot E^{-1}(s_b^* s_b^{*'}) = E^{-1}(s_b^* s_b^{*'}), \tag{6.11}$$

which agrees with (6.7).

If the observations with $y = 0$ are removed from the data set, we get a *truncated regression model* where the likelihood for one datum is the truncated density function

$$s^{-1}\phi\{(y - x'b)/s\}/\{1 - \Phi(-x'b/s)\} = s^{-1}\phi\{(y - x'b)/s\}/\Phi(x'b).$$

The denominator is the normalizing constant for $x'\beta + u > 0 \Leftrightarrow u > -x'\beta$. The log-likelihood and the gradient are

$$Q_N(b, s) = (1/N) \sum_i \{(-1/2) \cdot \ln(2\pi) - (1/2) \cdot \ln(s^2)$$

$$- (2s^2)^{-1}(y_i - x_i'b)^2 - \ln \Phi(x'b/s)\};$$

$$\partial Q_N(b, s)/\partial b = (1/N) \sum_i [(y_i - x_i'b)/s^2 - \phi(x'b/s)/\{\Phi(x'b/s) \cdot s\}]x_i;$$

$$\partial Q_N(b, s)/\partial s = (1/N) \sum_i [-1/s + (y - x'b)^2/s^3$$

$$+ \{(x'b) \cdot \phi(x'b/s)\}/\{\Phi(x'b/s) \cdot s^2\}]. \tag{6.12}$$

With this, implementing the truncated model MLE is straightforward.

Besides the MLEs, Heckman's (1979) two-stage estimator combining probit MLE and LSE is another parametric method good for the censored and truncated models, although it is less efficient than MLE. Since the above MLEs are easy to implement and since a better use of Heckman's method will be made in "selection models" given in Chapter 5, in the following we simply give the basic idea of the Heckman's method.

Under $u \cong N(0, \sigma^2)$, it can be shown that

$$E(y \mid x, y > 0) = x'\beta + E(u \mid x, y > 0) = x'\beta + E(u \mid x, u > -x'\beta)$$

$$= x'\beta + \sigma \cdot \{\phi(x'\beta/\sigma)/\Phi(x'\beta/\sigma)\}. \tag{6.13}$$

Defining $\alpha \equiv \beta/\sigma$, for y with $y > 0$,

$$y = E(y \mid x, y > 0) + y - E(y \mid x, y > 0) = x'\beta + \sigma \cdot \{\phi(x'\alpha)/\Phi(x'\alpha)\} + v, \tag{6.14}$$

where $E(v \mid x) = 0$; v is heteroskedastic by definition. Equation (6.14) is a nonlinear regression model in $(\beta', \sigma, \alpha')$ as such. The Heckman two-stage estimator is: first apply probit to the two groups $y > 0$ and $y = 0$ to get a_N, an estimate for α, then using the data with $y > 0$ only, do LSE of y on x and $\phi\{x'a_N\}/\Phi\{x'a_N\}$ to estimate β and σ.

4.7 Basic Duration Analysis

Suppose we want to explain duration y of a certain state with some explanatory variables x. The state can be being unemployed or being alive (with some disease). In the latter, the duration (or survival) is good, while it is bad in the former. Assuming that x_i is not time-varying, we can use cross-section data $\{(x_i', y_i)\}$, $i = 1, \ldots, N$, to examine duration. In this section, we study basic duration models, drawing upon Miller (1981), Kiefer (1988), and Lancaster (1992). We will use the duration of unemployment as our main example throughout.

Ignore x_i for a while, and let $F(t)$ denote the distribution function for y: $F(t) = P(y \le t)$. Also define the *survival function* $S(t) \equiv 1 - F(t)$ and the density function $f(t)$. A key concept for duration models is *hazard function*, or *hazard rate* $\lambda(t)$, which is defined by

$$\lambda(t) \equiv f(t)/S(t). \tag{7.1}$$

This can be interpreted as

$$\lambda(t) \cdot dt = P(t < y \le t + dt \mid y > t) = P(t < y \le t + dt, y > t \mid y > t) \tag{7.2}$$

$$= P(\text{leaving the state in } (t, t + dt] \mid \text{survived past } t). \tag{7.3}$$

Note that $P(t < y \le t + dt \mid y > t) > P(t < y \le t + dt)$ in view of (7.2). From (7.1) and $S(0) = 1$, we get

$$\lambda(t) = -d\{\ln S(t)\}/dt, \tag{7.4}$$

$$S(t) = \exp\left\{-\int_0^t \lambda(v)dv\right\} \equiv \exp\{-\Lambda(t)\}, \tag{7.5}$$

where $\Lambda(\tau) \equiv \int_0^t \lambda(v)dv$ is called the *integrated hazard*. The relation between $\lambda(t)$ and $S(t)$ is a two-way street: we can specify $S(t)$, which determines $\lambda(t)$, or we can specify $\lambda(t)$ first, which will then determine $S(t)$.

Although the normal distribution is a basic building block for MLE, it is not good for duration analysis, since duration should be nonnegative; also duration is often asymmetric with a long right tail. The basic distribution in duration analysis is exponential distribution indexed by one parameter θ. With y following the exponential distribution with θ ($\theta > 0$), we get the following:

(i) $f(t) = \theta \cdot \exp(-\theta t)$;

(ii) $S(t) = \exp(-\theta t)$, $F(t) = 1 - \exp(-\theta t)$;

(iii) $\lambda(t) = \theta$, $\Lambda(t) = \theta t$;   (7.6)

(iv) $E(y) = 1/\theta$, $V(y) = 1/\theta^2$;

(v) $E\{\ln(y)\} = -\ln(\theta) - 0.577, \quad V\{\ln(y)\} = 1.645.$

To understand (v), we introduce *Type I extreme value distribution* (or Gumbel distribution) with the parameter μ and ψ:

$$F(z) = \exp(-e^{-(z-\mu)/\psi}), \quad -\infty < z < \infty, \quad -\infty < \mu < \infty, \quad \psi > 0,$$
$$E(z) = \mu + \psi \cdot \gamma, \quad \gamma \text{ is the Euler's constant} \cong 0.577, \qquad (7.7)$$
$$V(z) = \psi^2 \pi^2/6.$$

Observe that, with y following the exponential distribution with θ,

$$P(-\ln(y) < t) = P(\ln(y) > -t) = P(y > e^{-t}) = \exp(-\theta \cdot e^{-t})$$

$$= \exp(-e^{-t} \cdot e^{\ln \theta}) = \exp(-e^{-(t-\ln \theta)}); \qquad (7.8)$$

that is, $-\ln(y)$ follows a type I extreme value distribution with $\mu = \ln \theta$ and $\psi = 1$. Hence,

$$E\{-\ln(y)\} = \ln \theta + 0.577, \quad V\{-\ln(y)\} = \pi^2/6 \cong 1.645, \qquad (7.9)$$

which implies (7.6)(v).

For regression analysis, usually we specify

$$\theta(x) = \exp(x'\beta) \Rightarrow E(y \mid x) = \exp(-x'\beta) \qquad (7.10)$$

as in the Poisson regression [$\exp(\cdot)$ guarantees $\theta(x) > 0$]. Using (7.6)(v),

$$\ln(y) = -x'\beta + u, \quad E(u \mid x) = -0.577, \quad V(u \mid x) = 1.645; \qquad (7.11)$$

$-u \mid x$ follows the type I extreme value distribution with $\mu = 0$ and $\psi = 1$, which is independent of x. The mode of u is 0; the density function of u is unimodal and asymmetric around 0.

As in the Poisson regression, the major problem with (7.11) is that $V(y \mid x)$ is decided as soon as $E(y \mid x)$ is set. In (7.11), we can absorb $E(u \mid x)$ into the intercept in $x'\beta$ to make $E(u \mid x) = 0$ and apply LSE. $V\{u \mid x\}$, however, is still a known constant. Also, the hazard function $\exp(-x'\beta)$ is not a function of time, not allowing the hazard rate to change across time. For instance, as unemployment duration goes up, the unemployed may be more willing to accept a job offer. Then the hazard rate will go up as y increases. The following Weibull distribution solves these problems in exponential distribution.

Weibull distribution with two parameters $\theta > 0$ and $\alpha > 0$ is

(i) $f(t) = \alpha \theta \cdot t^{\alpha-1} \exp(-\theta t^\alpha);$

(ii) $S(t) = \exp(-\theta t^\alpha), \quad F(t) = 1 - \exp(-\theta t^\alpha);$

(iii) $\lambda(t) = \alpha \theta \cdot t^{\alpha-1}, \quad \Lambda(t) = \theta t^\alpha; \qquad (7.12)$

(iv) $E(y^r) = \theta^{-r/\alpha}\Gamma(1 + r\alpha^{-1})$, where $\Gamma(w) = \int_0^\infty z^{w-1}e^{-z}dz$, $w > 0 \Rightarrow$
$E(y) = \theta^{-1/\alpha}\Gamma(1 + \alpha^{-1})$ and $V(y) = \theta^{-2/\alpha}\{\Gamma(1 + 2\alpha^{-1}) - \Gamma^2(1 + \alpha^{-1})\}$;

(v) $E\{\ln(y)\} = -(1/\alpha) \cdot \ln\theta - 0.577/\alpha$ and $V\{\ln(y)\} = 1.645/\alpha^2$.

For Weibull distribution, $\lambda(t)$ is increasing if $\alpha > 1$ and decreasing if $\alpha < 1$. The case in which $d\lambda(t)/dt < 0$ is called negative *duration dependence* [$d\lambda(t)/dt > 0$ is positive duration dependence]: as time progresses, it becomes less and less likely that the duration ends. Weibull distribution includes exponential distribution as a special case when $\alpha = 1$. In view of $S(t)$, Weibull distribution becomes exponential distribution by redefining t^α as t. If $\alpha > 1$, then time accelerates, which is equivalent to $\lambda(t)$ increasing over time. Owing to $S(\infty) = 0$, eventually everybody will leave unemployment. This however may be too restrictive; there may be "super-survivors" surviving until the end.

Assuming $\theta = \exp(x'\beta)$, we get

$$\ln(y) = x'(-\beta/\alpha) + u, \quad E(u \mid x) = -0.577/\alpha, \quad V(u \mid x) = 1.645/\alpha^2.$$
(7.13)

Unlike the exponential distribution, there is an unknown constant α in $V(u \mid x)$, which is not a function of x. Doing LSE of $\ln(y)$ on x, we can estimate $-\beta/\alpha$ with g_N, and

$$(1.645/\alpha^2) =^P s_N^2 \equiv (1/N)\sum_i\{\ln(y_i) - x'g_N\}^2.$$
(7.14)

From this, an estimate a_N for α is $(1.645)^{0.5}/s_N$. An estimate b_N for β is then obtained by $b_N = g_N(-a_N)$. Note that the intercept is still off the target due to $E(u \mid x) = -0.577/\alpha$, which, however, is not of main concern.

The only problem with the LSE for the linear model (7.13) is that, in many cases, the data are censored in duration analysis for various reasons. This may be due to stopping the study before all durations end, or individuals dropping out of the study (say, moving overseas). Even if there is no censoring, with the distribution of u known, LSE is inefficient compared with MLE. In the following, we examine MLE under Weibull specification allowing censoring.

Suppose we have (y_i, x_i') with a part of data censored; for the censored data, the only information is that the duration is greater than y_i. Then the log likelihood with a generic density $f(y \mid x)$ is

$$(1/N)\sum_i d_i \ln\{f(y_i \mid x_i; a, b)\} + (1/N)\sum_i(1 - d_i) \cdot \ln\{S(y_i \mid x_i; a, b)\},$$
(7.15)

where $d_i = 1$ if the ith observation is uncensored and 0 otherwise. Using $f = \lambda \cdot S$ and $\ln S = -\Lambda$, this can be rewritten as

$$(1/N)\sum_i d_i \ln\lambda(y_i, x_i; a, b) - (1/N)\sum_i \Lambda(y_i, x_i; a, b).$$
(7.16)

Using the Weibull hazard specification with $\theta = \exp(x'\beta)$, the likelihood function and the gradient become

$$Q(a,b) = (1/N) \sum_i d_i \ln(a) + d_i(a-1) \ln(y_i) + d_i x_i' b - y_i^a \exp(x_i'b); \quad (7.17)$$

$$\partial Q/\partial b = (1/N) \sum_i \{d_i - y_i^a \exp(x_i'b)\} \cdot x_i,$$

$$\hspace{6cm} (7.18)$$

$$\partial Q/\partial a = (1/N) \sum_i d_i/a + \{d_i - y_i^a \exp(x_i'b)\} \cdot \ln(y_i).$$

With this, the MLE can be implemented. In (7.18), $d_i - y_i^a \exp(x_i'b)$ may be regarded as an error term ε_i in the moment condition $(1/N) \sum_i \varepsilon_i x_i = 0$.

In the usual linear regression model, one way to view the error term is that it is a combination of omitted variables uncorrelated with the regressors. Suppose we include v_i in the regression function to reflect omitted variables uncorrelated with x_i [recall (7.12)(iii)]:

$$\lambda(x_i, y_i, v_i) = \exp(x_i'\beta + v_i) \cdot \alpha y_i^{\alpha-1}. \quad (7.19)$$

Then analogously to (7.13), we get

$$\ln(y_i) = -x_i'\beta\alpha - v_i/\alpha + u_i, \quad (7.20)$$

where $u_i - v_i/\alpha$ is the error term. This can be estimated with LSE if there is no censoring. If $V(v \mid x)$ is a function of x, then (7.20) has heteroskedasticity of an unknown form and is often called "unobserved heterogeneity."

If we want to apply MLE either for efficiency or for a censoring problem, the presence of v_i creates a difficulty. Since v_i is not observed, we need to assume a distribution for v_i. Also, in general, we need to assume that v_i is independent of x_i, not just uncorrelated with x_i. The likelihood function such as (7.17) should be integrated wrt v_i, for we only observe (x_i', y_i). Not only is this time consuming, but the estimation of α and β depends critically on the assumed distribution of v_i. Ignoring v_i in MLE will cause a downward bias in duration dependence estimation as shown in the following paragraph. Thus, if the censoring percentage is low, using the LSE for (7.20) rather than MLE may be a good idea.

Ignore x and imagine $\lambda(v) = v\theta$; that is, the hazard rate depends only on v and θ, which is a constant. Assume v takes 1 and 2 with equal probability. Then the half of the population has hazard θ (Group 1) and the other half has hazard 2θ (Group 2). Initially there are equal proportions of Group 1 and Group 2 people in the population, for $P(v=1) = P(v=2) = 0.5$. As time progresses, people in Group 2 with the higher hazard rate will leave the state, and the remaining population will have more and more Group 1 people. This scenario is indistinguishable from the situation where we have $\lambda(t)$ with $\lambda'(t) < 0$. Thus even when we have $\lambda'(t) = 0$ for all t, if we estimate $\lambda(t)$ ignoring v, we will end up $\lambda'(t) < 0$, which is a downward bias in duration dependence.

Recall the Weibull hazard with x: $\lambda(t, x) = \alpha t^{\alpha-1} \cdot e^{x'\beta}$ which is the product of $\alpha t^{\alpha-1}$ and $e^{x'\beta}$: $\lambda(t, x)$ is separated into a function of t and a function of x. In general,

$$\lambda(t, x) = \lambda_0(t) \cdot \phi(x, \beta) \tag{7.21}$$

is called a *proportional hazard function*, and $\lambda_0(t)$ is called the *base line hazard*. Equation (7.21) is a generalization of the Weibull hazard. A further generalization of (7.21) is a hazard function nonseparable in t and x.

Unlike the usual cross-section study, there are several different ways to collect samples for unemployment duration. One way is to fix an interval (say, a month) and sample those who become unemployed in the month to observe their unemployment period; this is called *flow sampling*. Another way is to set a date and sample unemployed people to ask them of their elapsed time as well as to observe their time unemployed; this is called *stock sampling*. Our exposition above is based on flow sampling. Stock sampling is subject to the so-called *length-biased sampling* or *inspection paradox*. Imagine two people with different hazard rates starting their employment at day 0. Suppose we select one person at day 100. Then it is more likely for the person with the lower hazard to be picked up, since the person with the higher hazard has a better chance to end the duration before day 100 and not to be sampled. Thus sampling at a given date results in data with lower hazard than average.

4.8 Chi-Square Specification Tests

A general approach to test model specifications is χ^2 tests, or often called "goodness of fit tests," suggested by Heckman (1984), which were then generalized by Andrews (1988a, 1988b). In this section, we review these tests that are widely applicable to most parametric models in econometrics.

Let $f(y \mid x, \beta)$ and $F(y \mid x, \beta)$ denote the density and the distribution function of $y \mid x$, respectively. Suppose β is estimated by b_N and

$$\sqrt{N}(b_N - \beta) =^p (1/\sqrt{N}) \sum_i \eta_i =^d N(0, \Omega). \tag{8.1}$$

Consider a partition of the range of y into J cells:

$$C_1 \equiv (-\infty, c_2), \quad C_2 \equiv [c_2, c_3) \dots C_J \equiv [c_J, \infty), \tag{8.2}$$

where c_j's are known constants. Define indicator functions d_{ij} as

$$\begin{aligned} d_{ij} &= 1 \quad \text{if } y_i \in C_j \\ &= 0 \quad \text{otherwise,} \end{aligned} \tag{8.3}$$

to get (analogously to ordered probit)

$$d_{ij} \equiv E(d_{ij} \mid x_i, \beta) + \varepsilon_{ij} = F(c_{j+1} \mid x_i, \beta) - F(c_j \mid x_i, \beta) + \varepsilon_{ij}. \quad (8.4)$$

Note that ε_{ij} is defined by this equation. Stacking up ε_{ij}'s, $j = 2, \ldots, J$, we get $\varepsilon_i \equiv (\varepsilon_{i2}, \ldots, \varepsilon_{iJ})'$, and $V(\varepsilon_i)$ is such that its diagonal terms are $E(d_{ij} \mid x_i, \beta) \cdot \{1 - E(d_{ij} \mid x_i, \beta)\}$, $j = 2, \ldots, J$, and the off-diagonal terms are $-E(d_{ij} \mid x_i, \beta) \cdot E(d_{im} \mid x_i, \beta)$, $j, m = 2, \ldots, J$.

Define

$$d_i \equiv (d_{i2} \ldots d_{iJ})', \quad E(d_i \mid x_i, \beta) \equiv (E(d_{i2} \mid x_i, \beta) \ldots E(d_{iJ} \mid x_i, \beta))'. \quad (8.5)$$

Heckman's (1984) χ^2 test statistic is

$$\left[(1/\sqrt{N}) \sum_i \{d_i - E(d_i \mid x_i, b_N)\} \right]'$$

$$\cdot W^{-1} \cdot \left[(1/\sqrt{N}) \sum_i \{d_i - E(d_i \mid x_i, b_N)\} \right] =^d \chi^2_{J-1}, \quad (8.6)$$

where W is the variance matrix of the adjacent vector. The test compares the observed frequency d_i with the "predicted frequency" $E(d_i \mid x_i, b_N)$ in each cell, because the two should be close if the model is correct. In the following, we derive W, which is affected by $b_N - \beta$.

Rewrite $(1/\sqrt{N}) \sum_i \{d_i - E(d_i \mid x_i, b_N)\}$ into

$$(1/\sqrt{N}) \sum_i \{d_i - E(d_i \mid x_i, \beta)\} - (1/\sqrt{N}) \sum_i \{d_i \mid x_i, b_N) - E(d_i \mid x_i, \beta)\}$$

$$= (1/\sqrt{N}) \sum_i \varepsilon_i - (1/\sqrt{N}) \sum_i \{E(d_i \mid x_i, b_N) - E(d_i \mid x_i, \beta)\}. \quad (8.7)$$

Applying the mean value theorem to the second term, it becomes $o_p(1)$ equivalent to [recall (8.1)]

$$(1/N) \sum_i \{\partial E(d_i \mid x_i, \beta)/\partial b\} \cdot \sqrt{N}(b_N - \beta)$$

$$=^p E\{\underbrace{\partial E(d_i \mid x_i, \beta)/\partial b}_{(J-1) \times k}\} \cdot (1/\sqrt{N}) \sum_i \eta_i. \quad (8.8)$$

Observe that

$$E\{\partial E(d_i \mid x_i, \beta)/\partial b\} = E_x \left[\partial \int \{d_i f(y_i \mid x_i, \beta) dy_i\} /\partial b \right]$$

$$= E_x \left[\int d_i \{\partial f(y_i \mid x_i, \beta)/\partial b\} \cdot dy_i \right]$$

$$= E_x \left[\int d_i \{\partial \ln(f(y_i \mid x_i, \beta))/\partial b\} \cdot f(y_i \mid x_i, \beta) dy_i \right]$$

$$= E[d_i \partial \ln\{f(y_i \mid x_i, \beta)\}/\partial b],$$

which is then used for the estimation of $E\{\partial E(d_i \mid x_i, \beta)/\partial b\}$. Substituting (8.8) into (8.7), (8.7) becomes

$$(1/\sqrt{N}) \sum_i [\varepsilon_i - E\{\partial E(d_i \mid x_i, \beta)/\partial b\} \cdot \eta_i] \equiv (1/\sqrt{N}) \sum_i \lambda_i. \qquad (8.9)$$

Then W can be estimated by

$$W_N \equiv (1/N) \sum_i \lambda_i \lambda_i' \qquad (8.10)$$

where the unknowns in λ_i should be replaced by their estimates.

Andrews (1988b) generalizes Heckman's (1984) test by allowing the partition to be random and to be on $z = (x', y')'$ space. The partition can be almost arbitrary, allowed to be data-dependent or estimate-dependent, so long as it is convergent to a fixed partition as $N \to \infty$. We will denote the cells in the partition as C_{Nj}, $j = 1, \ldots, J$. Assume that

$$C_{Nj} \to C_j \quad \text{as } N \to \infty, \quad j = 1, \ldots, J;$$

however, the number of cells J is assumed to be constant. There is a technical restriction on C_{Nj} but the restriction is satisfied in most of the applications [see Andrews (1988b, p. 1426)].

Suppose we partition the space for $z_i \equiv (x', y')'$ into J many cells. Define g_{Ni} and g_i as

$$\begin{aligned} g_{Ni} &\equiv (1[z_i \in C_{N1}], \ldots, 1[z_i \in C_{NJ}])', \\ g_i &\equiv (1[z_i \in C_1], \ldots, 1[z_i \in C_J])'; \end{aligned} \qquad (8.11)$$

that is, g_{Ni} is a $J \times 1$ vector of indicator functions showing to which cell z_i belongs among C_{Nj}'s, and g_i is the limiting version of g_{Ni}. Then the test statistic is

$$(1/\sqrt{N}) \sum_i \{g_{Ni} - E(g_{Ni} \mid x_i, b_N)\}' \cdot W^{-1}$$

$$\cdot (1/\sqrt{N}) \sum_i \{g_{Ni} - E(g_{Ni} \mid x_i, b_N)\}, \qquad (8.12)$$

where W is the variance of the adjacent vector and b_N is an estimate for β such that

$$\sqrt{N}(b_N - \beta) =^p (1/\sqrt{N}) \sum_i \eta_i =^d N(0, \Omega). \qquad (8.13)$$

The test statistic converges to a χ^2 distribution with its dof being the rank of W. The idea behind (8.12) is to compare the predicted frequency $E(g_{Ni} \mid x_i, b_N)$ with the observed frequency g_{Ni}. In the following, we discuss how to estimate W and get the rank of W.

In order to see how to estimate W, first we note that (8.12) is $o_p(1)$ equivalent to

$$(1/\sqrt{N}) \sum_i \{g_{Ni} - E(g_{Ni} \mid x_i, \beta) - A\eta_i\}' W^{-1}$$

$$\cdot (1/\sqrt{N}) \sum_i \{g_{Ni} - E(g_{Ni} \mid x_i, \beta) - A\eta_i\}, \qquad (8.14)$$

where

$$A \equiv E\{\partial E(g \mid x, \beta)/\partial b\} = E[g \cdot \{\partial \ln(f(y \mid x, \beta))/\partial b\}].$$

It can be shown that ["stochastic equicontinuity," see, e.g., Andrews (1994)]

$$(1/\sqrt{N}) \sum_i \{g_{Ni} - E(g_{Ni} \mid x_i, \beta)\} =^p (1/\sqrt{N}) \sum_i \{g_i - E(g_i \mid x_i, \beta)\}. \qquad (8.15)$$

Thus, instead of (8.14), we can examine

$$(1/\sqrt{N}) \sum_i \{g_i - E(g_i \mid x_i, \beta) - A\eta_i\}' W^{-1}$$

$$\cdot (1/\sqrt{N}) \sum_i \{g_i - E(g_i \mid x_i, \beta) - A\eta_i\}. \qquad (8.16)$$

There are two terms in the vector $\{g_i - E(g_i \mid x_i, \beta) - A\eta_i\}'$. The variance of the first term $g_i - E(g_i \mid x_i, \beta)$ is

$$E(gg') - E\{g \cdot E(g \mid x, \beta)'\} - E\{E(g \mid x, \beta) \cdot g'\} + E\{E(g \mid x, \beta) \cdot E(g \mid x, \beta)'\}$$

$$= E(gg') - E\{E(g \mid x, \beta) \cdot E(g \mid x, \beta)'\}, \qquad (8.17)$$

for $E\{g \cdot E(g \mid x, \beta)'\} = E\{E(g \mid x, \beta) \cdot E(g \mid x, \beta)'\}$. Thus, we get

$$(1/\sqrt{N}) \sum_i \{g_{Ni} - E(g_{Ni} \mid x_i, b_N)\} =^d N(0, W); \qquad (8.18)$$

$$W = E(gg') - E\{E(g \mid x, \beta) \cdot E(g \mid x, \beta)'\} + A \cdot E(\eta\eta') \cdot A'$$

$$- E[\{g - E(g \mid x, \beta)\} \cdot \eta'] \cdot A' - A \cdot E[\eta \cdot \{g - E(g \mid x, \beta)\}'].$$

W can be estimated in a number of ways. One easy way is to use the outer-product $W_N \equiv (1/N) \sum_i \delta_{Ni} \delta'_{Ni}$, where $\hat{\eta}_i =^p \eta_i$,

$$\delta_{Ni} \equiv \{g_{Ni} - E(g_{Ni} \mid x_i, b_N) - A_N \hat{\eta}_i\}',$$
$$A_N \equiv (1/N) \sum_i g_{Ni} \{\partial \ln(f(y_i \mid x_i, b_N))/\partial b\}. \qquad (8.19)$$

As mentioned subsequently, since the idea of the χ^2 test is comparing the conditional mean with the empirical conditional mean, the rank of W depends on how many categories we allow for x. If we partition the z space based on y [imagine drawing parallel lines in the two dimensional (x, y) plane], then there is only one category for x, and the rank of W is $J - 1$ as in the Heckman's χ^2 test. If we partition the x-space and the y-space separately, say 4 for x and 5 for y to get $4 \times 5 = 20$ cells, then

$$\text{rank}(W) = 4(5 - 1) = 20 - 4 = J - \text{number of } x\text{-categories.}$$

If we partition based on the residual $y_i - x_i'b_N$, then this is like drawing slanted parallel lines in the two-dimensional plane, which is little different from partitioning the z-space based on y only. Thus, in this case, $\text{rank}(W) = J - 1$. In all cases, since W is singular, a generalized inverse should be used unless we drop a number of cells and reduce the dimension of (8.11) to the rank of W. See Andrews (1987b) for more on test statistics with generalized inverses. Andrews (1988a) shows some guide lines of using the χ^2 test on specific examples, some of which are applied by Melenberg and Van Soest (1993, 1995a).

5

Parametric Estimators for Multiple Equations

5.1 Introduction

In Chapter 4, we studied MLE in general and introduced various single equation examples of MLE. In this chapter, our focus is on parametric estimators for multiple equations. We will discuss MLE, but we will also study other parametric estimators perhaps less efficient than MLE but more convenient and robust in practice. Since multiple equations are more difficult to handle than single equations, this chapter is, in general, more difficult than the preceding chapter.

In Section 2 the multinomial choice model is examined and the multinomial probit is introduced. In Section 3, as a practically easy alternative for multinomial probit, multinomial logit is examined. In Section 4, methods of simulated moments to overcome computational problems in multinomial probit are introduced. In Section 5, smooth simulators for multinomial probit are discussed and methods of simulated likelihood are examined. In Section 6, various selection models are discussed. Section 7, minimum distance estimation is introduced. In Section 8, model specification tests based upon the distances of estimators are studied.

5.2 Multinomial Choice Model

Ordered response models extend the binary response model "horizontally" by allowing more ordered values. Multinomial choice models extend the binary response model "vertically" by considering multiple binary response equations. Suppose we have N individuals each having J alternatives to choose from. Each person will choose one alternative that yields the highest utility, depending on his or her attributes and the characteristics of the alternatives.

One example is a transportation mode choice problem where the alternatives are bus, car, and train. Let i index the people and let j index the alternatives. Here, the explanatory variables are of three types: the first type varies across i and j (cost and time for each mode), the second type varies only across i (attributes of people such as income, sex, and race), and the third type varies only across j (some characteristics of mode common to all i such as whether the mode has a dining facility). Another example is industry choice for job where the alternatives are manufacturing, service, government, and the others. For a survey on theoretical background of multinomial choice, see, for instance, Anderson et al. (1992).

The following two models are popular in multinomial choice:

$$s_{ij} = x'_{ij}\delta + u_{ij}, \qquad (2.1)$$

$$s_{ij} = z'_i\eta_j + u_{ij}, \qquad (2.2)$$

where "s" in s_{ij} comes from "satisfaction" of the ith person in choosing j; note that x_{ij} may include variables that vary only across j and that η_j is choice-dependent. Combining (2.1) and (2.2), we get

$$s_{ij} = x'_{ij}\delta + z'_i\eta_j + u_{ij}. \qquad (2.3)$$

The term $z'_i\eta_j$ in (2.3) needs some care before it is used. Suppose we have $z'_i\eta$ in (2.3) instead of $z'_i\eta_j$, then the effect of z_i on all s_{ij}, $j = 1, \ldots, J$, is the same. This means that $z'_i\eta$ plays no role in the choice and so z_i drops out of the picture. If we use $z'_i\eta_j$ to include z_i in our analysis, we need to justify why z_i has different coefficients for the alternatives. For instance, if z_i is income, then $\eta_j = \partial s_{ij}/\partial z_i$; we should ask ourselves why a unit increase in income changes the utilities of different transportation modes in different ways. If this cannot be done, then z_i should be removed from (2.3). Also, even if $z'_i\eta_j$ is legitimate, still the identification of η can be fragile when there are no exclusion restrictions that some explanatory variables in s_{ij} are excluded from s_{im} for all $m \neq j$ (that is, the variables have the coefficients 0 in s_{im}); see Keane (1992).

Although (2.3) is more general than (2.1) and (2.2), in practice, we transform our model into one that looks like (2.1). Suppose $J = 3$ and consider the following differences of the regression functions:

$$
\begin{aligned}
2 \text{ and } 1: \ & x'_{i2}\delta + z'_i\eta_2 - (x'_{i1}\delta + z'_i\eta_1) \\
& = (x_{i2} - x_{i1})'\delta + z'_i(\eta_2 - \eta_1); \\
3 \text{ and } 1: \ & x'_{i3}\delta + z'_i\eta_3 - (x'_{i1}\delta + z'_i\eta_1) \\
& = (x_{i3} - x_{i1})'\delta + z'_i(\eta_3 - \eta_1).
\end{aligned}
\qquad (2.4)
$$

Define w_{i2}, w_{i3} and β as

$$w_{i2} \equiv (x'_{i2} - x'_{i1}, \ z'_i, \ 0_{\dim(z)'})',$$
$$w_{i3} \equiv (x'_{i3} - x'_{i1}, \ 0_{\dim(z)'}, \ z'_i)', \qquad (2.5)$$
$$\beta \equiv (\delta', \ \eta'_2 - \eta'_1, \ \eta'_3 - \eta'_1)',$$

where $0_{\dim(z)}$ is the zero vector with the same dimension as z. Then the above regression function differences in (2.4) can be written as

$$w'_{i2}\beta \quad \text{and} \quad w'_{i3}\beta,$$

which have the same form as (2.1). Define w_i as w_{ij}'s stacked for $j = 1 \ldots J$.
   Define y_{ij} as

$$y_{ij} = 1 \quad \text{if the } i\text{th person chooses } j$$
$$= 0 \quad \text{otherwise.}$$

The first alternative will be chosen, if (ignoring ties)

$$s_{i1} > s_{i2}, s_{i1} > s_{i3}, \ldots, s_{i1} > s_{iJ}.$$

Note that

$$\sum_{j=1}^{J} y_{ij} = 1, \ \sum_{j=1}^{J} P(y_{ij} = 1 \mid w_i) = 1 \quad \text{for all } i. \qquad (2.6)$$

Setting $J = 3$ and omitting i for simplicity, we get

$$P(y_1 = 1 \mid w) = P(s_1 > s_2, s_1 > s_3 \mid w)$$
$$= P(u_2 - u_1 < -w'_2\beta, u_3 - u_1 < -w'_3\beta \mid w),$$
$$P(y_2 = 1 \mid w) = P(u_2 - u_1 > -w'_2\beta, u_3 - u_2 < (w_2 - w_3)'\beta \mid w)$$
$$= P(u_2 - u_1 > -w'_2\beta, (u_2 - u_1) - (u_3 - u_1) > -(w_2 - w_3)'\beta \mid w),$$
$$P(y_3 = 1 \mid w) = P(u_3 - u_1 > -w'_3\beta, u_3 - u_2 > (w_2 - w_3)'\beta \mid w)$$
$$= P(u_3 - u_1) > -w'_3\beta, (u_3 - u_1) - (u_2 - u_1) > -(w_3 - w_2)'\beta \mid w). \ (2.7)$$

Here all choice probabilities are written in terms of the difference from the first alternative. Further define the following

$$v_2 \equiv u_2 - u_1, \ v_3 \equiv u_3 - u_1, \ \sigma^2 \equiv V(v_2), \qquad (2.8)$$

and rewrite (2.7) as (omit "$\mid w$" for now)

$$P(y_1 = 1) = P(v_2 < -w'_2\beta, v_3 < -w'_3\beta)$$
$$= P(v_2/\sigma < -w'_2(\beta/\sigma), v_3/\sigma < -w'_3(\beta/\sigma)),$$
$$P(y_2 = 1) = P(v_2 > -w'_2\beta, v_2 - v_3 > -(w_2 - w_3)'\beta)$$
$$= P(v_2/\sigma) > -w'_2(\beta/\sigma), (v_2 - v_3)/\sigma > -(w_2 - w_3)'(\beta/\sigma)),$$
$$P(y_3 = 1) = P(v_3 > -w'_3\beta, v_3 - v_2 > -(w_3 - w_2)'\beta)$$
$$= P(v_3/\sigma > -w'_3(\beta/\sigma), (v_3 - v_2)/\sigma > -(w_3 - w_2)'(\beta/\sigma)).$$

In view of this, the case with the generic J can be written as

$$P(y_1 = 1) = P(v_2/\sigma < -w_2'(\beta/\sigma), \ldots, v_J/\sigma < -w_J'(\beta/\sigma)),$$
$$P(y_2 = 1) = P(v_2/\sigma > -w_2'(\beta/\sigma), (v_2 - v_3)/\sigma > -(w_2 - w_3)'(\beta/\sigma),$$
$$\ldots, (v_2 - v_J)/\sigma > -(w_2 - w_J)'(\beta/\sigma)),$$

$$\vdots$$

$$P(y_J = 1) = P(v_J/\sigma > -w_J'(\beta/\sigma), (v_J - v_2)/\sigma > -(w_J - w_2)'(\beta/\sigma),$$
$$\ldots, (v_J - v_{J-1})/\sigma > -(w_J - w_{J-1})'(\beta/\sigma)) \tag{2.9}$$

where v_j and w_j for $j \geq 4$ are defined analogously to v_3 and w_3. The probabilities in (2.9) depend on the joint distribution of $(v_2/\sigma, \ldots, v_J/\sigma)$. The parameters to estimate are β/σ and the covariance matrix of $(v_2/\sigma, \ldots, v_J/\sigma)$.

To see better what is actually estimated in the covariance, consider the case $J = 3$. In the variance matrix of $(v_2/\sigma, v_3/\sigma)$, only

$$\sigma_{23} \equiv \text{COV}(v_2/\sigma, v_3/\sigma) \quad \text{and} \quad \sigma_3^2 \equiv V(v_3/\sigma)$$

are unknown $[V(v_2/\sigma) = 1$ by the definition of $\sigma]$. Thus, overall, the parameters to estimate when $J = 3$ are

$$\gamma \equiv (\beta'/\sigma, \sigma_{23}, \sigma_3^2)',$$

where β is identified only up to the scale σ as in the binary model. In practice, usually we do not estimate σ_{23} and σ_3^2 directly, since $\sigma_3^2 > 0$ is not guaranteed in numerical algorithms. Instead, we use a triangular decomposition of the variance matrix Σ of $u_{2/\sigma}, u_{3/\sigma})'$: set $\Sigma = AA'$ where A is lower-triangular. That is, for our case, consider a_{23} and a_{33} such that

$$\begin{bmatrix} 1 & 0 \\ a_{23} & a_{33} \end{bmatrix} \cdot \begin{bmatrix} 1 & a_{23} \\ 0 & a_{33} \end{bmatrix} = \begin{bmatrix} 1 & a_{23} \\ a_{23} & a_{23}^2 + a_{33}^2 \end{bmatrix} = \begin{bmatrix} 1 & \sigma_{23} \\ \sigma_{23} & \sigma_3^2 \end{bmatrix}. \tag{2.10}$$

So we estimate $(\beta'/\sigma, a_{23}, a_{33})'$ where $a_{23} = \sigma_{23}$ and $\sigma_3^2 = a_{23}^2 + a_{33}^2$. This way, the estimate for σ_3^2 is always positive. If $J = 4$, then γ becomes

$$\gamma \equiv (\beta'/\sigma, \sigma_{23}, \sigma_{24}, \sigma_3^2, \sigma_{34}, \sigma_4^2)'.$$

Again, we can use the triangular decomposition. In the generic case with J,

$$\gamma = (\beta'/\sigma, \sigma_{23}, \ldots, \sigma_{2J}, \sigma_3^2, \sigma_{34}, \ldots, \sigma_{3J}, \sigma_4^2, \sigma_{45}, \ldots, \sigma_{5J},$$
$$\cdots\cdots\cdots, \sigma_{J-1}^2, \sigma_{J-1,J}, \sigma_J^2). \tag{2.11}$$

Defining $P_{ij}(\gamma)$ as

$$P_{ij}(\gamma) = P(i \text{ chooses } j \mid w_i), \tag{2.12}$$

the log-likelihood function to be maximized wrt g is

$$(1/N) \sum_{i=1}^{N} \sum_{j=1}^{J} y_{ij} \ln P_{ij}(g). \tag{2.13}$$

The first derivative is

$$(1/N)\sum_{i=1}^{N}\sum_{j=1}^{J}y_{ij}\{\partial P_{ij}(g)/\partial g\}/P_{ij}(g). \tag{2.14}$$

The score function is $\sum_{j=1}^{J}y_{ij}\{\partial P_{ij}(g)/\partial g\}/P_{ij}(g)$, and $\sqrt{N}(g_N - \gamma)$ follows $N(0, I_f^{-1})$, where I_f is estimated by the outer-product of the score:

$$(1/N)\sum_{i=1}^{N}\left[\sum_{j=1}^{J}y_{ij}\{\partial P_{ij}(g)/\partial g\}/P_{ij}(g)\right]$$

$$\cdot\left[\sum_{j}y_{ij}\{\partial P_{ij}(g)/\partial g\}/P_{ij}(g)\right]'.$$

Differentiating the second equation of (2.6), we have

$$\sum_{j=1}^{J}P_{ij}(g)\cdot\{\partial P_{ij}(g)/\partial g\}/P_{ij}(g) = \sum_{j=1}^{J}\partial P_{ij}(g)/\partial g = 0. \tag{2.15}$$

Subtract this from the first-order condition (2.14) to get

$$(1/N)\sum_{i=1}^{N}\sum_{j=1}^{J}\{y_{ij} - P_{ij}(g)\}\cdot\{\partial P_{ij}(g)/\partial g\}/P_{ij}(g)$$

$$= (1/N)\sum_{i=1}^{N}\sum_{j=1}^{J}\{y_{ij} - P_{ij}(g)\}\cdot\partial\ln P_{ij}(g)/\partial g = 0. \tag{2.16}$$

This can be viewed as a moment condition where $y_{ij} - P_{ij}(g)$ is the error term and $\partial\ln P_{ij}(g)/\partial g$ is the instrument.

If we assume that u_1 to u_J are jointly normally distributed, then v_2 to v_J are also jointly normally distributed. So we get *multinomial probit* where $P_{ij}(g)$ is obtained by integrating a $J - 1$ dimensional multivariate normal density function [recall (2.9)]. Numerical integration in high dimensions (say, greater than 4) is computationally burdensome and not reliable. Also getting $\partial P_{ij}(g)/\partial g$ is somewhat cumbersome. In the next two sections, we discuss how to avoid these problems in multinomial probit.

5.3   Multinomial Logit

Multinomial probit has two problems. One is the numerical integration already mentioned, and the other is estimating the covariance matrix. Although the covariance matrix is estimable, in practice the likelihood function is often rather flat for the elements of the covariance matrix. One way to avoid the problem in multinomial probit is multinomial logit [see Maddala (1983) and McFadden (1984)].

If $-u_{i1}, \ldots, -u_{iJ}$ are iid with the type I extreme value distribution, then we get the following *multinomial logit* specification:

$$P_{ij}(\beta) = \exp(x'_{ij}\delta + z'_i\eta_j) \Big/ \sum_{j=1}^{J} \exp(x'_{ij}\delta + z'_i\eta_j). \qquad (3.1)$$

The assumption on $(u_{i1} \ldots u_{iJ})$ is restrictive, for the error terms have the same variance and are uncorrelated with one another. This is essentially throwing away the covariance matrix estimation problem in multinomial probit. With (3.1), the ratio of the probabilities P_{ij} and P_{im}, $j \neq m$, is [recall (2.5) for the notations]

$$P_{ij}(\beta)/P_{im}(\beta) = \exp(x'_{ij}\delta + z'_i\eta_j - x'_{im}\delta - z'_i\eta_m) = \exp\{(w_{ij} - w_{im})'\beta\};$$

availability of the other alternatives does not play any role in this ratio so long as the attributes of the other alternatives are not in $w_{ij} - w_{im}$. This feature, known as *independence of irrelevant alternatives* (IIA), is unlikely to hold in real life when some other alternatives are similar to either j or m in their attributes. A well-known example against IIA is three commuting alternatives: a blue bus, a red bus, and a car, where the probability ratio of red bus to car should depend on the availability of a blue bus, for a blue bus will take away one half of the red bus commuters.

To see a more restrictive nature of multinomial logit [see Anderson et al. (1992, p. 44) for the following], define

$$\lambda_{ij} \equiv x'_{ij}\delta + z'_i\eta_j, \quad \lambda_{im} \equiv x'_{im}\delta + z'_i\eta_m$$

to get

$$P_{ij}(\beta) = \exp(\lambda_{ij})\Big/\sum_k \exp(\lambda_{ij}), \quad P_{im}(\beta) = \exp(\lambda_{im})\Big/\sum_j \exp(\lambda_{ij}).$$

Omitting i and β, we have

$$\partial P_j/\partial\lambda_j = P_j(1 - P_j), \quad \partial P_j/\partial\lambda_m = -P_jP_m,$$

$$\partial \ln P_j/\partial \ln \lambda_j = (1 - P_j)\lambda_j, \quad \partial \ln P_j/\partial \ln \lambda_m = -P_m\lambda_m.$$

In the last equality, the cross elasticities wrt λ_m are the same for all j. This highly restrictive feature can be shown to be due to IIA. If the alternative m is dropped from the choice set, then the increase in the choice probability in the remaining alternatives is the old choice probability times a factor common to all alternatives. To be specific,

$$\text{new prob.} = \text{old prob.} \cdot \sum_j \exp(\lambda_{ij})\Big/\sum_{j\neq m} \exp(\lambda_{ij});$$

the second term in the right-hand side (rhs) is greater than 1.

Divide the numerator and the denominator of $P_{i1} \ldots P_{iJ}$ by P_{i1} for a normalization (this does not change the choice probabilities) to get

$$P_{i1} = 1 \Big/ \left\{ 1 + \sum_{j=2}^{J} \exp(w'_{ij}\beta) \right\},$$

$$P_{ij} = \exp(w'_{ij}\beta) \Big/ \left\{ 1 + \sum_{j=2}^{J} \exp(w'_{ij}\beta) \right\}, \quad j = 2 \ldots J. \tag{3.2}$$

Recall the first derivative (2.14) with g replaced by b:

$$(1/N) \sum_{i=1}^{N} \sum_{j=1}^{J} y_{ij} \{\partial P_{ij}(b)/\partial b\}/P_{ij}(b). \tag{3.3}$$

Getting $\partial \ln P_{ij}(b)/\partial b = \{\partial P_{ij}(b)/\partial b\}/P_{ij}(b)$,

$$\partial \ln P_{i1}(b)/\partial b = - \left(\sum_{j=2}^{J} w_{ij} P_{ij} \right),$$

$$\partial \ln P_{ij}(b)/\partial b = w_{ij} - \left(\sum_{j=2}^{J} w_{ij} P_{ij} \right), \quad j = 2 \ldots J. \tag{3.4}$$

Substituting this into (3.3), (3.3) becomes

$$(1/N) \sum_{i=1}^{N} \left\{ -y_{i1} \left(\sum_{j=2}^{J} w_{ij} P_{ij} \right) + \sum_{j=2}^{J} y_{ij} w_{ij} - \sum_{j=2}^{J} y_{ij} \left(\sum_{j=2}^{J} w_{ij} P_{ij} \right) \right\}$$

$$= (1/N) \sum_{i=1}^{N} \left\{ \sum_{j=2}^{J} y_{ij} w_{ij} - \sum_{j=2}^{J} w_{ij} P_{ij} \right\} \quad \left(\text{for } \sum_{j=1}^{J} y_{ij} = 1 \right)$$

$$= (1/N) \sum_{i=1}^{N} \sum_{j=2}^{J} w_{ij}(y_{ij} - P_{ij}). \tag{3.5}$$

The last line may be viewed as a moment condition analogously to (2.16), with $y_{ij} - P_{ij}$ being the error term.

The likelihood function is well behaving and multinomial logit is computationally attractive despite its rigid specification. The computational advantage is too great to ignore, particularly in dynamic discrete choice problems [see Eckstein and Wolpin (1989) for a survey] where the regression functions cannot be written in a closed form (they involve a "value function" in dynamic programming). See Rust (1987, 1994) for dynamic multinomial logit.

One way to correct for IIA in multinomial logit is "nested (multinomial) logit," which allows error term correlations in subsets of similar alternatives, but the nested logit does not behave well computationally with many local maxima. Also "simulated methods of moments" discussed in the next section solve the computational problem in multinomial probit. Hence, the nested logit is not as attractive as it used to be, and we omit any discussion of it.

5.4   Methods of Simulated Moments (MSM)

In this section, we introduce a method of simulated moments in McFadden (1989), which solves the computational problem in multinomial probit. The idea of *method of simulated moments* (MSM) is to recast multinomial probit in a method-of-moments framework, and estimate $P(y_{ij} = 1)$ using simulated random variables (rv's) drawn from the same distribution as that of u_{ij}'s. This section serves as an introduction to the idea of using "simulators," which will be further discussed in the following section.

Recall (2.1) [dealing with (2.3) is similar]:

$$s_{ij} = x'_{ij}\beta + u_{ij}. \tag{4.1}$$

A slightly different specification, a *random coefficient model*, is

$$s_{ij} = x'_{ij}\alpha_i \equiv x'_{ij}(\beta + \Gamma e_i) = x'_{ij}\beta + x'_{ij}\Gamma e_i, \tag{4.2}$$

where e_i is a $J \times 1$ error vector and Γ is a matrix defined conformably. Defining $x'_{ij}\Gamma e_i$ as u_{ij}, (4.2) becomes (4.1) except that now u_{ij} has a known form of heteroskedasticity: with $e_i =^d N(0, I_J)$,

$$u_{ij} \mid x_{ij} = N(0, x'_{ij}\Gamma\Gamma'x_{ij}) \equiv N(0, x'_{ij}Gx_{ij}), \quad G \equiv \Gamma\Gamma'.$$

More generally, we may consider

$$s_{ij} = x'_{ij}\alpha_i + \zeta_{ij} = x'_{ij}(\beta + \Gamma e_i) + \zeta_{ij} = x'_{ij}\beta + (x'_{ij}\Gamma e_i + \zeta_{ij}), \tag{4.3}$$

which includes both (4.1) and (4.2). McFadden (1989) uses (4.2). Models (4.1) and (4.2) differ in simulating the error terms and getting the derivatives of P_{ij}. We will accommodate both models.

Denote the estimable parameter as a $k \times 1$ vector γ. Recall the method-of-moments interpretation of multinomial probit in (2.16):

$$(1/N) \sum_{i=1}^{N} \sum_{j=1}^{J} (y_{ij} - P_{ij}) \cdot \partial \ln P_{ij}/\partial g = 0. \tag{4.4}$$

Regard the true parameter as the one satisfying the population version of this moment condition; there may be many such parameters, which is an inherent problem once we turn a MLE into a method of moments. Define the instrument vector

$$z_{ij} \equiv \partial \ln P_{ij}/\partial g = \{\partial P_{ij}/\partial g\}/P_{ij},$$

which depends on the unknown parameters, differently from the usual case of instruments being observed variables. Further define y_i, P_i, and z_i as [dimension is in $(\cdot)$]

$$
\begin{aligned}
y_i(J \times 1) &\equiv (y_{i1}, \ldots, y_{iJ})' & P_i(J \times 1) &\equiv (P_{i1}, \ldots, P_{iJ})', \\
u_i(J \times 1) &\equiv (u_{i1}, \ldots, u_{iJ})' & z_i(k \times J) &\equiv (z_{i1}, \ldots, z_{iJ}),
\end{aligned}
\tag{4.5}
$$

to rewrite (4.4) as

$$(1/N) \sum_{i=1}^{N} z_i(y_i - P_i) = 0. \tag{4.6}$$

Alternatively, define $Z_j(N \times k)$, $Y_j(N \times 1)$, and $P_j(N \times 1)$ to rewrite (4.4) as

$$\sum_{j=1}^{J} Z_j'(Y_j - P_j)/N = 0; \tag{4.7}$$

(4.6) may be more convenient for theory and (4.7) may be better for computation.

The key step in the MSM is simulating $P_i(g)$. Let n be the simulated (or generated) sample size. Then the MSM can be done as follows.

(i) Generate $\{\varepsilon_i(t)\}_{t=1}^n$ for each i such that $\varepsilon_i(t)$, $t = 1 \ldots n$, are iid following $N(0, I_J)$. $\{\varepsilon_i(t)\}_{t=1}^n$ is independent of $\{\varepsilon_m(t)\}_{t=1}^n$ for all $m \neq i$. Overall, $N \cdot J \cdot n$ rv's are generated.

(ii) For the random coefficient model (4.2), fix a matrix Γ (recall $\Gamma\Gamma' = G$) and obtain $\eta_{ij}(t)$, a "pseudo u_{ij}," by

$$(\eta_{i1}(t) \ldots \eta_{iJ}(t)) \equiv (x_{i1}' \ldots x_{iJ}')\Gamma\varepsilon_i(t), \quad \text{for } t = 1 \ldots n. \tag{4.8}$$

For (4.1), let $\Omega \equiv V[(u_{i1} \ldots u_{iJ})']$ and fix A such that $AA' = \Omega$. Then

$$(\eta_{i1}(t) \ldots \eta_{iJ}(t))' = A\varepsilon_i(t), \quad \text{for } t = 1 \ldots n.$$

(iii) Fix b and generate $s_{ij}(t)$:

$$s_{ij}(t) = x_{ij}'b + \eta_{ij}(t) \quad \text{for } t = 1 \ldots n.$$

(iv) Find the relative frequency that the ith person chooses j:

$$f_{ij}(g) \equiv (1/n) \sum_{t=1}^{n} 1[s_{ij}(t) > s_{im}(t), \ m \neq j],$$

which is a "simulated estimate" for $P_{ij}(g)$. Stack up $f_{ij}(g)$ to get the $J \times 1$ simulated estimate vector $f_i(g)$.

(v) To get $f_i(g)$ for different values of g, simply repeat (ii) through (iv) with the same $\varepsilon_i(t)$'s obtained in (i).

(vi) With $f_i(g)$ replacing $P_i(g)$, iterate until convergence with

$$g_1 = g_0 + \left[(1/N) \sum_i z_i(g_0)\{y_i - P_i(g_0)\}\{y_i - P_i(g_0)\}' z_i(g_0)' \right]^{-1}$$
$$\cdot (1/N) \sum_i z_i(g_0)\{y_i - P_i(g_0)\}.$$

In the iteration formula, it is explicit that both P_i and z_i depend on g. The formula is derived from "Gauss–Newton" algorithm (to be seen in the following chapter) whose idea is simple: replace P_i in (4.6) with $\{\partial P_i(g_0)/\partial g\} \cdot (g_1 - g_0)$ and solve (4.6) for g_1. This is a linearization of the nonlinear function $P_i(g)$. Since $z_{ij}(g)$ is also a nonlinear function of g, we may want to linearize $z_{ij}(g)$ too. But if $g_0 \cong \gamma$, the term

$$(1/N) \sum_i \sum_j \{y_{ij} - P_{ij}(g_0)\} \cdot \{\partial z_{ij}(g_0)/g\}$$

is negligible, which is why $\partial z_{ij}/\partial g$ does not appear in the iteration scheme. This idea of omitting negligible terms when $g_0 \cong \gamma$ was already seen in the method of scoring.

In order to get z_{ij}, we need $\partial P_{ij}/\partial g$, which can be obtained analytically as in McFadden (1989) and Hajivassiliou and Ruud (1984). But the derivative is somewhat complicated. Instead, we can use a numerical derivative

$$\partial P_{ij}(g)/\partial g_m \cong \{P_{ij}(g + he_m) - P_{ij}(g - he_m)\}/(2h), \quad m = 1, \ldots, k,$$

where e_m is a $k \times 1$ zero vector with its mth component replaced by 1, and h is a small positive scalar close to 0. Since z_i should be orthogonal to $y_i - P_i$, the generated ε_{it}'s already used in $y_i - P_i$ are not good for simulating z_i. Instead, generate new error vectors and use them to simulate z_i through a procedure analogous to (i)–(v). Since the simulator $f_{ij}(g)$ has an indicator function, the numerical derivatives may be difficult to get; better simulators replacing the indicator function with a smooth function are shown in the following section.

In the remainder of this section, we derive the asymptotic distribution of the MSM. The reader may skip the details to (4.19).

If P_i and z_i are obtained without simulation, then we can get a method-of-moments estimator g_{mme} satisfying (4.4); g_{mme} has the same asymptotic variance as the MLE:

$$\sqrt{N}(g_{mme} - \gamma) =^d N\left(0, \left[(1/N)\sum_i z_i(y_i - P_i)(y_i - P_i)'z_i'\right]^{-1}\right). \quad (4.9)$$

The MSM g_{msm} will be shown to have a variance slightly larger than this due to the simulation error. As is the case usually, the error in estimating the instrument z_i does not affect the variance of g_{msm}, so we will treat z_i as if it is an observed variable. The asymptotic distribution of $g_{msm} = g_N$ can be derived from the moment condition

$$(1/\sqrt{N})\sum_i z_i\{y_i - f_i(g_N)\} = (1/\sqrt{N})\sum_i z_i[\{y_i - f_i(\gamma)\}$$

$$- \{P_i(g_N) - P_i(\gamma)\} + \{f_i(\gamma) - P_i(\gamma)\} - \{f_i(g_N) - P_i(g_N)\}] = 0. \quad (4.10)$$

Using "stochastic equicontinuity" [see Pollard (1984) and Pakes and Pollard (1989) for instance] and $E\{f_i(\gamma)\} = P_i(\gamma)$ [this unbiasedness of $f_i(\gamma)$ for $P_i(\gamma)$ holds due to a LLN and the fact that $f_i(\gamma)$ is an average of the simulated samples], we can show that

$$(1/\sqrt{N})\sum_i z_i[\{f_i(\gamma) - P_i(\gamma)\} - \{f_i(g_N) - P_i(g_N)\}] =^p 0. \quad (4.11)$$

Then, (4.10) becomes

$$(1/\sqrt{N})\sum_i z_i\{y_i - f_i(\gamma)\} - (1/\sqrt{N})\sum_i z_i\{P_i(g_N) - P_i(\gamma)\} =^p 0. \quad (4.12)$$

Apply the mean value theorem to $P_i(g_N)$ around γ to get

$$P_i(g_N) - P_i(\gamma) = \{\partial P_i(g^*)/\partial g'\} \cdot (g_N - \gamma).$$

Substitute this into (4.12) to get

$$(1/\sqrt{N})\sum_i z_i\{y_i - f_i(\gamma)\} - \left\{(1/N)\sum_i z_i\partial P_i/\partial g'\right\}\sqrt{N}(g_N - \gamma) =^p 0. \quad (4.13)$$

$$\Leftrightarrow \sqrt{N}(g_N - \gamma) =^p \left\{(1/N)\sum_i z_i\partial P_i/\partial g'\right\}^{-1} \cdot (1/\sqrt{N})\sum_i z_i\{y_i - f_i(\gamma)\}$$

$$=^p -C^{-1} \cdot (1/\sqrt{N})\sum_i z_i\{y_i - f_i(\gamma)\}, \quad (4.14)$$

where

$$C \equiv -(1/N) \sum_i z_i \partial P_i(\gamma)/\partial g' = (1/N) \sum_i z_i(y_i - P_i)(y_i - P_i)'z_i'.$$

Owing to $E\{f_i(\gamma)\} = P_i(\gamma) = E(y_i)$, we get $E(y_i - f_i) = 0$ and so

$$(1/\sqrt{N}) \sum_i z_i\{y_i - f_i(\gamma)\} = N\left(0, (1/N) \sum_i z_i(y_i - f_i)(y_i - f_i)'z_i'\right).$$
(4.15)

Subtracting and adding P_i to y_i, the variance in (4.15) becomes

$$C + (1/N) \sum_i z_i(y_i - P_i)(P_i - f_i)'z_i' + (1/N) \sum_i z_i(P_i - f_i)(y_i - P_i)'z_i'$$

$$+ (1/N) \sum_i z_i(P_i - f_i)(P_i - f_i)'z_i'.$$
(4.16)

By a LLN, the second (and third) term converges to 0:

$$E\{z_i(y_i - P_i)(P_i - f_i)'z_i'\} = E[z_i \cdot \{E_{y|z_i}(y_i - P_i)\} \cdot (P_i - f_i)'z_i']$$

$$= E[z_i(P_i - P_i) \cdot (P_i - f_i)'z_i'] = 0.$$
(4.17)

By a LLN, the fourth term converges to (ε is the simulation error)

$$E[z_i \cdot E_\varepsilon\{(P_i - f_i)(P_i - f_i)'\} \cdot z_i'].$$
(4.18)

$E_\varepsilon\{(P_i - f_i)(P_i - f_i)'\}$ is the variance of f_i as a simulation estimator for P_i which can be also written as $E\{(P_i - y_i)(P_i - y_i)'\}/n$. Hence (4.16) is $C + (1/n) \cdot C = \{1 + (1/n)\} \cdot C$, and so

$$\sqrt{N}(g_{msm} - \gamma) =^d N(0, C^{-1} \cdot \{1 + (1/n)\} \cdot C \cdot C^{-1}) = N(0, \{1 + (1/n)\}C^{-1}).$$
(4.19)

If $n \to \infty$, then we get $\sqrt{N}(g_{msm} - \gamma) =^d N(0, C^{-1})$; since C is the information matrix, and MSM is as efficient as MLE when $n \to \infty$.

5.5   Smooth Simulators and Methods of Simulated Likelihood

In this section, first we introduce simulators improving upon the frequency simulator f_{ij} in (4.8). Then we discuss "methods of simulated likelihood."

The simulator f_{ij} in (4.8) has the following three disadvantages. First, f_{ij} is not a smooth function due to the indicator function. Second, f_{ij} can be 0, a problem for $z_{ij} = \{\partial P_{ij}/\partial g\}/P_{ij}$. Third, it may take too many draws (too large n) to estimate a small P_{ij}. Replacing the indicator function with a smooth function taking a value in $(0,1)$ can solve these problems. In the following, we introduce two smoothing ideas in Stern (1992) and Börsch–Supan and Hajivassiliou (1993); see McFadden (1989) and Hajivassiliou and Ruud (1994) for more.

To see the idea of Stern (1992), observe that the event $s_{ij} > s_{im}$, $m \neq j$, for $f_{ij}(\gamma)$ in (4.8) can be written as [we omit t in $s_{ij}(t)$ here, and use γ instead of g in $f_{ij}(\cdot)$ to prevent notational confusion]

$$a_i < A_j v_i / \sigma \tag{5.1}$$

for a $(J-1) \times (J-1)$ matrix A_j, a $(J-1) \times 1$ vector $a_i \equiv (a_{i2} \ldots a_{iJ})'$, and $v_i \equiv (v_{i2} \ldots v_{iJ})'$; recall the definition of $v_{ij} \equiv u_{ij} - u_{i1}$ for $j \geq 2$ in Section 2. For instance, for $J = 3$ and $j = 3$, recalling (2.8) to (2.9),

$$a_i = \begin{bmatrix} -w'_{i3}(\beta/\sigma) \\ -(w_{i3} - w_{i2})'(\beta/\sigma) \end{bmatrix} \quad A_3 = \begin{bmatrix} 0 & 1 \\ -1 & 1 \end{bmatrix}. \tag{5.2}$$

Since $V(u_i) = \Omega$ and $v_i = Bu$ for a constant matrix B composed of 0 and ± 1, $V(v_i/\sigma) = B\Omega B'/\sigma^2 \equiv \Sigma$; we estimate $(\beta/\sigma$ and$)$ Σ, not Ω. Then $V(A_j v_i/\sigma) = A_j \Sigma A_j' \equiv M$. Using the additivity of normal rv's, decompose $A_j v_i/\sigma$ as

$$A_j v_i/\sigma = \zeta_D + \zeta_E, \quad \zeta_D \cong N(0, D), \quad \zeta_E \cong N(0, M - D), \tag{5.3}$$

where D is a diagonal matrix $\mathrm{diag}(d_2 \ldots d_J)$, $M - D$ is p.d., $\zeta_D = (\zeta_{D2} \ldots \zeta_{DJ})'$, and $\zeta_E = (\zeta_{E2} \ldots \zeta_{EJ})'$; see Stern (1992) for how to choose D. Note that ζ_D and ζ_E should carry the subscript ij, which we omit however. Then

$$P_{ij} = P(a_i < A_j v_i/\sigma) = P(a_i - \zeta_E < \zeta_D)$$

$$= \int P(a_i - \zeta_E < \zeta_D \mid \zeta_E) \cdot f(\zeta_E) d\zeta_E = \int H(\zeta_E) \cdot f(\zeta_E) d\zeta_E, \tag{5.4}$$

where $f(\zeta_E)$ is the density function for ζ_E, and

$$H(\zeta_E) = \prod_{j=2}^{J} \Phi\{(\zeta_{Ej} - a_{ij})/d_j\}.$$

Drawing $\{\zeta_E(t)\}_{t=1}^{n}$ from $N(0, M - D)$, $P_{ij}(\gamma)$ can be simulated by

$$(1/n) \sum_{t=1}^{n} H\{\zeta_E(t)\}. \tag{5.5}$$

Stern's (1992) simulator, which is unbiased, has two advantages over the frequency simulator. First, it is differentiable wrt the parameters and the derivatives of the simulator are unbiased. Second, the variance of the simulator is smaller than that of the frequency simulator. This is because only ζ_E is simulated and $V(\zeta_E)$ is smaller than M.

Turning to Börsch–Supan and Hajivassiliou (1993), assume $J = 3$ and $j = 3$ as in the preceding paragraph for ease of exposition, which makes $A_j v_i$ two dimensional. Let $M = TT'$, where T is the lower triangular Cholesky factor ($\tau_{ii} = \sqrt{m_{ii}}$ where $M = [m_{ij}]$),

$$T = \begin{bmatrix} \tau_{22} & 0 \\ \tau_{32} & \tau_{33} \end{bmatrix}. \tag{5.6}$$

Let $e \equiv (e_2, e_3)$ where e_2 and e_3 are iid $N(0, 1)$. Then $Te \cong N(0, M)$. Consider $a_i < Te$ instead of $a_i < A_j v_i / \sigma$. The inequality $a_i < Te$ is

$$a_{i2}/\tau_{22} < e_2 \quad \text{and} \quad (a_{i3} - \tau_{32} e_2)/\tau_{33} < e_3. \tag{5.7}$$

Then

$$P_{ij}(\gamma) = P\{(a_{i2}/\tau_{22}) < e_2\} \cdot P\{(a_{i3} - \tau_{32} e_2)/\tau_{33} < e_3\}$$

$$= \Phi(-a_{i2}/\tau_{22}) \cdot \Phi\{(-a_{i3} + \tau_{32} e_2)/\tau_{33}\} \equiv Q_2 \cdot Q_3(e_2), \tag{5.8}$$

and $P_{ij}(\gamma)$ can be simulated by

$$(1/n) \sum_{t=1}^{n} \Phi(-a_{i2}/\tau_{22}) \cdot \Phi\{(-a_{i3} + \tau_{32} e_2)/\tau_{33}\}$$

$$= (1/n) \sum_{t=1}^{n} Q_2 \cdot Q_3(e_{2t}). \tag{5.9}$$

For $J = 4$ and $j = 4$, there will be one more inequality in (5.7) for e_4 being greater than a function of e_2 and e_3, and the simulator will look like $(1/n) \sum_{t=1}^{n} Q_2 \cdot Q_3(e_{2t}) \cdot Q_4(e_{2t}, e_{3t})$, from which we can infer that the simulator for the general case looks like

$$(1/n) \sum_{t=1}^{n} \{Q_2 \cdot Q_3(e_{2t}) \cdot Q_4(e_{2t}, e_{3t}) \cdots Q_J(e_{2t} \ldots e_{J-1,t})\}. \tag{5.10}$$

As for simulating $e_{2t} \cong N(0, 1)$ truncated by a_{i2}/τ_{22}, we can use

$$\Phi^{-1}[\{1 - \Phi(a_{i2}/\tau_{22})\} \cdot \xi + \Phi(a_{i2}/\tau_{22})], \tag{5.11}$$

where $\zeta \cong U[0, 1]$; to see this, imagine $P((5.11) < s)$ and invert Φ^{-1}. More generally, $e \cong N(0, 1)$ subject to $c_1 < e < c_2$ can be simulated by

$$\Phi^{-1}[\{\Phi(c_2) - \Phi(c_1)\} \cdot \xi + \Phi(c_1)]. \tag{5.12}$$

The simulator in (5.10) is unbiased and likely to have a smaller variance than the Stern's simulator when M has large off-diagonal terms.

To introduce methods of simulated likelihood (MSL), consider a duration problem where $\{(y_i, x_i')\}_{i=1}^N$ are observed and y_i is the survival time. Suppose $y_i \mid (x_i', \varepsilon_i)$ follows a Weibull distribution, where ε is an unobserved "error" term (the so-called "heterogeneity" factor) following $N(0, 1)$. As in the usual linear model, ε can be regarded as a collection of unobserved variables affecting y_i. The hazard function of $y \mid (x', \varepsilon)$ is (recall our discussion in the preceding chapter)

$$\lambda(y, x, \varepsilon) = \alpha y^{\alpha-1} \cdot \exp(x'\beta + \varepsilon). \tag{5.13}$$

Since ε is not observed, we need the hazard of $y \mid x$, not of $y \mid (x', \varepsilon)$. Denote the $N(0, 1)$ density as ϕ and integrate (5.13) wrt ε:

$$\lambda(y, x) = \int \alpha y^{\alpha-1} \cdot \exp(x'\beta + \varepsilon)\phi(\varepsilon)d\varepsilon. \tag{5.14}$$

Then the maximand for the MLE is ($d_i = 1$ if uncensored and 0 otherwise, and Λ is the integrated hazard)

$$(1/N) \sum_i d_i \ln \lambda(y_i, x_i) - (1/N) \sum_i \Lambda(y_i, x_i)$$

$$= (1/N) \sum_i d_i \ln \left\{ \int a y_i^{a-1} \exp(x_i'b + \varepsilon)\phi(\varepsilon)d\varepsilon \right\}$$

$$- (1/N) \sum_i \int y_i^a \exp(x_i'b + \varepsilon)\phi(\varepsilon)d\varepsilon \tag{5.15}$$

to be maximized wrt a and b. Here the integration can be done numerically. But we can use a simulator as well: generate n $N(0, 1)$ random variables for each (y_i, x_i') and (a, b') to get

$$\lambda_n(y_i, x_i) \equiv (1/n) \sum_{t=1}^n a y_i^{a-1} \cdot \exp(x_i'b + \varepsilon(t)),$$

$$\Lambda_n(y_i, x_i) \equiv (1/n) \sum_{t=1}^n y_i^a \cdot \exp(x_i'b + \varepsilon(t)). \tag{5.16}$$

Then we can maximize (5.15) with λ and Λ replaced by the simulators.

More generally, consider a log-likelihood function for $\{z_i\}_{i=1}^N$

$$(1/N) \sum_i \ln\{f(z_i, b)\}. \tag{5.17}$$

The estimation can be done iteratively with (omit z in f from now on)

$$b_1 = b_0 + \left\{ (1/N) \sum_i \{\nabla f(b)/f(b)\}\{\nabla f(b)/f(b)\}' \right\}^{-1} (1/N) \sum_i \nabla f(b)/f(b),$$

$$(5.18)$$

where $\nabla f(b) \equiv \partial f(b)/\partial b$. Suppose $\ln\{f(b)\}$ is an integrated quantity and we want to apply MSL. Then we need to simulate $\nabla f(b)$ as well as $f(b)$. In MSM, we showed $V[\sqrt{N}(g_{msm} - \gamma)] = (1 + n^{-1})C^{-1}$, where C^{-1} is the variance with no simulation error; that is, the simulation error results in one additional term C^{-1}/n. In MSL, however, it is difficult to get the variance with a finite n. To see the difficulty, examine the following which the MSL estimator b_{msl} should satisfy:

$$\sqrt{N}(b_{msl} - \beta) =^p \left\{ (1/N) \sum_i \{\nabla f_n(\beta)/f_n(\beta)\}\{\nabla f_n(\beta)/f_n(\beta)\}' \right\}^{-1}$$

$$\cdot (1/\sqrt{N}) \sum_i \{\nabla f_n(\beta)/f_n(\beta)\}, \qquad (5.19)$$

where f_n denotes a simulator for f. It can be shown that the inverted matrix converges to the information matrix I_f. If we have an unbiased simulator for the score function, that is, if

$$E\{\nabla f_n(\beta)/f_n(\beta)\} = \nabla f(\beta)/f(\beta), \qquad (5.20)$$

then we can rewrite $(1/\sqrt{N}) \sum_i \{\nabla f_n(\beta)/f_n(\beta)\}$ in (5.19) as

$$(1/\sqrt{N}) \sum_i \{\nabla f_n(\beta)/f_n(\beta)\} = (1/\sqrt{N}) \sum_i \{\nabla f(\beta)/f(\beta)\}$$

$$+ (1/\sqrt{N}) \sum_i [\nabla f_n(\beta)/f_n(\beta)\} - E\{\nabla f_n(\beta)/f_n(\beta)\}]. \qquad (5.21)$$

The first term on the rhs yields the variance of the MLE for (5.17), while a CLT can be applied to the second term, which reflects the pure simulation error. Thus, if (5.20) holds, $V[\sqrt{N}(b_{msl} - \beta)]$ will be a sum of two terms analogously to MSM. However, (5.20) does not hold in general: for two rv's λ_1 and λ_2, $E(\lambda_1/\lambda_2) \neq E(\lambda_1)/E(\lambda_2)$ in general. It can be shown that, if $n/\sqrt{N} \to \infty$, then the second term in (5.21) is negligible (that is, MSL becomes as efficient as the MLE) under certain conditions; for instance, see Propositions 4 and 5 in Hajivassiliou and Ruud (1994).

In short, although we can simulate unbiased estimates for f and ∇f separately, it is difficult to simulate an unbiased estimate for the score function. Since $\nabla f/f = \partial \ln f/\partial b$, if we can simulate $\ln\{f(b)\}$ which is differentiable wrt b, then it may be possible to get an unbiased simulator for the score function. To see that this idea will not work however, recall the duration example (5.14) to (5.15): what we need is $\ln\{E_\varepsilon \lambda(y, x, \varepsilon)\}$, the log of the likelihood for the observation (y, x'), not $E_\varepsilon \ln\{\lambda(y, x, \varepsilon)\}$; note that $E_\varepsilon \ln(\cdot) \leq \ln E_\varepsilon(\cdot)$. MSM, by using the moment condition, avoids this problem associated with MSL. In MSM, the simulation takes place in the error term [see $y_i - P_i$ in (4.6)], and the simulation error appears linearly rather than log-linearly. In MSL, the denominator of the score function is simulated, which is the source of the trouble. However, one should not take MSM as a panacea; as already mentioned, due to the potential multiple solutions in MSM, it is not clear how to establish identification in MSM, which is automatic in MSL. Readers with further interest in simulation-based methods can refer to Gourieroux and Monfort (1993) in addition to aforementioned papers.

5.6   Selection Models

Let s be a binary rv taking 1 and 0. Suppose we observe (x', y) only when $s = 1$. Our interest is still in $E(y \mid x)$, but what we can get from the observations is $E(y \mid x, s = 1)$. Note that

$$E(y \mid x) = E(y \mid x, s = 1) \cdot P(s = 1 \mid x) + E(y \mid x, s = 0) \cdot P(s = 0 \mid x).$$
(6.1)

If $E(y \mid x, s = 0) = E(y \mid x, s = 1)$, then $E(y \mid x) = E(y \mid x, s = 1)$. But, otherwise, we cannot get $E(y \mid x)$ from the selectively observed (x', y).

Suppose we take $E(\cdot \mid x, s = 1)$ on the linear model $y = x'\beta + u$ to get

$$E(y \mid x, s = 1) = x'\beta + E(u \mid x, s = 1).$$
(6.2)

So long as $E(u \mid x, s = 1)$ is not zero but a function of x, β is not $\partial E(y \mid x)/\partial x$, and applying LSE to (x', y) results in omitted variable bias [$E(u \mid x, s = 1)$ is omitted]. This bias is also called a *selection bias*. So the important factor is whether the event $s = 1$ changes $E(u \mid x) = 0$. While the source of the selection problem is simple, correcting for it requires specific knowledge on the event $s = 1$.

As the simplest example, consider the truncation of $y = x'\beta + u$ at 0 from below. In this case, $s = 1$ if $y > 0 \Leftrightarrow u > -x'\beta$, and

$$E(y \mid x, y > 0) = x'\beta + E(u \mid x, u > -x'\beta).$$
(6.3)

If u is independent of x, then $E(u \mid x, u > -x'\beta) = E(u \mid u > -x'\beta)$, which is a (decreasing) function of $x'\beta$. To be more specific, suppose $\varepsilon =^d N(0, 1)$. Then it can be shown that

$$E(\varepsilon \mid \epsilon > c) = \phi(c)/\Phi(-c) \quad \text{and} \quad E(\varepsilon \mid \varepsilon < c) = -\phi(c)/\Phi(c). \quad (6.4)$$

Assuming $u =^d N(0, \sigma^2)$, which is independent of x, we have $u = \varepsilon\sigma$ and

$$E(u \mid u > -x'\beta) = E(\sigma\varepsilon \mid \sigma\varepsilon > -x'\beta) = \sigma \cdot E(\varepsilon \mid \varepsilon > -x'\beta/\sigma)$$

$$= \sigma \cdot \phi(-x'\beta/\sigma)/\{1 - \Phi(-x'\beta/\sigma)\} = \sigma \cdot \phi(x'\beta/\sigma)/\Phi(x'\beta/\sigma). \quad (6.5)$$

Likewise,

$$E(u \mid u < -x'\beta) = -\sigma \cdot \phi(x'\beta/\sigma)/\Phi(-x'\beta/\sigma). \quad (6.6)$$

If u is not independent of x, say heteroskedasticity, then $E(u \mid x, u > -x'\beta)$ will not be as simple as (6.5). In the remainder of this section, we go over various selection model examples, increasingly more complicated.

Suppose we want to know what determines the earnings from getting a college diploma. If we collect a data set on earnings from college graduates (y_1) and their individual characteristics (x) to run LSE, then we will incur a selection problem, since those who tend to benefit more from college education will go to a college. To see this point better, denote the earnings from college education and no college education, respectively, as y_1 and y_2 and set up

$$y_1 = x'\beta_1 + u_1, \quad y_2 = x'\beta_2 + u_2. \quad (6.7)$$

College education is chosen if $y_1 > y_2 \Leftrightarrow u_1 - u_2 > -x'(\beta_1 - \beta_2)$. Thus,

$$\begin{aligned} s = 1[u_1 - u_2 > -x'(\beta_1 - \beta_2)] &= 1[(u_1 - u_2)/\sigma > -x'(\beta_1 - \beta_2)/\sigma] \\ &= 1[u > -x'(\beta_1 - \beta_2)/\sigma], \end{aligned} \quad (6.8)$$

where $\sigma^2 \equiv V(u_1 - u_2)$ and $u \equiv (u_1 - u_2)/\sigma$ [note that $V(u) = 1$ by definition].

Assume (u_1, u_2) follows $N(0, \Omega)$, which is independent of x, to get

$$E(u_1 \mid x, y_1 > y_2) = E(u_1 \mid u > -x'(\beta_1 - \beta_2)/\sigma). \quad (6.9)$$

Since (u_1, u) follows a joint normal distribution, we get, using the linear projection,

$$E(u_1 \mid u) = \{\text{COV}(u_1, u)/V(u)\} \cdot u = [\{E(u_1^2) - E(u_1 u_2)\}/\sigma] \cdot u$$

$$= \{(\sigma_1^2 - \sigma_{12})/\sigma\} \cdot u. \quad (6.10)$$

Consequently,

$$E(u_1 \mid u > -x'(\beta_1 - \beta_2)/\sigma) = \{(\sigma_1^2 - \sigma_{12})/\sigma\} \cdot E(u \mid u > -x'(\beta_1 - \beta_2)/\sigma). \quad (6.11)$$

Define $\alpha \equiv (\beta_1 - \beta_2)/\sigma$ and use (6.5) to get

$$E(u_1 \mid u > -x'\alpha) = \{(\sigma_1^2 - \sigma_{12})/\sigma\} \cdot \phi(x'\alpha)/\Phi(x'\alpha);$$

if $\text{cov}(u_1, u) = 0$, then the selection bias is zero. Therefore,

$$E(y_1 \mid x, y_1 > y_2) = x'\beta_1 + \{(\sigma_1^2 - \sigma_{12})/\sigma\} \cdot \phi(x'\alpha)/\Phi(x'\alpha). \quad (6.12)$$

If we knew α, then β_1 and $\{(\sigma_1^2 - \sigma_{12})/\sigma\}$ could be estimated by regressing y_1 on x and $\phi(x'\alpha)/\Phi(x'\alpha)$.

In the last example, we had observations only when $s = 1$. Generalizing this further, suppose we observe (x_1', y_1) if $s = 1$ and (x_2', y_2) if $s = 0$. Specifically,

$$s = 1[x_0'\beta_0 + u_0 > 0], \quad y_1 = x_1'\beta_1 + u_1, \quad y_2 = x_2'\beta_2 + u_2, \qquad (6.13)$$

where $(u_0, u_1, u_2)'$ follows $N(0, \Omega)$ independent of x with $V(u_0) = 1$ (normalization for the binary model). Here a "regime" is chosen by the criterion function s. This kind of model is called a *switching regression model*. If u_0 is correlated with u_1 and u_2, then the model is called *endogenous switching model*. If u_0 is not correlated with u_1 and u_2, then we have an *exogenous switching model*. Another classification for switching regression is whether sample separation is known or not. In the college example, sample separation was known. One example of unknown sample separation is a market disequilibrium model where $\min(y_1, y_2)$ is observed with $y_1 = $ demand and $y_2 = $ supply; that is, we only observe $\min(y_1, y_2)$ without knowing which one is the smaller.

The college example is an endogenous switching model with the same regressor x in both regimes. As an example of endogenous switching regression model with different regressors, consider a women thinking about working. She makes two decisions: one is whether to work or not, and the other is how much to work if she works. The first decision depends on her "reservation wage" (the minimum wage that she is willing to take) and offered wage: she will work if her offered wage $w = m'\delta_w + u_w$ is larger than her reservation wage $r = m'\delta_r + u_r$ where m is a regressor vector:

$$w > r \Leftrightarrow m'(\delta_w - \delta_r) + e > 0, \quad e \equiv u_w - u_r. \qquad (6.14)$$

Let $y = x'\beta + u$ be her working hour equation. Assume that $(e, u)'$ follows $N(0, \Omega)$ independent of x and m. Then, with $\alpha \equiv (\delta_w - \delta_r)/\sigma_e$,

$$E(y \mid x, m, e/\sigma_e > -m'\alpha) = x'\beta + E(u \mid e/\sigma_e > -m'\alpha)$$

$$= x'\beta + \text{COV}(u, e/\sigma_e) \cdot E(e/\sigma_e \mid e/\sigma_e > -m'\alpha)$$

$$= x'\beta + (\sigma_{eu}/\sigma_e) \cdot \phi(m'\alpha)/\Phi(m'\alpha). \qquad (6.15)$$

This includes (6.12) as a special case when $m = x$.

In the job choice example, selection is made by the subject, which may be called "self-selection." There are cases where the selection is made by somebody else, and in those cases there may or may not be a selection bias problem. Consider a job-training program for the poor where the administrator selects the trainees. To simplify this discussion, assume that those selected cannot refuse to take the training. The concern is on whether the job-training program is effective or not. If we do an experiment, an ideal way to test the program effectiveness is to select a number of people for the program and divide them randomly into two groups such that one group receives the training and the other group does not. Then looking at the difference of the (post-training) earnings of the two groups, we can see if the program is effective, without being misled by the selection bias.

The above scenario, however, is not attractive for we dont' want to send a group of people chosen for the program back home simply for the experiment; even when the social experiment is feasible, there are several contentious issues including "randomization bias" and "substitution bias" (Heckman and Smith, (1995). What is usually done in practice is comparing the job success of the trainees with that of those not selected for the program. But this does not tell much about the program effectiveness, since the program administrator may select people of high ability who have a better chance of higher earnings with or without the training than the nonselected people. This may result in a selection bias. To see this, assume that

$$y_i = x_i'(\beta + \alpha \cdot s_i) + u_i, \tag{6.16}$$

where y is earnings, α is a $k \times 1$ parameter vector and $s = 1$ if the subject was in the program and 0 otherwise. Note that this model is a special case of (6.7) with $u_1 = u_2$, $\beta = \beta_2$, and $\beta + \alpha = \beta_1$. We want to know if $\alpha = 0$, which implies the ineffectiveness of the program.

Observe that

$$E(y \mid x, s = 1) = x'(\beta + \alpha) + E(u \mid x, s = 1) \quad \text{for trainees;}$$

$$E(y \mid x, s = 0) = x'\beta + E(u \mid x, s = 0) \quad \text{for nontrainees.}$$

Subtracting the latter from the former, we get

$$E(y \mid x, s = 1) - E(y \mid x, s = 0) = x'\alpha + \{E(u \mid x, s = 1) - E(u \mid x, s = 0)\}. \tag{6.17}$$

The expression in $\{\cdot\}$ is the selection bias term. If we had selected both groups from the same population and had discarded one group randomly, then s would be independent of u and the bias term would be zero. If $\{\cdot\} \neq 0$, a positive left-hand side (lhs) does not necessarily imply $x'\alpha > 0$ (program effectiveness). Note that if we use (6.7), which is more general than (6.16), then (6.17) will look like

$$E(y_1 \mid x, s = 1) - E(y_2 \mid x, s = 0) = x'(\beta_1 - \beta_2)$$

$$+ \{E(u_1 \mid x, s = 1) - E(u_2 \mid x, s = 0)\}. \tag{6.18}$$

The selection bias depends on how the administrator makes his decision. If the selection depends only on x and some other variable z such that $E(u \mid x, z) = 0$, then there will be no selection bias. Suppose the selection is made if $z'\gamma + v > 0$, and v is not observed to the researcher (so not included in x) but is correlated with u, then $\text{COV}(u, v) \neq 0$ making $\{\cdot\}$ in (6.17) nonzero. If $(u, v)'$ follows $N(0, \Omega)$ independent of x, then the bias term in (6.17) can be derived analogously to (6.15).

In reality, we have situations more complicated than the above training program example. Usually, the admission process starts with applications from qualified poor people. Then the administrator admits if $z'\gamma + v > 0$. In this case, the event of being in the program is the intersection of two events {application} and {admission}. The event of application is similar to $\{y_1 > y_2\} = \{u_1 - u_2 > -x'(\beta_1 - \beta_2)\}$ in (6.7) where y_1 and y_2 are, respectively, the earnings with and without the program. Then the selection bias term depends on the joint distribution of (u_1, u_2, v). See Maddala (1983, 1986) for more on selection and switching regression models, and Heckman and Hotz (1989) and Heckman and Smith (1995) for training program evaluation.

5.7   Two-Stage Estimation for Selection Models

Consider the following (generic) selection model:

$$s = 1[x_0'\beta_0 + u_0 > 0], \quad y_1 = x_1'\beta_1 + u_1, \quad y_2 = x_2'\beta_2 + u_2. \tag{7.1}$$

Regime 1 (2) is observed if $s = 1$ (0),

$$u \equiv (u_0, u_1, u_2)' \cong N(0, \Omega) \text{ independent of } x, \ V(u_0) = 1.$$

Let $f_{012}(u_0, u_1, u_2)$ denote the density function for u. Also let f_{01} and f_{02} denote the density function, respectively, for $(u_0, u_1)'$ and $(u_0, u_2)'$. Define R_1 and R_2 as the regime 1 and 2, respectively. If the sample separation is known, then the likelihood function for one individual is

$$L_1 \equiv \int_{-\infty}^{\infty} \int_{-x_0'\beta_0}^{\infty} f_{012}(u_0, y_1 - x_1'\beta_1, u_2) \cdot du_0 du_2$$

$$= \int_{-x_0'\beta_0}^{\infty} f_{01}(u_0, y_1 - x_1'\beta_1) \cdot du_0 \quad \text{if the individual is in } R_1,$$

$$L_2 \equiv \int_{-\infty}^{\infty} \int_{\infty}^{-x_0'\beta_0} f_{012}(u_0, u_1, y_2 - x_2'\beta_2) \cdot du_0 du_1 \tag{7.2}$$

$$= \int_{-\infty}^{-x_0'\beta_0} f_{02}(u_0, y_2 - x_2'\beta_2) \cdot du_0 \quad \text{if the individual is in } R_2.$$

If the sample separation is unknown, then the likelihood is

$$L \equiv L_1 + L_2. \tag{7.3}$$

The MLE maximizing (7.2) or (7.3) is cumbersome. But, there is an easy two-stage estimation method (Heckman, 1979). In the following, we discuss the selection model (6.15) and its two stage estimation.

Recall (6.15): with $y = x'\beta + u$ and $s = 1[e/\sigma_e > -m'\alpha]$,

$$E(y \mid x, e/\sigma_e > -m'\alpha) = x'\beta + \gamma \cdot \lambda(m'\alpha), \qquad (7.4)$$

$$\gamma \equiv \sigma_{eu}/\sigma_e, \quad \lambda(m'\alpha) \equiv \phi(m'\alpha)/\Phi(m'\alpha).$$

Let T denote the total number of observations combining the two regimes and N denote the number of observations in the first regime for $y = x'\beta + u$. The Heckman's two-stage estimation goes as follows. First estimate α by a_T using the probit with observations on (s, m'). Second, for the subsample with y observed, do LSE of y on x and $\lambda(m'a_T)$ to estimate β and γ. If γ is not significantly different from 0, then there is no selection bias. Define

$$z_i \equiv (x_i', \lambda(m_i\alpha))' \quad \text{and} \quad z_{Ti} \equiv (x_i', \lambda(m_i a_T))'.$$

The sample moment condition for b_N and g_N in the second stage is

$$(1/T) \sum_{i=1}^{T} s_i\{y_i - x_i'b_N - g_N \cdot \lambda(m_i'a_T)\} \cdot z_{Ti} = 0; \qquad (7.5)$$

this is a two-stage problem with a nuisance parameter α as discussed in Chapter 3. Since the first-stage estimation error in z_{Ti} does not affect the second stage, instead of (7.5), consider

$$(1/T) \sum_{i=1}^{T} s_i\{y_i - x_i'b_N - g_N \cdot \lambda(m_i'a_T)\} \cdot z_i = 0. \qquad (7.6)$$

Define

$$d_N \equiv (b_N', g_N')' \quad \delta \equiv (\beta', \gamma)' \quad \varepsilon_i \equiv y_i - x_i'\beta - \gamma \cdot \lambda(m_i'\alpha),$$

and observe

$$\sqrt{T}(a_T - \alpha) =^p (1/\sqrt{T}) \sum_{i=1}^{T} I_f^{-1} m_i(s_i - \Phi_i)\phi_i/\{\Phi_i(1 - \Phi_i)\} \equiv (1/\sqrt{T}) \sum_{i=1}^{T} \eta_i,$$

where I_f is the probit information matrix, $\phi_i \equiv \phi(m_i'\alpha)$ and $\Phi_i \equiv \Phi(m_i'\alpha)$. Follow the two-stage estimation discussion in Chapter 3 to get

$$\sqrt{T}(d_N - \delta) =^p E^{-1}(szz') \cdot (1/\sqrt{T}) \sum_i \{s_i\varepsilon_i z_i + A \cdot \eta_i\} \qquad (7.7)$$

where $A \equiv -\gamma \cdot E\{s\lambda'(m'\alpha)zm'\}$ and $\lambda' \equiv d\lambda(a)/da = -a\lambda - \lambda^2$. Then

$$\sqrt{T}(d_N - \delta) =^d N(0, \Omega), \qquad (7.8)$$

$$\Omega = E^{-1}(szz') \cdot E\{(s\varepsilon z + A\eta)(s\varepsilon z + A\eta)'\} \cdot E^{-1}(szz').$$

The term $A\eta$ is the effect of the first stage on the second. An estimate for Ω can be obtained by replacing the parameters with their estimates; note that $E(\cdot)$ should be replaced by $(1/T)\sum_{i=1}^{T}$ not by $(1/N)\sum_{i=1}^{N}$. It is interesting to see that the covariance matrix $E(s\varepsilon z\eta')A'$ between $s\varepsilon z$ and $A\eta$ is zero for [recall $E(\varepsilon \mid m, x, s = 1) = 0$ by definition of ε]

$$E(s\varepsilon z\eta') = E\{E(s\varepsilon z\eta' \mid m, x, s = 1)\} = E\{E(\varepsilon z\eta' \mid m, x, s = 1)\}$$

$$= E[I_f^{-1}m(1 - \Phi)\phi\{\Phi(1 - \Phi)\}^{-1} \cdot z \cdot E(\varepsilon \mid m, x, s = 1)] = 0. \quad (7.9)$$

One serious problem in testing selectivity bias with $H_0 : \gamma = 0$ in (7.4) is that it is impossible in general to separate the selection effect from a regression function misspecification. For instance, if $(x'\beta)^2$ is omitted in the regression function of the second stage, then γ can be significantly different from 0 in (7.4) due to the correlation between $(x'\beta)^2$ and $\lambda(m'\alpha)$. In this case, $\lambda(m'\alpha)$ simply picks up the presence of the omitted variable. The only way to separate the true selection effect from regression function misspecifications is a prior exclusion restriction: there should be at least one important variable in m which is not included in x.

It is known that the second-stage estimator in Heckman's two stage estimator is sensitive to the assumption of normality (the first-stage probit estimator seems to be rather robust to violations of normality assumption). One way to get a robust estimator for selection models is as follows. Suppose $E(y \mid x) = x'\beta$ and y is observed if $m'\alpha + (e/\sigma_e) > 0$, where m includes at least one significant variable not in x. Assume that u and e are independent of x and m, and that

$$E(y \mid x, m, m'\alpha + (e/\sigma_e) > 0) = x'\beta + E(u \mid m'\alpha + (e/\sigma_e) > 0)$$
$$= x'\beta + \gamma_1(m'\alpha) + \gamma_2(m'\alpha)^2. \quad (7.10)$$

Here we assume that $E(u \mid m'\alpha + (e/\sigma_e) > 0)$ depends on x and m only through the linear index $m'\alpha$, and that $E(u \mid m'\alpha + (e/\sigma_e) > 0)$ can be approximated by a quadratic function in $m'\alpha$; extension to higher-order polynomials is straightforward. For (7.10), define

$$\begin{aligned}
z_i &\equiv (x_i', (m_i'\alpha), (m_i'\alpha)^2)', \\
z_{Ti} &\equiv (x_i', (m_i'a_T), (m_i'a_T)^2)', \\
\delta &\equiv (\beta', \gamma_1, \gamma_2)', \quad d_N \equiv (b_N', g_{N1}, g_{N2})', \\
\varepsilon_i &\equiv y_i - x_i'\beta - \gamma_1(m_i'\alpha) - \gamma_2(m_i'\alpha)^2.
\end{aligned} \quad (7.11)$$

Then we can show that δ is estimated by the second stage LSE d_N with

$$\sqrt{T}(d_N - \delta) =^d N(0, \Omega_q), \quad (7.12)$$

$$\Omega_q \equiv E^{-1}(szz') \cdot E\{(s\varepsilon z + A_q\eta)(s\varepsilon z + A_q\eta)'\} \cdot E^{-1}(szz'),$$

$$A_q \equiv -E[s\{\gamma_1 + 2\gamma_2(m'\alpha)\}zm'].$$

5.8 Minimum Distance Estimation (MDE)

Consider a parameter β and a function $R(\beta)$ that is estimable by R_N. Then minimum distance estimation (MDE), in a wide sense, estimates β by minimizing a distance between R_N and $R(\beta)$. Namely,

$$b_{\text{mde}} = \text{argmin}_b \|R_N - R(b)\|, \tag{8.1}$$

where $\|\cdot\|$ is a norm. For instance, $\|\cdot\|$ can be the Euclidean norm if $R(b)$ is a vector, or it can also be the sum of the absolute values of the components in the vector; (8.1) includes many estimators as special cases. In a narrow sense, MDE uses the Euclidean norm, in which case MDE is also called "minimum χ^2 estimation." In this chapter, we will use the term MDE in the narrow sense. MDE is useful in estimating simultaneous equations [see L.F. Lee (1992), Lee (1995a) and the references therein] and panel data models (Chamberlain, 1982).

MDE has a long history; see Malinvaud (1970), Rothenberg (973), and Chamberlain (1982) among many others and the references therein. The main problem that MDE answers is how to optimally impose overidentifying restrictions. Consider a $k \times 1$ parameter vector β and K ($K \geq k$) restrictions $\theta = \psi(\beta)$, where $\psi(\beta)$ has the continuous first derivative. Suppose θ can be estimated with an estimator θ_N. Then $\theta = \psi(\beta)$ can be written as

$$\theta_N - \psi(\beta) = \theta_N - \theta. \tag{8.2}$$

Now choose an estimator b_N for β such that $\theta_N - \psi(b_N)$ is as small as possible. Since $\theta_N - \psi(b_N)$ is a $K \times 1$ vector, we turn it into a scalar using a quadratic norm.

An (inefficient) MDE for β is obtained by minimizing

$$\{\theta_N - \psi(b)\}' \cdot W^{-1} \cdot \{\theta_N - \psi(b)\} \tag{8.3}$$

wrt b, where W is a $K \times K$ symmetric p.d. matrix (for instance $W = I_K$). The efficient MDE is obtained by setting

$$W = V\{\sqrt{N} \cdot (\theta_N - \psi(\beta))\} = V\{\sqrt{N} \cdot (\theta_N - \theta)\}.$$

Since W is unknown, W should be estimated, for which we need a $\sqrt{N}$-consistent estimate for β. The inefficient estimator with $W = I_K$ in (8.3) is sufficient for this purpose. Often the efficient MDE is simply called (the) MDE. Denoting the MDE as b_N, it can be shown that

$$\{\theta_N - \psi(b_N)\}' \cdot [V\{\sqrt{N} \cdot (\theta_N - \theta)\}]^{-1} \cdot \{\theta_N - \psi(b_N)\} \Rightarrow \chi^2_{K-k}. \tag{8.4}$$

This can be used as a test for the restriction $\theta = \psi(\beta)$.

It is illuminating to examine MDE with a linear restriction:

$$\theta = \Psi \cdot \beta, \quad \Psi \text{ is } K \times k, \quad K \geq k, \quad \text{rank}(\Psi) = k. \tag{8.5}$$

First consider $K = k$. Then $\beta = \Psi^{-1}\theta$, and an estimator for β is

$$b_N = \Psi^{-1}\theta_N, \tag{8.6}$$

where $\sqrt{N}(\theta_N - \theta) =^d N(0, C)$. Since $\theta_N =^p \theta$, $b_N =^p \Psi^{-1}\theta = \beta$; that is, $b_N =^p \beta$. Also the distribution of $\sqrt{N}(b_N - \beta)$ is straightforward:

$$\sqrt{N}(b_N - \beta) =^d N(0, \Psi^{-1} \cdot C \cdot \Psi^{-1'}). \tag{8.7}$$

Now consider $K > k$. For a $K \times K$ p.d. matrix W, $\Psi \cdot \beta = \theta$ implies

$$\Psi'W^{-1}\Psi \cdot \beta = \Psi'W^{-1} \cdot \theta. \tag{8.8}$$

Similarly to (8.6), we have

$$b_N = (\Psi'W^{-1}\Psi)^{-1}\Psi'W^{-1} \cdot \theta_N. \tag{8.9}$$

This b_N actually minimizes (8.3) with $\psi(\beta) = \Psi \cdot \beta$, and the distribution is

$$\sqrt{N}(b_N - \beta) =^d N(0, (\Psi'W^{-1}\Psi)^{-1}\Psi'W^{-1}CW^{-1}\Psi(\Psi'W^{-1}\Psi)^{-1}). \tag{8.10}$$

The covariance matrix attain its minimum when $W = C$ (as in GMM). Thus, the efficient MDE is

$$b_N = (\Psi'C^{-1}\Psi)^{-1}\Psi'C^{-1} \cdot \theta_N, \quad \sqrt{N}(b_N - \beta) =^d N(0, (\Psi'C^{-1}\Psi)^{-1}). \tag{8.11}$$

Returning to the case $K = k$, we can use W as in the case $K > k$. But since there is only one b_N satisfying $\Psi b_N = \theta_N$, b_N is the same regardless of W when $K = k$. The variance matrix $(\Psi'C^{-1}\Psi)^{-1}$ in (8.11) is the same as $\Psi^{-1} \cdot C \cdot \Psi^{-1}$, for Ψ is invertible when $K = k$.

Consider a slightly more complicated situation where the linear restriction Ψ is a function of θ. That is,

$$\theta = \Psi(\theta) \cdot \beta, \quad \Psi(\theta) \text{ is } K \times k, \quad K \geq k, \quad \text{rank}(\Psi(\theta)) = k. \tag{8.12}$$

Rewrite (8.12) as

$$\theta_N - [\theta_N - \theta - \{\Psi(\theta_N) - \Psi(\theta)\} \cdot \beta] = \Psi(\theta_N) \cdot \beta. \tag{8.13}$$

Defining an $K \times 1$ error vector ε as

$$\varepsilon \equiv \theta_N - \theta - \{\Psi(\theta_N) - \Psi(\theta)\} \cdot \beta,$$

and denoting $\Psi(\theta_N)$ as Ψ_N, (8.13) becomes

$$\theta_N = \Psi_N \cdot \beta + \varepsilon. \tag{8.14}$$

This equation looks like the usual linear model with $\theta_N = y$ and $\Psi_N = x$. We can estimate β $\sqrt{N}$-consistently with LSE b_c:

$$b_c = (\Psi'_N\Psi_N)^{-1}\Psi'_N\theta_N. \tag{8.15}$$

Since $V[\varepsilon]$ may not be diagonal, we can do better with the generalized LSE (GLS). If we can estimate $V(\varepsilon) \equiv \Omega$ with Ω_N, then the GLS is feasible:

$$b_N = (\Psi'_N \Omega_N^{-1} \Psi_N)^{-1} \Psi'_N \Omega_N^{-1} \theta_N, \quad \sqrt{N}(b_N - \beta) =^d N(0, (\Psi'_N \Omega_N^{-1} \Psi_N)^{-1}).$$
(8.16)

This asymptotic distribution is the same as $N(0, (\Psi' \Omega^{-1} \Psi)^{-1})$. In practice, getting Ω_N can be difficult. If $K = k$, then (8.15) = (8.16). The reader may wonder how we can apply LSE or GLS with only K many "observations." The answer is that the error term ε is not the usual error term; ε has a degenerate distribution converging to 0. Thus (8.14) behaves like a K linear deterministic equation.

Further generalizing MDE, consider [see L.F. Lee (1992)]

$$f(\underset{q \times 1}{\theta}, \underset{K \times 1}{\beta}) = 0 \atop k \times 1$$

where f is twice continuously differentiable, $\sqrt{N}(\theta_N - \theta) =^d N(0, C)$, and $k \le q \le K$. By the δ-method,

$$\sqrt{N}\{f(\theta_N, \beta) - f(\theta, \beta)\} =^d N(0, \{\partial f(\theta, \beta)/\partial\theta\} \cdot C \cdot \{\partial f(\theta, \beta)/\partial\theta'\}). \quad (8.17)$$

Consider a (generalized) MDE b_N for β by minimizing

$$f(\theta_N, b)' \cdot \Omega_N^{-1} \cdot f(\theta_N, b) \tag{8.18}$$

wrt b, where $\Omega_N =^p \Omega \equiv \{\partial f(\theta, \beta)/\partial\theta\} \cdot C \cdot \{\partial f(\theta, \beta)/\partial\theta'\}$. Then

$$\sqrt{N}(b_N - \beta) =^d N(0, [\{\partial f(\theta, \beta)/\partial\beta\} \cdot \Omega^{-1} \cdot \{\partial f(\theta, \beta)/\partial\beta'\}]^{-1}), \quad (8.19)$$

$$N \cdot f(\theta_N, b_N)\Omega_N^{-1} f(\theta_N, b_N)' =^d \chi_{q-k}^2. \tag{8.20}$$

By equating $f(\theta, \beta)$ to $\theta - \psi(\beta)$ or $\theta - \Psi\beta$, or $\theta - \Psi(\theta)\beta$, this result includes all the preceding cases with $q = K$ and $\partial f(\theta, \beta)/\partial\theta = I_K$.

If θ_N is a MLE and $\theta = \Psi \cdot \beta$, then there are two choices in estimating β. The first is MDE as we did above: get θ_N first by maximizing $(1/N)\sum_i \ln f(\theta)$ and then use the MDE to find β. The second is to maximize $(1/N)\sum_i \ln f(\Psi \cdot b)$ wrt b directly. Now we will show that the two methods are equivalent. Denote the score vector for θ as s_θ and define I_θ as $E(s_\theta s'_\theta)$. Consider the MDE b_N:

$$b_N = (\Psi' I_\theta \Psi)^{-1} \Psi' I_\theta \cdot \theta_N, \quad \sqrt{N}(b_N - \beta) =^d N(0, (\Psi' I_\theta \Psi)^{-1}). \quad (8.21)$$

Now if we maximize $(1/N)\sum_i \ln f(\Psi \cdot b)$ directly wrt b, then the score function for b is (by the chain rule) $\Psi' s_\theta$. Thus the information matrix is $E(\Psi' s_\theta s'_\theta \Psi) = \Psi' I_\theta \Psi$, and this direct MLE has the same variance as the MDE. In short, if θ_N is a MLE, then MDE is the efficient way to impose the restrictions.

5.9 Specification Tests Based on Difference of Estimators

The Wald test for the linear hypothesis $R\beta = r$ looks at the difference between the two estimators Rb_N and r for the parameter $R\beta$. Rb_N is consistent for $R\beta$ under both H_0 and H_a, whereas r is consistent for $R\beta$ only under H_0. Thus if H_0 holds, $Rb_N - r =^p 0$; otherwise $Rb_N - r \neq^p 0$. Extending this idea, we can test the model assumption by comparing two estimators that are supposed to be close if the assumptions are right.

Consider two estimators a_N and b_N for β, where b_N is $\sqrt{N}$-consistent under both H_0 and H_a and a_N is $\sqrt{N}$-consistent only under H_0. Assume that both estimators follow asymptotic normal distributions. Then for a C,

$$\sqrt{N}(a_N - b_N) = \sqrt{N}(a_N - \beta) - \sqrt{N}(b_N - \beta) =^d N(0, C). \qquad (9.1)$$

Then

$$\sqrt{N}(a_N - b_N)' \cdot C^{-1} \cdot \sqrt{N}(a_N - b_N) = N \cdot (a_N - b_N)' \cdot C^{-1} \cdot (a_N - b_N) = \chi^2_{\text{rank}(C)}. \qquad (9.2)$$

If

$$\sqrt{N}(a_N - \beta) =^p (1/\sqrt{N}) \sum_i v_i, \quad \sqrt{N}(b_N - \beta) =^p (1/\sqrt{N}) \sum_i w_i, \qquad (9.3)$$

then

$$\sqrt{N}(a_N - b_N) = \sqrt{N}(a_N - \beta) - \sqrt{N}(b_N - \beta) = (1/\sqrt{N}) \sum_i (v_i - w_i).$$

Thus,

$$V[\sqrt{N}(a_N - b_N)] = C =^p (1/N) \sum_i (v_i - w_i)(v_i - w_i)'. \qquad (9.4)$$

In practice, the most cumbersome step is in getting (9.3).

Define

$$A \equiv E(vv') = V[\sqrt{N}(a_N - \beta)], \quad B \equiv E(ww') = V[\sqrt{N}(b_N - \beta)].$$

In addition to the assumptions on a_N and b_N, further assume that a_N is efficient under H_0 while b_N is not. Then, (9.2) with C replaced by $B - A$

$$N \cdot (a_N - b_N)'(B - A)^{-1}(a_N - b_N) \Rightarrow \chi^2_{\text{rank}(B-A)} \qquad (9.5)$$

is called a *Hausman test* (Hausman, 1978) in econometrics; we will show why $B - A$ is p.s.d. below. There are many examples of Hausman tests:

(i) Recall the tests in Chapter 2 on instruments legitimacy.

(ii) Let $\beta \equiv (\gamma', \eta')'$ and $H_0 \colon \eta = 0$. The restricted MLE $a_N \equiv (g_N', 0')'$ is efficient under H_0 and inconsistent under H_a. The unrestricted MLE $b_N \equiv (g_N', h_N')'$ is consistent under both H_0 and H_a, but inefficient under H_0.

(iii) Let a_N be a multinomial logit MLE when the number of alternative is J. Let b_N be the multinomial logit for a subset of data where the number of alternatives is smaller than J. The independence of the irrelevant alternative has be tested (Hausman and McFadden, 1984).

(iv) Let a_N be an ordered probit MLE when y takes four ordered values 0, 1, 2, and 3. By grouping 0 and 1 into 0, and 2 and 3 into 1, we can collapse the ordered response into binary response. Let b_N be the binary probit. Then we get a Hausman (type) test analogous to (iii). This test has been suggested by Lee (1993b).

Usually in an Hausman test, H_0 and H_a are not specific. H_0 is the set of model assumptions that makes a_N efficient and consistent and b_N inefficient but consistent, while H_a is the set of model assumptions that makes a_N inconsistent and b_N consistent. This raises a question of what Hausman's test really tests; when we reject H_0, we do not know which model assumption in H_0 is violated. It may be the likelihood or the linear model or something else.

In practice, the sample estimate for $B - A$ may not be invertible, nor p.s.d. even if invertible. If the number of samples is small, then a poor estimate for the covariance matrix may be the reason for this problem. However, if the number of samples is large, the problem should be taken as rejecting H_0, since $B - A$ being p.s.d. is valid only under H_0. One way to avoid these problems is to use (9.4).

Turning to the question on why $V[\sqrt{N}(b_N - a_N)] = B - A$, recall (9.3). If a_N is the efficient estimator, then under certain regularity conditions,

$$\sqrt{N}(b_N - \beta) = (1/\sqrt{N}) \sum_i w_i = (1/\sqrt{N}) \sum_i (v_i + \eta_i) \qquad (9.6)$$

where $E(v\eta') = 0$. That is, an inefficient asymptotically normal $\sqrt{N}$-consistent estimator is a convolution of $(1/\sqrt{N}) \sum_i v_i$ (from the efficient estimator) and $(1/\sqrt{N}) \sum_i \eta_i$ orthogonal to $(1/\sqrt{N}) \sum_i v_i$. This has been known since 1950's; see, for instance, Bickel et al. (1993). Intuitively, if $E(v\eta') \neq 0$, then we can get the following estimator more efficient than a_N by projecting v on η:

$$(1/\sqrt{N}) \sum_i \{v_i - E(v\eta') \cdot E^{-1}(\eta\eta')\eta_i\}. \qquad (9.7)$$

This contradicts the efficiency of a_N, and so $E(v\eta')$ must be zero. From (9.6), $\sqrt{N}(b_N - a_N) = (1/\sqrt{N}) \sum_i \eta_i$ and $B = E(ww') = A + E(\eta\eta')$, which implies $V[\sqrt{N}(b_N - a_N)] = E(\eta\eta') = B - A$.

The idea of comparing two estimators is old, and $V[\sqrt{N}(b_N - a_N)] = B - A$ under the efficiency of a_N was known well before Hausman (1978). Also in practice, frequently one has to use (9.4), for the sample version of $B - A$ does not work. In view of these observations, the term "Hausman test" is somewhat overused in econometrics.

Although different from Wald-type tests, it is possible to test model specification using variance matrix estimates. Consider two estimators Ω_{N1} and Ω_{N2} for $\Omega = V[\sqrt{N}(b_N - \beta)]$. Under H_0, both Ω_{N1} and Ω_{N2} are consistent for Ω. Under H_a, only Ω_{N1} is consistent for Ω. In the following, we introduce two tests based on this idea.

White's heteroskedasticity test (White, 1980) for the linear model $y = x'\beta + u$ uses a heteroskedasticity consistent variance matrix and the variance matrix under homoskedasticity. The test looks at the difference

$$E^{-1}(xx') \cdot E(xx'u^2) \cdot E^{-1}(xx') - \sigma^2 E^{-1}(xx'). \tag{9.8}$$

Stacking up some elements of the sample version of (9.8), we get a vector which follows a normal distribution with the mean 0 when normed by $\sqrt{N}$; the matrix in (9.8) is symmetric and it has $k(k+1)/2$ distinct elements at maximum. Then a Wald test statistic can be formed. This test is somewhat cumbersome to implement; later we will see an artificial regression version of the White test which is much easier to implement.

Another specification test based on covariance matrix comparison is the *information matrix test* by White (1982). Under the model assumption of a MLE, the second-order matrix times minus one should be equal to the sum of the outer-products of the score functions. That is, if the MLE specification is correct, the following k by k matrix moment condition holds:

$$E\{\partial^2 \ln f/\partial b \partial b' + (\partial \ln f/\partial b)\partial \ln f/\partial b'\} = 0. \tag{9.9}$$

As in the White heteroskedasticity test, select a few components out of (9.9) and construct a χ^2 test statistic. The information matrix test (and the heteroskedasticity test) also suffers from the same drawback as Hausman's test: Rejection of H_0 does not indicate which assumption of the model is violated. Also deriving the second and third derivatives of the likelihood is difficult; the third derivatives are needed for the asymtotic distribution of the information matrix test.

6

Nonlinear Models and Generalized Method of Moments

6.1 Introduction

Consider a nonlinear regression model

$$y = r(x, \beta) + u, \quad E(u \mid x) = 0, \tag{1.1}$$

where β is a $k \times 1$ vector and the form of $r(\cdot)$ is known. In contrast to the linear model, the dimension of x is not necessarily the same as that of β. Depending on cases, we may omit either x or β in $r(x, \beta)$. A model more general than (1.1) is

$$\rho(y, x, \beta) = u, \quad E(u \mid x) = 0, \tag{1.2}$$

which includes (1.1) as a special case when $\rho(y, x, \beta) = y - r(x, \beta)$.

Since (1.1) includes the linear model as a special case when $r(x, \beta) = x'\beta$, it seems natural to estimate β by minimizing

$$(1/N) \sum_i \{y_i - r(x_i, b)\}^2 \tag{1.3}$$

with respect to (wrt) b. The estimator is called a *nonlinear least squares estimator* (NLS). The first-order condition is

$$(1/N) \sum_i -2 \cdot \{y_i - r(x_i, b)\} \cdot r_b(x_i, b) = 0, \tag{1.4}$$

where $r_b = \partial r / \partial b$. With this, NLS may be viewed as a method-of-moments estimator (MME) with the population moment condition

$$E[\{(y - r(x, \beta)\} \cdot r_b(x, \beta)] = E\{u \cdot r_b(x, \beta)\} = 0. \qquad (1.5)$$

Differently from LSE, however, there may be many solutions for (1.5) even when (1.4) has a unique minimizer. Still, regarding NLS as MME has its advantage: if some regressors in x are correlated with u, then we can use an instrument. As in Chapter 2, this observation and (1.5) brings up a host of questions. First, since $E(u \mid x) = 0$ implies $E\{u \cdot g(x)\} = 0$ for any function $g(x)$, how do we choose $g(x)$? Second, although (1.5) suggests $r_b(x, \beta)$ be used as an instrument, can we do better with other instruments? Third, if some variables in x are correlated with u, what kinds of and how many instruments should we use? These questions will be answered in this chapter.

The rest of this chapter is organized as follows. In Section 2, we show various nonlinear models and study NLS. In Section 3, testing with NLS is examined; the three classical tests in Chapter 4 are discussed in detail. In Section 4, the Gauss–Newton algorithm is introduced, which encompasses numerical iterative algorithms using only gradients. In Section 5, the basics of GMM with nonlinear moment conditions is studied. In Section 6, GMM for simultaneous (non)linear equation system is examined. In Section 7, three tests with GMM and GMM for dependent observations are introduced. In Section 8, we introduce a LM test for linear regression function specification against nonlinear alternatives.

6.2   Nonlinear Models and Nonlinear LSE

There are many ways to generalize the linear model to nonlinear models while retaining the linearity to a certain extent. Two well-known ways are "index models" and "transformation of variables." In a (multiple) *index model*, x affects $E(y \mid x)$ through J number of indices $x'_{(j)}\beta_{(j)}$, $j = 1 \ldots J$:

$$y = r\{x'_{(1)}\beta_{(1)}, \ldots, x'_{(J)}\beta_{(J)}\} + u,$$

where $x_{(j)}$ is a subset of x (overlaps in $x_{(j)}$ and $x_{(m)}$ are allowed). Special (practical) cases of this are

$$y = \sum_{j=1}^{J} r_j\{x'_{(j)}\beta_{(j)}\} + u \quad \text{or} \quad y = \sum_{j=1}^{J} r\{x'_{(j)}\beta_{(j)}\} + u. \qquad (2.1)$$

The simplest index model is a *single index model*:

$$y = s(x'\beta) + u; \qquad (2.2)$$

the effect of x_k on y is measured by

$$\partial s(x'\beta)/\partial x_k = \beta_k \cdot s'(x'\beta), \quad \text{where } s'(x'\beta) \equiv \partial s(x'\beta)/\partial(x'\beta).$$

Thus, although $\beta_k \cdot s'(x'\beta)$ depends on $x'\beta$, the relative effect

$$\{\partial s(x'\beta)/\partial x_j\}/\{\partial s(x'\beta)/\partial x_k\} = \beta_j/\beta_k, \quad j \neq k$$

does not. Sometimes only strictly monotonic $s(\cdot)$'s are called single index models. An example of (2.2) is the probit written as a nonlinear model:

$$y = E(y \mid x) + v = P(y = 1 \mid x) + v = \Phi(x'\beta) + v. \tag{2.3}$$

Another example is the Poisson MLE where $s(x'\beta) = \exp(x'\beta)$.

A *transformation of variable model* is

$$h(y) = g_1(x_1) \cdot \beta_1 + \cdots + g_k(x_k) \cdot \beta_k + u. \tag{2.4}$$

As in the single index model, often we restrict the transformation to be (strictly) monotonic. A well-known transformation is the *Box–Cox transformation* (Box and Cox, 1964): for $x_k > 0$,

$$\begin{aligned} g_k(x_k) &= (x_k^{\alpha_k} - 1)/\alpha_k \quad &&\text{if } \alpha_k \neq 0; \\ g_k(x_k) &= \ln(x_k) \quad &&\text{if } \alpha_k = 0. \end{aligned} \tag{2.5}$$

Using the L'Hopital's rule, we get $(x_k^{\alpha_k} - 1)/\alpha_k \to \ln(x_k)$ as $\alpha_k \to 0$. We estimate α_k's as well as β to find the "best" transformation. Note that y can also be transformed as in (2.5) so long as $y > 0$. To relax the restriction of $x_k > 0$ in (2.5), Bickel and Doksum (1981) suggest the following transformation: for $\alpha_k > 0$,

$$g_k(x_k) = \{|x_k|^{\alpha_k} \cdot \text{sgn}(x_k) - 1\}/\alpha_k, \tag{2.6}$$

which is convex if $x_k > 0$ and concave if $x_k < 0$. Also available is "shifted power transformation":

$$g_k(x_k) = (x_k - \mu_k)^{\alpha_k}. \tag{2.7}$$

See Caroll and Ruppert (1988) for more on transformation, and Breiman and Friedman (1985) and Tibshirani (1988) on "optimal transformations". Also see MacKinnon and Magee (1990) for another alternative to (2.5).

Since the set of all polynomial functions on $[a_1, a_2]$ with rational coefficients can approximate any continuous function on $[a_1, a_2]$ arbitrarily well [see, for instance, Luenberger (1969, p. 43)], if $s(\cdot)$ is unknown in (2.2), then we can consider a polynomial in $x'\beta$ as in

$$y = \sum_{m=1}^{M} \gamma_m \cdot (x'\beta)^m + u \tag{2.8}$$

for a given M. Allowing transformation in x as in (2.5), we get

$$y = \sum_{m=1}^{M} \gamma_m \cdot \{g(x)'\beta\}^m + u, \tag{2.9}$$

where $g(x) \equiv (g_1(x_1)\ldots g_k(x_k))'$.

It is possible to further classify the nonlinear models. For instance, $r(x) = \sum_{j=1}^{k} r_j(x_j)$ is an "additive model"; $r(x) = r_1(x_1,\ldots,x_j)+r_2(x_{j+1},\ldots,x_k)$ is a "partially additive model." Since it is cumbersome to treat these models one by one, we will discuss only the general models (1.1) and (1.2), which include the above models as special cases. As in the previous chapters, MME will be the key concept in estimating (1.1) and (1.2).

Suppose we allow transformation of y in (2.9) to get

$$h(y) = \sum_{m=1}^{M} \gamma_m \cdot \{g(x)'\beta\}^m + u. \tag{2.10}$$

One example is a CES production function $y^\alpha = \sum_{j=1}^{k} \beta_j x_j^\alpha + u$. Assuming $h(\cdot)$ is strictly monotonic, we can invert $h(\cdot)$ to get

$$y = h^{-1}\left[\sum_{m=1}^{M} \gamma_m \cdot \{g(x)'\beta\}^m + u\right]. \tag{2.11}$$

Then

$$E(y \mid x) = \int h^{-1}\left[\sum_{m=1}^{M} \gamma_m \cdot \{g_k(x)'\beta\}^m + u\right] \cdot dF_{u|x}(u) \equiv R(x;\beta). \tag{2.12}$$

Thus we get an equation that looks like (1.1):

$$y = R(x;\beta) + v, \tag{2.13}$$

where $v \equiv y - R(x;\beta)$. If $F_{u|x}$ is unknown, then the form of $R(\cdot)$ is unknown; unknown regression functions will not be discussed in this chapter.

Returning to NLS in (1.3), let $Q(b)$ denote the population version of (1.3) times $1/2$. Assume that r is twice continuously differentiable wrt b. The first two derivatives of $Q(b)$ are

$$Q_b(b) \equiv -E[\{y - r(b)\} \cdot r_b(b)];$$
$$Q_{bb}(b) \equiv E\{r_b(b) \cdot r_b(b)'\} - E[\{y - r(b)\} \cdot r_{bb}(b)], \tag{2.14}$$

where $r_{bb} \equiv \partial r/\partial b\partial b'$. The following three conditions together are sufficient for the identification of β in NLS:

$$Q_b(\beta) = 0, \; Q_{bb}(\beta) \text{ is p.d. and } Q_{bb}(b) \text{ is p.s.d. for any } b. \tag{2.15}$$

The first two conditions together make β a local minimum, and the last condition ensures that β is the unique global minimum. Observe that

$$Q_b(\beta) = -E\{u \cdot r_b(\beta)\} = 0, \quad Q_{bb}(\beta) = E\{r_b(\beta) \cdot r_b(\beta)'\} \text{ is p.s.d.} \quad (2.16)$$

Assuming $E\{r_b(\beta) \cdot r_b(\beta)'\}$ is of full rank analogously to the assumption that $E(xx')$ is of full rank, $Q_{bb}(\beta)$ is p.d. Hence, the first two conditions in (2.15) are easily satisfied, and β is a local minimum. The third in (2.15) appears difficult to check unless $r(b)$ is specified.

The asymptotic distribution of NLS is straightforward:

$$\sqrt{N}(b_{\text{NLS}} - \beta) =^d N(0, E^{-1}(r_b r_b') \cdot E(r_b r_b' \cdot u^2) \cdot E^{-1}(r_b r_b')). \quad (2.17)$$

If $E(u^2 \mid x) = \sigma^2$,

$$\sqrt{N}(b_{\text{NLS}} - \beta) =^d N(0, \sigma^2 \cdot E^{-1}(r_b r_b')). \quad (2.18)$$

With $r_b = x$, (2.17) and (2.18) become the asymptotic variance for LSE under heteroskedasticity and homoskedasticity, respectively.

In the probit model (2.3), $y = \Phi(x'\alpha) + v$, where $v \equiv y - \Phi(x'\alpha)$. By definition, $E(v \mid x) = 0$. As for the variance,

$$V(v \mid x) = E_{y|x}\{y - \Phi(x'\alpha)\}^2 = E_{y|x}\{y^2 - 2y \cdot \Phi(x'\alpha) + \Phi(x'\alpha)^2\}$$

$$= E_{y|x}(y^2) - 2E_{y|x}(y) \cdot \Phi(x'\alpha) + \Phi(x'\alpha^2) = \Phi(x'\alpha) \cdot \{1 - \Phi(x'\alpha)\}, \quad (2.19)$$

which is natural, for $y \mid x$ is a binary random variable (rv) with $P(y = 1 \mid x)$ being $\Phi(x'\alpha)$. Thus, v has heteroskedasticity of a known form. Suppose we apply NLS to $y = \Phi(x'\alpha) + v$ and get the NLS estimator a_0. Then we can estimate $V(v \mid x_i)$ with $w_i^2 \equiv \Phi(x_i'a_0)\{1 - \Phi(x_i'a_0)\}$. Transform $y_i = \Phi(x_i'\alpha) + v_i$ into

$$y_i/w_i = \Phi(x_i\alpha)/w_i + v_i/w_i. \quad (2.20)$$

Applying NLS to (2.20), we get a more efficient estimator. This two-stage procedure is a weighted NLS (WNLS). The variance matrix turns out to be the same as that of probit.

NLS is not applicable to models where the dependent variable is an unknown transformation of y. To see this, consider minimizing $E\{h(y, a) - x'b\}^2$ wrt a and b. The first-order condition at the true values α and β is

$$E\{u \cdot \partial h(y, \alpha)/\partial a\} = 0, \quad E(ux) = 0. \quad (2.21)$$

Since $\partial h(y, \alpha)/\partial a$ is a function of y, the first equation is unlikely to hold. But instrumental variable estimators (IVE) with instruments x and its functions are still applicable.

6.3   Three Classical Tests with NLS

Testing a linear hypothesis $H_0: R\beta = r$ where $\text{rank}(R) = g \leq k$ with NLS is similar to that with MLE: there are three kinds of tests corresponding to Wald, Lagrangian multiplier (LM), and likelihood ratio (LR) tests. Let $V\{\sqrt{N}(b_N - \beta)\} = C$, where $b_N = $ NLS. The Wald test statistic is

$$N(Rb_N - r)'\{RCR'\}^{-1}(Rb_N - r) =^d \chi_g^2; \qquad (3.1)$$

C can be estimated by its sample analog

$$\left\{(1/N)\sum_i r_b(b_N)r_b(b_N)'\right\}^{-1} \cdot (1/N)\sum_i [\{y_i - r(b_N)\}^2 r_b(b_N)r_b(b_N)']$$

$$\cdot \left\{(1/N)\sum_i r_b(b_N)r_b(b_N)'\right\}^{-1}. \qquad (3.2)$$

If $E(u^2 \mid x) = \sigma^2$, (3.1) becomes

$$N(Rb_N - r)'\{R \cdot E^{-1}(r_b r_b') \cdot R'\}^{-1}(Rb_N - r)/\sigma^2 =^d \chi_g^2; \qquad (3.3)$$

$E^{-1}(r_b r_b')$ and σ^2 can be estimated, respectively, by the first matrix in (3.2) and $(1/N)\sum\{y_i - r(b_N)\}^2$. In the rest of this section, we go over LM and LR type tests under homoskedasticity assumption, which will give results analogous to those in tests for MLE.

Define NLS with restriction by

$$b_{Nr} \equiv \text{argmax}\, Q_{Nr}(b, \lambda) \equiv (1/N)\sum_i \{y_i - r(b)\}^2 + 2 \cdot \lambda'(Rb - r), \quad (3.4)$$

where λ is a $g \times 1$ Lagrangian multiplier. In order to derive the LM test, differentiate $Q_{Nr}(b, \lambda)$ wrt b and λ to get

$$-(1/N)\sum_i [r_b(b_{Nr}) \cdot \{y_i - r(b_{Nr})\}] + R'\lambda = 0, \qquad (3.5)$$

$$Rb_{Nr} - r = 0. \qquad (3.6)$$

Solving these for b_{Nr} and λ, we will get the LM test (3.14) below; the details up to (3.14) may be skipped.

Define B (from b in r_b) as

$$B \equiv E(r_b r_b'). \qquad (3.7)$$

Multiply (3.5) by RB^{-1} to get

$$\lambda = (RB^{-1}R')^{-1}RB^{-1}\left[(1/N)\sum_i r_b(b_{Nr})\{u_i + r(\beta) - r(b_{Nr})\}\right]. \qquad (3.8)$$

Expanding $r(\beta)$ into $r(b_{Nr}) + r_b(b^*)'(\beta - b_{Nr})$ where $b^* \in (\beta, b_{Nr})$, we get

$$\lambda = (RB^{-1}R')^{-1}RB^{-1}(1/N)\sum_i \{r_b(b_{Nr})u_i + r_b(b_{Nr})r_b(b^*)'(\beta - b_{Nr})\}.$$
(3.9)

From this,

$$\left[\sqrt{N}\cdot\lambda - (RB^{-1}R')^{-1}RB^{-1}\left\{(1/\sqrt{N})\sum_i r_b(b_{Nr})u_i\right\}\right]$$

$$= (RB^{-1}R')^{-1}RB^{-1}\left\{(1/N)\sum_i r_b(b_{Nr})r_b(b^*)'\right\}\sqrt{N}(\beta - b_{Nr}) \quad (3.10)$$

$$=^p (RB^{-1}R')^{-1}\cdot\sqrt{N}(R\beta - Rb_{Nr}) = 0 \quad (3.11)$$

for $R\beta = r = Rb_{Nr}$ under H_0 due to (3.6). Hence,

$$\sqrt{N}\cdot\lambda =^p (RB^{-1}R')^{-1}RB^{-1}(1/\sqrt{N})\sum_i r_b(\beta)u_i. \quad (3.12)$$

From (3.12), $\sqrt{N}\cdot\lambda =^d N(0, C_\lambda)$ where

$$C_\lambda \equiv (RB^{-1}R')^{-1}RB^{-1}\cdot E(u^2 r_b r_b')\cdot B^{-1}R'(RB^{-1}R')^{-1} = \sigma^2(RB^{-1}R')^{-1}. \quad (3.13)$$

Note that the assumed homoskedasticity was not invoked until C_λ. Convert $\sqrt{N}\cdot\lambda =^d N(0, C_\lambda)$ into a χ^2 test statistic:

$$\sqrt{N}\cdot\lambda' C_\lambda^{-1}\sqrt{N}\cdot\lambda = N\cdot\lambda'R\cdot E^{-1}(r_b r_b')\cdot R'\lambda/\sigma^2$$

$$= N\cdot\left[(1/N)\sum_i r_b(b_{Nr})\{y_i - r(b_{Nr})\}\right]\cdot E^{-1}(r_b r_b') \quad (3.14)$$

$$\cdot\left[(1/N)\sum_i r_b(b_{Nr})\{y_i - r(b_{Nr})\}\right]/\sigma^2 =^d \chi_g^2$$

due to (3.5). Compare this to the LM test statistic in MLE.

In general, the idea of LM test (or score test) is the following. Suppose that H_0 implies two sets of restrictions S_1 and S_2, and β can be estimated using only S_1. Then the validity of H_0 can be checked by examining if the estimate obtained using only S_1 can satisfy S_2. If S_2 is a moment condition, then the LM test becomes a special case of method of moments tests.

In order to see the relation b_{Nr} and b_N [the reader may skip to (3.20)], expand $r(b_{Nr})$ around b_N and substitute it into (3.5) to get

$$\sqrt{N}\cdot R'\lambda = (1/\sqrt{N})\sum_i r_b(b_{Nr})\{y_i - r(b_N) - r_b(b^*)'(b_{Nr} - b_N)\}$$

$$=^p (1/\sqrt{N}) \sum_i r_b(b_{Nr})\{y_i - r(b_N)\} - B \cdot \sqrt{N}(b_{Nr} - b_N) \quad (3.15)$$

where $b^* \in (b_{Nr}, b_N)$. We will show that the first term is $o_p(1)$. By the definition of b_N, it satisfies $(1/\sqrt{N}) \sum_i r_b(b_N)\{y_i - r(b_N)\} = 0$, the first-order condition. Subtracting this from the first term of (3.15),

$$(1/\sqrt{N}) \sum_i \{y_i - r(b_N)\}\{r_b(b_{Nr}) - r_b(b_N)\}$$

$$= (1/N) \sum_i \{y_i - r(b_N)\}r_{bb}(b^{**}) \cdot \sqrt{N}(b_{Nr} - b_N) \quad (3.16)$$

$$=^p (1/N) \sum_i \{y_i - r(\beta)\}r_{bb}(\beta) \cdot O_p(1) = o_p(1) \cdot O_p(1) = o_p(1).$$

Hence the first term of (3.15) is $o_p(1)$ and so (3.15) becomes

$$\sqrt{N} \cdot R'\lambda = B \cdot \sqrt{N}(b_N - b_{Nr}) + o_p(1). \quad (3.17)$$

Multiply both sides of (3.17) by RB^{-1} to get

$$\begin{aligned}\sqrt{N} \cdot \lambda &= \sqrt{N}(RB^{-1}R')^{-1}R(b_N - b_{Nr}) + o_p(1) \\ &= \sqrt{N}(RB^{-1}R')^{-1}(Rb_N - r) + o_p(1), \quad \text{for } Rb_{Nr} = r.\end{aligned} \quad (3.18)$$

Substituting (3.18) into (3.17),

$$R'(RB^{-1}R')^{-1}(Rb_N - r) = B(b_N - b_{Nr}) + o_p(1/\sqrt{N}). \quad (3.19)$$

Solving this for b_{Nr} and using $R\beta = r$,

$$b_{Nr} = b_N - B^{-1}R'(RB^{-1}R')^{-1}(Rb_N - R\beta) + o_p(1/\sqrt{N}) \quad (3.20)$$

which is equivalent to

$$\sqrt{N}(b_{Nr} - \beta) = \{I_k - B^{-1}R'(RB^{-1}R')^{-1}R\} \cdot \sqrt{N}(b_N - \beta) + o_p(1). \quad (3.21)$$

Using (3.20), we can obtain b_{Nr} easily from b_N.

Turning to the LR test for NLS, expand $N \cdot Q_N(b_{Nr})$ around b_N to the second order and use the fact $\partial Q_N(b_N)/\partial b = 0$ to get

$$N \cdot \{Q_N(b_{Nr}) - Q_N(b_N)\}$$

$$= (1/2)\sqrt{N}(b_{Nr} - b_N)' \left[(1/N) \sum_i r_b(b_N^*)r_b(b_N^*)'\right] \sqrt{N}(b_{Nr} - b_N)$$

$$=^p (1/2)\sqrt{N}(b_{Nr} - b_N)'B\sqrt{N}(b_{Nr} - b_N). \quad (3.22)$$

Define

$$W_N(b) \equiv 2N \cdot Q_N(b) = \sum_i \{y - r(b)\}^2, \quad (3.23)$$

so that

$$W_N(b_{Nr}) - W_N(b_N) = \sqrt{N}(b_{Nr} - b_N)'B\sqrt{N}(b_{Nr} - b_N). \qquad (3.24)$$

Substitute (3.20) into (3.24) to get

$$N(Rb_N - r)'(RB^{-1}R')^{-1}(Rb_N - r). \qquad (3.25)$$

Owing to (3.3), the LR test statistic for NLS is

$$(3.25)/\sigma^2 = \{W_N(b_{Nr}) - W_N(b_N)\}/\sigma^2 \Rightarrow \chi_g^2. \qquad (3.26)$$

There are two differences between (3.26) and the LR statistic for MLE, which is $2 \cdot [\ln\{f(b_N)\} - \ln\{f(b_{Nr})\}]$. One is that the number "2" is canceled by "2" in $\sum_i\{y - r(b)\}^2$ when this is differentiated. The other is that the LR test in MLE does not have σ^2 as in (3.26), since the variance matrix in MLE is the inverse of the expected outer-product of the first-order vector; in NLS, the variance matrix is the inverse of the expected outer-product of the first order vector times σ^2.

In the linear model with $u =^d N(0, \sigma^2)$ independent of x, there is an exact (not asymptotic) F test:

$$\{(N - k)/g\} \cdot [\{W_N(b_{Nr}) - W_N(b_N)\}/W_N(b_N)] = F(g, N - k), \quad (3.27)$$

(3.26) is $o_p(1)$ equal to

$$N \cdot \{W_N(b_{Nr}) - W_N(b_N)\}/W_N(b_N). \qquad (3.28)$$

Observe that $(3.27)^*g \cong (3.28)$ and $g \cdot F(g, \infty) =^d \chi_g^2$; that is, $g \cdot F(g, N - k) =^d \chi_g^2$ as $N \to \infty$. Hence, under homoskedasticity, the exact F-test under the normality assumption is asymptotically equivalent to the Wald test.

6.4 Gauss–Newton Algorithm and One-Step Efficient Estimation

One well-known way to implement NLS is the *Gauss–Newton* algorithm. Taylor expand $r(b)$ around b_0, an initial estimate:

$$r(b) \cong r(b_0) + r_b(b_0)'(b - b_0). \qquad (4.1)$$

Substitute this into $r(b)$ and minimize the following wrt b:

$$(1/2) \cdot (1/N) \sum_i \{y_i - r(b_0) - r_b(b_0)'(b - b_0)\}^2. \qquad (4.2)$$

The first-order condition is

$$-(1/N) \sum_i r_b(b_0)\{y_i - r(b_0) - r_b(b_0)'(b - b_0)\} = 0. \qquad (4.3)$$

Solve this for b and denote the solution by b_1:

$$b_1 = b_0 - \left\{(1/N) \sum_i r_b(b_0)r_b(b_0)'\right\}^{-1} (1/N) \sum_i -\{y_i - r(b_0)\}r_b(b_0). \quad (4.4)$$

Repeat this until a stopping criterion is met.

The Gauss–Newton method yields a new perspective for NLS. Replace b by β in (4.1) and substitute (4.1) into $y = r(\beta) + u$ to get

$$y - r(b_0) + r_b(b_0)'b_0 = r_b(b_0)'\beta + u. \qquad (4.5)$$

Treat the left-hand side as a new dependent variable and $r_b(b_0)$ as the regressor. Applying LSE to this model, we obtain b_1 in (4.4). That is, the Gauss–Newton algorithm is equivalent to applying LSE to the linearized version of the nonlinear model.

Gauss–Newton algorithm was also used in MLE. Recall the Newton–Rhapson algorithm with the second-order matrix approximated by the outer-product of the score function. This version has the same format as (4.4); namely,

$$\text{new estimate} = \text{old estimate} - (-\text{outer product})^{-1} \cdot (\text{gradient}). \qquad (4.6)$$

If the approximation of the second-order matrix by the outer-product is not good in practice, (4.6) may oscillate near the peak of the maximand.

Sometimes we may get an initial $\sqrt{N}$-consistent estimate easily, and start (4.6) from the estimate. In this case, applying (4.6) only once is often asymptotically as good as doing (4.6) many times. To see this in MLE, let I_f denote the information matrix and s_i denote the score vector for i. Take one Gauss–Newton step from a $\sqrt{N}$-consistent estimator b_0:

$$b_N = b_0 - \left\{-(1/N) \sum_i s_i(b_0)s_i(b_0)'\right\}^{-1} \cdot (1/N) \sum_i s_i(b_0)$$

$$\Rightarrow \sqrt{N}(b_N - \beta) = \sqrt{N}(b_0 - \beta) + \left\{(1/N) \sum_i s_i(b_0)s_i(b_0)'\right\}^{-1}$$

$$\cdot (1/\sqrt{N}) \sum_i s_i(b_0). \qquad (4.7)$$

Let $w_i \equiv \partial s/\partial b$. Apply the mean value theorem to $s_i(b_0)$ around β to get

$$\sqrt{N}(b_N - \beta) =^p \sqrt{N}(b_0 - \beta) + I_f^{-1} \cdot (1/\sqrt{N}) \sum_i \{s_i(\beta) + w_i(b^*)(b_0 - \beta)\}$$

$$\Rightarrow \sqrt{N}(b_N - \beta) =^p I_f^{-1}(1/\sqrt{N}) \sum_i s_i(\beta)$$

$$+ \sqrt{N}(b_0 - \beta)\{I_k + I_f^{-1}(1/N) \sum_i w_i(b^*)\}. \tag{4.8}$$

But, owing to $E\{w(\beta)\} = -I_f$,

$$I_k + I_f^{-1}(1/N) \sum_i w_i(b^*) =^p I_k - I_k = 0.$$

Therefore $\sqrt{N}(b_N - \beta) =^p I_f^{-1}(1/\sqrt{N}) \sum_i s_i(\beta) =^d (0, I_f^{-1})$, and b_N in (4.7) is as efficient as MLE, without no further gain in repeating (4.7).

Suppose we have two estimators b_1 and b_2 for β and

$$\sqrt{N}(b_1 - \beta) =^p (1/\sqrt{N}) \sum_i \lambda_{1i} + (1/N) \sum_i \delta_{1i},$$

$$\sqrt{N}(b_2 - \beta) =^p (1/\sqrt{N}) \sum_i \lambda_{2i} + (1/N) \sum_i \delta_{2i}. \tag{4.9}$$

Then the asymptotic variances of $\sqrt{N}(b_1 - \beta)$ and $\sqrt{N}(b_2 - \beta)$ depend only on $V(\lambda_1)$ and $V(\lambda_2)$, respectively. If $V(\lambda_1) = V(\lambda_2)$, then b_1 and b_2 have the same *first-order efficiency*. The terms δ_1 and δ_2 will determine the "second-order efficiency." More generally, if there are more terms in (4.9), we can consider higher-order efficiencies. Unless otherwise mentioned, we mean the first-order efficiency whenever we discuss efficiency. The above result, that taking one step from a $\sqrt{N}$-consistent estimate is enough for MLE, is also based on the first-order efficiency; repeating (4.7) may raise higher-order efficiencies.

6.5   Basics of GMM in Nonlinear Models

In Chapter 2, we introduced GMM for linear models where GMM improves upon LSE in two aspects: taking heteroskedasticity of unknown form into account, and optimally combining more than enough instruments. GMM is also applicable to nonlinear models and includes NLS as a special case. In nonlinear models, GMM has the following additional important advantage (compared with NLS). In many optimization problems, there are first-order conditions that can be turned into moment conditions. Often the conditions are not solvable explicitly for the variable of interest (y). In such cases, NLS is not applicable while GMM is.

Let

$$E\psi(z; \beta) = 0$$

be a $s \times 1$ population moment condition where β is a $k \times 1$ vector. For instance, in NLS for $y = r(x, \beta) + u$,

$$E\psi(z, \beta) = 0 \Leftrightarrow E[\{y - r(x, \beta)\} \cdot r_b(x, \beta)] = 0. \qquad (5.1)$$

The sample version is

$$(1/N) \sum_i \psi(z_i, \beta) = 0 \Leftrightarrow (1/N) \sum_i \{y_i - r(x_i, \beta)\} \cdot r_b(x_i, \beta) = 0. \qquad (5.2)$$

As mentioned in Chapter 2, regardless of $\psi(\beta)$ being linear on nonlinear, if $\{z_i\}$ is iid the most efficient way to combine s moment restrictions $E\psi(z, \beta) = 0$ is to minimize the quadratic norm:

$$Q_N(b) = \left\{ (1/N) \sum_i \psi(b)' \right\} \left\{ (1/N) \sum_i \psi(\beta)\psi(\beta)' \right\}^{-1} \left\{ (1/N) \sum_i \psi(b) \right\} \qquad (5.3)$$

wrt b, with the middle weighting matrix being the inverse of $V[(1/\sqrt{N}) \sum_i \psi(\beta)]$. Intuitively, whatever metric we may use on R^s, only its quadratic approximation matters for the variance. Hence, GMM in Chapter 2 indexed by the weighting matrix W is a large enough class for the moment condition under iid assumption. In practice, since β in the middle weighting matrix of $Q_N(b)$ is unknown, we need to replace β by an initial consistent estimate for which NLS or some IVE may be used. As long as the initial estimate is $\sqrt{N}$-consistent, the asymptotic distribution of the GMM estimator minimizing the feasible version is the same as that of the nonfeasible GMM, which would minimize (5.3).

The asymptotic distribution of GMM minimizing (5.3) cannot be obtained from the theorem for the extremum estimators. The reason is that the minimand is not of the form $(1/N) \sum_i q(z_i, b)$. Still, a Taylor's expansion of the first-order condition yields the asymptotic distribution. For a general weighting matrix W^{-1}, not just $\{E\psi(\beta)\psi(\beta)'\}^{-1}$, the first-order condition is

$$(1/N) \sum_i \psi_b(b_N) \cdot W^{-1} \cdot (1/\sqrt{N}) \sum_i \psi(b_N) = 0, \qquad (5.4)$$

where $\psi_b \equiv \partial\psi/\partial b$. Expand $(1/\sqrt{N}) \sum_i \psi(b_N)$ around β to get

$$(1/\sqrt{N}) \sum_i \psi(\beta) + (1/N) \sum_i \psi_b(b_N^*)' \cdot \sqrt{N}(b_N - \beta).$$

Substitute this into the first-order condition to get

$$(1/N) \sum_i \psi_b(b_N) \cdot W^{-1} \cdot \left\{ (1/\sqrt{N}) \sum_i \psi(\beta) \right.$$

$$+ (1/N) \sum_i \psi_b(b_N^*)' \sqrt{N}(b_N - \beta) \bigg\} = 0. \tag{5.5}$$

Hence,

$$\sqrt{N}(b_N - \beta) =^p -(E\psi_b W^{-1} E\psi_b')^{-1} \cdot E\psi_b W^{-1} \cdot (1/\sqrt{N}) \sum_i \psi(\beta)$$

$$=^d N(0, (E\psi_b W^{-1} E\psi_b')^{-1} \cdot E\psi_b W^{-1} E\psi\psi' W^{-1} E\psi_b'$$

$$\cdot (E\psi_b W^{-1} E\psi_b')^{-1}). \tag{5.6}$$

Choosing $W = E\psi\psi'$ simplifies the variance matrix, yielding the most efficient one for the GMM class. Therefore,

$$\sqrt{N}(b_N - \beta) = N(0, \{E\psi_b \cdot E^{-1}(\psi\psi') \cdot E\psi_b'\}^{-1}). \tag{5.7}$$

If $E\psi_b$ is invertible with $s = k$, then the variance matrix in (5.6) becomes that in (5.7) regardless of W. Hence, in this case, using the Euclidean metric with $W = I_k$ is optimal; also, the asymptotic distribution of b_N in (5.7) is the same as the one which is obtained by applying the mean value theorem to $(1/\sqrt{N}) \sum_i \psi(b_N) = 0$. This result is analogous to GMM becoming LSE when $s = k$ in the linear model; that is, if $s = k$, there is nothing to gain by weighting the moment conditions. For the nonlinear model $y = r(\beta) + u$ with the $k \times 1$ moment condition $E\{r_b(\beta)u\} = 0$, NLS becomes GMM, for the variance matrix of $\sqrt{N}(b_{\text{NLS}} - \beta)$ given in (2.17) can be obtained by subsituting $\psi = r_b(\beta) \cdot \{y - r(\beta)\}$ and $E\psi_b = E\{r_b(\beta)r_b(\beta)'\}$ into (5.7). Hence, the relationship between GMM and LSE in the linear model holds between GMM and NLS in the nonlinear model.

GMM is implemented by taking one step from an initial $\sqrt{N}$-consistent estimator. Substituting the initial estimate b_0 into β in (5.6), we can rewrite (5.6) (with $W = E\psi\psi$) as

$$b_N = b_0 - \left[\sum_i \psi_b(b_0) \cdot \left\{ \sum_i \psi(b_0)\psi(b_0)' \right\}^{-1} \cdot \sum_i \psi_b(b_0)' \right]^{-1}$$

$$\cdot \sum_i \psi_b(b_0) \cdot \left\{ \sum_i \psi(b_0)\psi(b_0)' \right\}^{-1} \cdot \sum_i \psi(b_0). \tag{5.8}$$

If there is no such initial consistent estimate, then GMM estimation is implemented iteratively by starting with an arbitrary b_0 and updating b_0 until a convergence criterion is met.

When $s > k$, if the moment conditions are correct, then GMM should not just minimize $Q_N(b)$ but also make $Q_N(b) =^p 0$. Hansen (1982) suggests a *GMM specification test* examining whether the moment conditions more than k can be satisfied by only k many parameters. The test statistic is obtained by norming $Q_N(b)$ properly:

$$(1/\sqrt{N}) \sum_i \psi(b_N)' \cdot \left\{ (1/N) \sum_i \psi(b_N)\psi(b_N)' \right\}^{-1}$$

$$\cdot (1/\sqrt{N}) \sum_i \psi(b_N) =^d \chi^2_{s-k}. \qquad (5.9)$$

Earlier in Chapter 3, we showed that, for an extreme estimator b_{opt} maximizing $(1/N) \sum_i q(z_i, b)$, $V[\sqrt{N}(b_{\text{opt}} - \beta)]$ can be written as

$$[E(s_b q_b') \cdot E^{-1}(q_b q_b') \cdot E(q_b s_b')]^{-1}, \qquad (5.10)$$

where s_b is the score function for β and $q_b \equiv \partial q / \partial b$. Through the same steps we went through for (5.10), GMM variance matrix can be rewritten as

$$[E(s_b \psi') \cdot E^{-1}(\psi\psi') \cdot E(\psi s_b')]^{-1}. \qquad (5.11)$$

This is the inverse of the part of $E(s_b s_b')$ explained by ψ. If ψ is s_b, then the GMM is MLE; if ψ is close to s_b, the GMM will be almost as efficient as MLE. Since $E(s_b s_b')$ is larger than the part explained by ψ, GMM is less efficient than MLE.

Let $z = (x', y)'$. If the moment condition we have is $E\{\psi(y, x) \mid x\} = 0$, not $E\psi(y, x) = 0$, then GMM is not efficient. The reason is that $E\{\psi(y, x) \mid x\}$ is much stronger than $E\psi(y, x) = 0$: $E\{\psi(y, x) \mid x\} = 0$ implies $E\{g(x) \cdot \psi(y, x)\} = 0$ for any function $g(x)$, rendering infinitely many unconditional moment conditions. The variance matrix of the efficient estimator (or the "efficiency bound") under $E\{\psi(y, x) \mid x\} = 0$ is (Chamberlain, 1987)

$$E_x^{-1}\{E(\psi_b \mid x) \cdot E^{-1}(\psi\psi' \mid x) \cdot E(\psi_b' \mid x)\}. \qquad (5.12)$$

If $g(x)$ is continuous, it can be well approximated by polynomial functions of x. Then $E(g \cdot \psi) = 0$ for an arbitrary $g(x)$ is equivalent to $E(\zeta_j(x) \cdot \psi) = 0$, $j = 1 \ldots J$, where $\zeta_j(x)$ are polynomial functions of x. The GMM with these unconditional moment conditions attains the bound (5.12) as $J \to \infty$ (Chamberlain, 1987).

As an example of (5.12), suppose $y = x'\beta + u$ with $E(\psi \mid x) = 0$ where $\psi(z, \beta) = y - x'\beta u$. Then (5.12) becomes

$$E_x^{-1}\{x \cdot E^{-1}(u^2 \mid x) \cdot x'\} = E^{-1}\{xx'/V(u \mid x)\}, \qquad (5.13)$$

the variance matrix of the generalized LSE (GLS). If $E(u^2 \mid x) = \sigma^2$, then the efficiency bound is $\sigma^2 \cdot E^{-1}(xx')$ [we already know that LSE is efficient under $E(ux) = 0$]. Therefore, for the linear model,

(i) if $E(u \mid x) = 0$ and $E(u^2 \mid x) = \sigma^2$, then LSE is efficient;

(ii) if $E(u \mid x) = 0$ and $E(u^2 \mid x) = h(x)$ whose form is known, then GLS attains the efficiency bound (5.13);

(iii) if $E(u \mid x) = 0$, whether homoskedasticity holds or not, there exists a GMM which is efficient attaining (5.13).

One relevant question for (ii) is: with $\psi = (y - x'\beta, (y - x'\beta)^2 - h(x))$, can we get an estimator more efficient than GLS? The answer is yes if $E(u^3 \mid x) \neq 0$ or β enters the $h(x)$; see Newey (1993, p. 427).

6.6   GMM for Linear Simultaneous Equations and Nonlinear Models

In this section, we apply GMM to linear simultaneous equations, and then examine GMM for multiple nonlinear equations.

Consider a three linear simultaneous equation system

$$\begin{bmatrix} y_1 - \alpha_{12}y_2 - \alpha_{13}y_3 - x_1'\beta_1 \\ y_2 - \alpha_{21}y_1 - \alpha_{23}y_3 - x_2'\beta_2 \\ y_3 - \alpha_{31}y_1 - \alpha_{32}y_2 - x_3'\beta_3 \end{bmatrix} = \begin{bmatrix} u_1 \\ u_2 \\ u_3 \end{bmatrix}. \tag{6.1}$$

Define $y \equiv (y_1, y_2, y_3)'$, $u \equiv (u_1, u_2, u_3)'$, $x \equiv x_1 \cup x_2 \cup x_3$, and γ as the system structural form (SF) coefficients:

$$\gamma \equiv (\alpha_{12}, \alpha_{13}, \beta_1', \alpha_{21}, \alpha_{23}, \beta_2', \alpha_{31}, \alpha_{32}, \beta_3')'. \tag{6.2}$$

Define further K and k as the row dimension of x and γ, respectively,

$$K \equiv \text{row}(x), \quad k \equiv \text{row}(\gamma) \tag{6.3}$$

where $\text{row}(\cdot)$ is the row dimension. Then (6.1) can be written as

$$y - w'\gamma = u, \tag{6.4}$$

where $(xj \equiv x_j)$

$$\begin{matrix} w' \\ 3 \times k \end{matrix} \equiv \begin{bmatrix} y_2 & y_3 & x_1' & 0 & 0 & 0'_{\text{row}(x2)} & 0 & 0 & 0'_{\text{row}(x3)} \\ 0 & 0 & 0'_{\text{row}(x1)} & y_1 & y_3 & x_2' & 0 & 0 & 0'_{\text{row}(x3)} \\ 0 & 0 & 0'_{\text{row}(x1)} & 0 & 0 & 0'_{\text{row}(x2)} & y_1 & y_2 & x_3' \end{bmatrix}. \tag{6.5}$$

Let $\otimes$ denote the Kronecker product. The moment condition is (the "subline" shows the dimensions of the terms above)

$$\underset{(3K) \times 1}{E(u \otimes x)} = \underset{(3K) \times 3}{E\{(I_3 \otimes x)} \cdot \underset{3 \times 1}{u\}} = \{E(x'u_1), E(x'u_2), E(x'u_3)\}' = 0; \tag{6.6}$$

the error term in each equation has zero covariance with x. Define

$$\underset{3K \times 3}{z_i} \equiv I_3 \otimes x_i \tag{6.7}$$

to express the moment condition as

$$E\{z \cdot (y - w'\gamma)\} = 0. \tag{6.8}$$

Then an instrumental variable estimator g_{ive} for γ is

$$g_{\text{ive}} = \left[\sum_i w_i z_i' \left\{ \sum_i z_i z_i' \right\}^{-1} \sum_i z_i w_i' \right]^{-1}$$

$$\cdot \sum_i w_i z_i' \left\{ \sum_i z_i z_i' \right\}^{-1} \sum_i z_i y_i. \tag{6.9}$$

Note that $z_i z_i' = (I_3 \otimes x_i) \cdot (I_3 \otimes x_i') = I_3 \otimes x_i x_i'$, and that $w_i z_i'$ is of dimension $(k \times 3) \cdot (3 \times 3K) = k \cdot 3K$. So each component of $w_i z_i'$ is actually a sum of three terms. For a single equation case, we called a version similar to (6.9) "Best IVE (BIV)," for the version was efficient under homoskedasticity. But since (6.9) is not efficient even under homoskedasticity for (6.1), we call (6.9) simply an IVE.

The GMM version for (6.8) is obtained by taking one step from g_{ive}. Define the residuals $\hat{u}_i \equiv y_i - w_i' g_{\text{ive}}$ to get

$$g_{\text{gmm}} = \left[\sum_i w_i z_i' \cdot \left\{ \sum_i z_i \hat{u}_i \hat{u}_i' z_i' \right\}^{-1} \cdot \sum_i z_i w_i' \right]^{-1}$$

$$\cdot \sum_i w_i z_i' \cdot \left\{ \sum_i z_i \hat{u}_i \hat{u}_i' z_i' \right\}^{-1} \cdot \sum_i z_i y_i; \tag{6.10}$$

$$\sqrt{N}(g_{\text{gmm}} - \gamma) =^d N \left(0, \left[\sum_i w_i z_i'/N \cdot \left\{ \sum_i z_i \hat{u}_i \hat{u}_i z_i'/N \right\}^{-1} \right. \right.$$

$$\left. \left. \cdot \sum_i z_i w_i'/N \right]^{-1} \right).$$

Differently from a single equation case, b_{gmm} does not include b_{ive} as a special case when the error term vector is homoskedastic; b_{gmm} becomes b_{ive} only when $E(uu')$ is a scalar matrix.

Turning to nonlinear equations, note that we did not try to solve the linear simultaneous equations for y in (6.1), and this makes extension to nonlinear models easy. The only complication is that the GMM in nonlinear equations requires an iterative scheme.

Suppose we have a simultaneous nonlinear equation system

$$\rho(\underset{s \times 1}{y}, \underset{s \times 1}{x}, \underset{K \times 1}{\gamma}) = \underset{s \times 1}{u}, \qquad (6.11)$$

where y is the endogenous variables and x is the exogenous variables of the system. This includes (6.4) as a special case with $s = 3$.

Assume

$$\underset{(sK) \times 1}{E(x \otimes u)} (= \underset{(sK) \times s}{E\{(I_s \otimes x)} \cdot \underset{s \times 1}{u\})} = 0, \qquad (6.12)$$

which is $s \cdot K$ moment conditions of the system where the error term in each equation has zero covariance with x. As in (6.7), define

$$\underset{sK \times s}{z_i} \equiv I_s \otimes x_i \qquad (6.13)$$

to rewrite the sK moment condition as [recall (6.8)]

$$E(z \cdot u) = 0 \Leftrightarrow E\{z \cdot \rho(y, x, \gamma)\} = 0. \qquad (6.14)$$

From now on, we will drop the adjective "simultaneous," since the following is applicable to any nonlinear equations satisfying (6.14).

Consider a nonlinear IVE (NIV) minimizing

$$\left[(1/N) \sum_i z_i \rho(y_i, x_i, g) \right]' \cdot \left\{ (1/N) \sum_i z_i z_i' \right\}^{-1}$$

$$\cdot \left[(1/N) \sum_i z_i \rho(y_i, x_i, g) \right] \qquad (6.15)$$

wrt g. This was first suggested by Amemiya (1974) under a different name. The form of NIV is similar to BIV in the linear model. Since ρ is nonlinear, NIV is obtained with an iterative method. Observe that

$$\rho(g_1) \cong \rho(g_0) + \rho_g(g_0)'(g_1 - g_0), \qquad \underset{k \times s}{\rho_g} \equiv \partial \rho / \partial g. \qquad (6.16)$$

Substituting this into (6.15) and solving the first-order condition for minimization wrt g_1, we obtain the following iterative scheme for NIV:

$$g_1 = g_0 - \left[\sum_i \rho_g(g_0) z_i' \cdot \left\{ \sum_i z_i z_i' \right\}^{-1} \cdot \sum_i z_i \rho_g(g_0)' \right]^{-1}$$

$$\cdot \sum_i \rho_g(g_0)z_i' \cdot \left\{\sum_i z_i z_i'\right\}^{-1} \cdot \sum_i z_i \rho(g_0). \tag{6.17}$$

GMM for (6.11) minimizes

$$\left[(1/N)\sum_i z_i \cdot \rho(y_i, x_i, g)\right]' \cdot W^{-1} \cdot \left[(1/N)\sum_i z_i \cdot \rho(y_i, x_i, g)\right], \tag{6.18}$$

where W is the $sK \times sK$ variance matrix of $(1/\sqrt{N})\sum_i z_i u_i$:

$$W = E(zuu'z') = E\{(I_s \otimes x)uu'(I_s \otimes x')\}.$$

With $\hat{u}_i \equiv \rho(y_i, x_i, g_{\mathrm{niv}})$, GMM is obtained as $(g_0 = g(\mathrm{niv})$

$$g_{\mathrm{gmm}} = g_0 - \left[\sum_i \rho_g(g_0)z_i' \cdot \left\{\sum_i z_i \hat{u}_i \hat{u}_i' z_i'\right\}^{-1} \cdot \sum_i z_i \rho_g(g_0)'\right]^{-1}$$

$$\cdot \sum_i \rho_g(g_0)z_i' \cdot \left\{\sum_i z_i \hat{u}_i \hat{u}_i' z_i'\right\}^{-1} \cdot \sum_i z_i \rho(g_0). \tag{6.19}$$

The asymptotic variance matrix for GMM in (6.19) is

$$\left[E(\rho_g z') \cdot \{E(zuu'z')\}^{-1} \cdot E(z\rho_g')\right]^{-1}, \tag{6.20}$$

where the dimension of $\rho_g z'$ is $k \times (sK)$. With $\rho_g = w$, this includes the linear simultaneous equation GMM in (6.10) as a special case.

Suppose we have the moment condition $E(u \mid x) = 0$, not just $E(x \otimes u) = 0$. Then (5.12) indicates that GMM is not efficient, but we can get an efficient estimator by augmenting the moment conditions as shown in the paragraph following (5.12). Newey (1988, 1993) suggests another way to get an efficient estimator for the model $\rho(y, x, \gamma) = u$. Consider a $r \times s$ $(r \geq k)$ known instrument matrix $A(x)$ and the moment condition

$$E\{A(x)\rho(y, x, \gamma)\} \equiv E\psi(z, \gamma) = 0. \tag{6.21}$$

Applying GMM to this, the asymptotic variance is

$$\left[E\{\rho_g \cdot A(x)'\} \cdot E^{-1}\{A(x)uu'A(x)'\} \cdot E\{A(x) \cdot \rho_g'\}\right]^{-1}. \tag{6.22}$$

If we set $r = k$ and

$$A(x) = E(\rho_g \mid x) \cdot E^{-1}(uu' \mid x), \tag{6.23}$$

then (6.22) becomes (5.12); that is, if we can get (6.23), we can attain the efficiency bound (5.12) with the GMM.

To get (6.23), we need $E(\rho_g \mid x)$ and $E(uu' \mid x)$. If homoskedasticity is assumed, then $E(uu' \mid x)$ is a constant matrix and can be estimated with the residuals from an initial estimator. Getting $E(\rho_g \mid x)$ is more problematic, for we need $f_{y|x}$ and the integration $\int \rho_g(y, x, \gamma) f_{y|x}(y) dy$ to get $E(\rho_g \mid x)$ in general. But as to be shown later, using nonparametric techniques, we can get (6.23). See Robinson (1991) and Newey (1993) who also shows several examples; Newey (1990b) treats the homoskedastic case only.

Finally in this section, we note that IVE (including GMM) is not valid for nonlinear errors-in-variable models, while IVE is applicable to errors-in-variable problems in linear models. To see this, consider $y = r(x^*) + u$ but we observe only $x = x^* + v$. Then, for a $x^{**} \in (x^*, x)$,

$$y = r(x) + r'(x^{**})(x^* - x) + u = r(x) + \{u - r'(x^{**})v\}. \tag{6.24}$$

Even if there is an instrument z that is correlated with x and uncorrelated with u and v, the new error term $u - r'(x^{**})v$ is correlated with z through $r'(x^{**})$. This problem occurs even in polynomial regression models that are linear in parameters. See Hausman et al. (1995) for more.

6.7 Three Tests in GMM and GMM for Dependent Observations

In this section, we examine two topics: three tests in GMM, and adjustment in GMM for dependent observations.

There are GMM analogs for Wald, LR, and LM tests. Recall the moment condition along with $H_0: h(\beta) = 0$:

$$\underset{s \times 1}{E\{\psi(z, \ \beta)\}} = 0, \quad \underset{g \times 1}{H_0: h(\beta) = 0.} \tag{7.1}$$

If $H_0: h(\beta) = r$, redefine $h(\beta)$ as $h(\beta) - r$ to get $H_0: H(\beta) = 0$. Define the efficient GMM as b_N, and its restricted version as b_{Nr}, which minimizes

$$Q_N(b) \equiv \left\{ (1/N) \sum_i \psi(b)' \right\} \cdot \left[V \left\{ (1/\sqrt{N}) \sum_i \psi(b) \right\} \right]^{-1}$$

$$\cdot \left\{ (1/N) \sum_i \psi(b) \right\} \tag{7.2}$$

subject to $h(b) = 0$. Let W_N denote an estimate for $V\{(1/\sqrt{N}) \sum_i \psi(\beta)\}$, and define (with $\psi_b \equiv \partial \psi / \partial b$)

$$\Omega_N(b_N) \equiv (1/N) \sum_i \psi_b(b_N) \cdot W_N^{-1} \cdot (1/N) \sum_i \psi_b(b_N)'. \tag{7.3}$$

Then we have the following test statistics, all following χ_g^2 [Newey and West (1987b); Newey and McFadden (1994) show even more]

$$\text{Wald} = N \cdot h(b_N)' \cdot \{\partial h(b_N)/\partial b \cdot \Omega_N^{-1}(b_N) \cdot \partial h(b_N)'/\partial b\}^{-1} \cdot h(b_N),$$

$$\text{LR} = N\{Q_N(b_{Nr}) - Q_N(b_N)\},$$

$$\text{LM} = N \cdot \left\{ (1/N) \sum_i \psi_b(b_{Nr}) \cdot W_N^{-1} \cdot (1/N) \sum_i \psi(b_{Nr}) \right\}' \cdot \Omega_N(b_{NR})^{-1}$$

$$\cdot \left\{ (1/N) \sum_i \psi_b(b_{Nr}) \cdot W_N^{-1} \cdot (1/N) \sum_i \psi(b_{Nr}) \right\}. \qquad (7.4)$$

Using the same estimator W_N in all cases is required; for instance, there is no guarantee for $LR > 0$ unless this condition is met. Under this condition, the observations practicality made for Wald, LR and LM tests in regard to MLE and NLS also hold for the GMM tests. If $\psi(z, \beta)$ and $h(\beta)$ are linear in β in (7.4), Wald = LR = LM numerically.

If we have a dependent data set, then the optimal weighting matrix in GMM requires some adjustment, although estimation of $E\psi_b(z, \beta)$ can be done in the same way as with iid data by $(1/N) \sum_i \psi_b(z_i, b_N)$. Recall that the optimal weighting matrix is the inverse of $W = V\{(1/\sqrt{N}) \sum_i \psi(z_i, \beta)\}$. By the definition of variance,

$$W = E\left[(1/N) \sum_i \sum_j \psi(z_i, \beta) \cdot \psi(z_j\beta)' \right]$$

$$= (1/N) \sum_i \sum_j E\{\psi(z_i, \beta) \cdot \psi(z_j, \beta)'\}. \qquad (7.5)$$

If $\{z_i\}$ are iid, all cross products in (7.5) disappear and W is estimated by $(1/N) \sum_i \psi(z_i, b_N)\psi(z_i, b_N)'$.

It is important to realize that (7.5) cannot be estimated by

$$(1/N) \sum_i \sum_j \psi(z_i, b_N) \cdot \psi(z_j, b_N)'$$

$$= (1/\sqrt{N}) \sum_i \psi(z_i, b_N) \cdot (1/\sqrt{N}) \sum_j \psi(z_j, b_N)' \qquad (7.6)$$

because $(1/\sqrt{N}) \sum_i \psi(z_i, b_N) =^p 0$ by the first-order condition. White and Domowitz (1984) suggests an estimator for (7.5): omitting b_N,

$$(1/N) \sum_i \psi(z_i)\psi(z_i)' + (1/N) \sum_{i=1}^m \sum_{j=i+1}^N \{\psi(z_j)\psi(z_{j-i})'$$

$$+ \psi(z_{j-i})\psi(z_j)'\}, \tag{7.7}$$

where $m < N - 1$. If $m = N - 1$, then this is the same as (7.6). Hence by removing some terms in (7.6), we get (7.7). The two terms in $\{\cdot\}$ guarantees the symmetry of the estimator. In (7.7), we attempt to make (7.6) nonzero by limiting the dependence over time:

$$E\{\psi(z_j, \beta) \cdot \psi(z_{j-i}, \beta)'\} = 0 \quad \text{for all } i > m. \tag{7.8}$$

However there is no good practical guideline on how to select the truncation number m. See White and Domowitz (1984, p. 153–154) for more.

Newey and West (1987a) modify (7.8) further which guarantees that the estimate is p.s.d. with a given N. Their estimate is

$$W_N = (1/N) \sum_i \psi(z_i)\psi(z_i)'$$

$$+ (1/N) \sum_{i=1}^{m} \{1 - i/(m+1)\} \sum_{j=i+1}^{N} \{\psi(z_j)\psi(z_{j-i})' + \psi(z_{j-i})\psi(z_j)'\}; \tag{7.9}$$

see Andrews (1991b) for more on estimating W. For instance, suppose we choose $m = 2$. Then (7.9) is

$$(3/3) \cdot (1/N) \sum_{j=1}^{N} \psi(z_j)\psi(z_j)'$$

$$+ (2/3) \cdot (1/N) \sum_{j=2}^{N} \{\psi(z_j)\psi(z_{j-1})' + \psi(z_{j-1})\psi(z_j)'\}$$

$$+ (1/3) \cdot (1/N) \sum_{j=3}^{N} \{\psi(z_j)\psi(z_{j-2})' + \psi(z_{j-2})\psi(z_j)'\}; \tag{7.10}$$

smaller weights are given to the terms with more lags. If $m = 1$, use the first two of (7.10) with the weights being 2/2 and 1/2, respectively.

6.8   A LM Test for Linear Regression Function

In this section, we propose a linear regression function specification test where the alternative models are single index and power transformation models. The two models give "directions" against which linear regression model can be tested. The reader not interested in the derivation of the test can skip to (8.11).

Recall the Box–Cox (power) transformation model:

$$\begin{aligned} y_i &= x_i(\alpha)'\beta + u_i \\ &\equiv \beta_1 + x_{i2}(\alpha_2) \cdot \beta_2 + \cdots + x_{ik}(\alpha_k) \cdot \beta_k + u_i, \end{aligned} \tag{8.1}$$

where

$$x_{ij}(\alpha_j) = \begin{matrix} (x_{ij}^{\alpha_j} - 1)/\alpha_j & \text{if } \alpha_j \neq 0, \\ \ln(x_{ij}) & \text{if } \alpha_j = 0, \ j = 2, \ldots, k; \end{matrix} \qquad (8.2)$$

if x_{ij} can be equal to or less than 0, set $x_{ij}(\alpha_j) = x_{ij}$. We can generalize (8.1) further by allowing the single index specification:

$$y_i = s\{x_i(\alpha)'\beta\} + u_i. \qquad (8.3)$$

We may impose the Box–Cox transformation also on y, but this will require a lower bound on y ($y > 0$). Instead, consider a strictly monotonic function $\tau(\cdot)$ such that

$$\tau(y_i) = s\{x_i(\alpha)'\beta\} + u_i \Leftrightarrow y_i = \tau^{-1}[s\{x_i(\alpha)'\beta\} + u_i]. \qquad (8.4)$$

Then

$$E(y_i \mid x_i) = \int \tau^{-1}[s\{x_i(\alpha)'\beta\} + u_i] \cdot dF_{u|x_i}(u_i) \equiv w\{x_i(\alpha)'\beta\}, \qquad (8.5)$$

where $F_{u|x}$ is the distribution function of $u \mid x$. Thus, defining $v_i \equiv y_i - w\{x_i(\alpha)'\beta\}$, y_i can be rewritten as

$$y_i = w\{x_i(\alpha)'\beta\} + v_i; \qquad (8.6)$$

(8.6) is more general than (8.3), for (8.6) allows transforming y.

Approximate (8.6) with the following polynomial: for some M,

$$y = \sum_{m=1}^{M} \gamma_m \cdot \{x(\alpha)'\beta\}^m + u. \qquad (8.7)$$

Suppose we estimate $\alpha = (\alpha_2, \ldots, \alpha_k)'$, β and $\gamma = (\gamma_2, \ldots, \gamma_M)'$ in (8.7) using NLS by maximizing ($\alpha_1 = 1$ and $\gamma_1 = 1$)

$$Q_N \equiv (1/N) \sum_{i=1}^{N} -(1/2) \cdot \left[y_i - \sum_{m=1}^{M} g_m \cdot \{x_i(a)'b\}^m \right]^2 \qquad (8.8)$$

wrt a, b, and g. The gradient evaluated at the true value is (assume that all regressors are transformed, which does not affect our test):

$$\partial Q_N/\partial g_m = (1/N) \sum_i u_i \{x_i(\alpha)'\beta\}^m, \quad m = 2, \ldots, M;$$

$$\partial Q_N/\partial a_j = (1/N) \sum_i u_i \beta_j \{\alpha_j \cdot x_{ij}^{\alpha_j} \ln(x_{ij}) - x_{ij}^{\alpha_j} + 1\}$$

$$\cdot \sum_{m=1}^{M} \gamma_m m \{x_i(\alpha)'\beta\}^{m-1}, \quad j = 2, \ldots, k; \qquad (8.9)$$

$$\partial Q_N/\partial b_1 = (1/N)\sum_i u_i \cdot \sum_{m=1}^{M} \gamma_m m(x_i(\alpha)'\beta)^{m-1};$$

$$\partial Q_N/\partial b_j = (1/N)\sum_i u_i \cdot x_{ij}(\alpha_j) \cdot \sum_{m=1}^{M} \gamma_m m(x_i(\alpha)'\beta)^{m-1}, \quad j = 2,\ldots k.$$

Suppose that the linear model is true: $\alpha = (1,\ldots,1)$, $\gamma_1 = 1$, $\gamma_2 = \cdots = \gamma_k = 0$. Substitute these into the gradient to set the gradient equal to 0. Remove redundant terms to get the following moment conditions:

(i) $(1/N)\sum_i u_i(x_i'\beta)^m = 0, \quad m = 2,\ldots, M;$

(ii) $(1/N)\sum_i u_i\{x_{ij}\beta_j \cdot \ln(x_{ij})\} = 0, \quad j = 2,\ldots, k;$ (8.10)

(iii) $(1/N)\sum_i u_i x_i = 0.$

A LM type linear regression function specification is possible, for LSE is obtained with (iii). The test examines if LSE also satisfies (i) and (ii). The test procedure is

Step 1: Estimate the linear model with LSE;

Step 2: Set up the following artificial regression model

$$y = x'\beta + \sum_{m=2}^{M} \delta_m(x'b_N)^m + \sum_{j=2}^{k} \theta_j\{x_j \cdot \ln(x_j)\} + \varepsilon \qquad (8.11)$$

to estimate β, $\delta_m's$ and $\theta_j's$ with LSE.

Step 3: Test $[H_0\colon \delta_m's = 0$ and $\theta_j's = 0]$ in the artificial model using a heteroskedasticity-consistent covariance matrix for

$$\{E(zz')\}^{-1} \cdot E(\varepsilon^2 zz') \cdot \{E(zz')\}^{-1}$$

where z denotes the regressors in Step 2.

Typically $M = 2$ or 3 with a number of $x_j \cdot \ln(x_j)$ will be enough. Without the terms in $\sum_{j=2}^{k}$, the test becomes the RESET test of Ramsey (1969).

7
Nonparametric Density Estimation

7.1 Introduction

In the linear model $y = x'\beta + u$ where x is a regressor vector and $E(u \mid x) = 0$, we estimate β in $E(y \mid x) = x'\beta$. However, the assumption of the linear model, or any nonlinear model for that matter, is a strong one. In nonparametric regression, we try to estimate $E(y \mid x)$ without specifying the functional form. Since

$$E(y \mid x) = \int y \cdot f(y \mid x) dy = \int y \cdot \{f(y,x)/f(x)\} dy, \qquad (1.1)$$

if we can estimate $f(y,x)$ and $f(x)$, we can also estimate $E(y \mid x)$. In this chapter, we study nonparametric density estimation for x as a prelude to nonparametric regression in the next chapter. We will assume that x has a continuous density function $f(x)$. If x is discrete, one can estimate $P(x = x_0)$ either with the same estimation method used for the continuous case or with the number of observations with $x_i = x_0$. There are several nonparametric density estimation methods available. The most popular is "kernel density estimation method" which we explore mainly. Other methods will be examined briefly in the last section. See Prakasa Rao (1983), Silverman (1986), Izenman (1991), Rosenblatt (1991) and Scott (1992) for more on nonparametric density estimation in general.

The rest of this chapter is organized as follows. In Section 2, kernel is introduced and a kernel density estimator is defined. In Sections 3 and 4, bias and variance of the kernel density estimator are derived, respectively. Combining the two, consistency is proved in Section 4. Strengthening the consistency, the uniform consistency is introduced in Section 5. In Section 6, we address the question of choosing a kernel and a "smoothing parameter." In Section 7, the asymptotic distribution of the kernel estimator is studied. Finally in Section 8, other nonparametric density estimators are examined.

7.2   Kernel Density Estimation

Suppose we have N observations x_i, $i = 1 \ldots N$, where x_i is a $k \times 1$ vector. If our interest is in $P(x \leq x_0) \equiv F(x_0)$, $P(x \leq x_0)$ can be estimated by $(1/N) \sum_{i=1}^{N} 1[x_i \leq x_0]$, the *empirical distribution function*. Although this converges to $F(x_0)$ in various senses, it is not differentiable while $F(x_0)$ is. Hence, we cannot estimate $f(x_0)$ by differentiating the empirical distribution function. Since $f(x_0) = dF(x_0)/dx$, it is conceivable to estimate $f(x_0)$ by approximating $dF(x_0)$ and dx.

Let h be a small positive number. Then, $dx \cong h^k$ and

$$dF(x_0) = P(x_0 < x < x_0 + h) \cong (1/N) \sum_{i=1}^{N} 1[x_0 < x_i < x_0 + h]$$

$$\cong (1/N) \sum_i (1/2) \cdot 1[x_0 - h < x_i < x_0 + h].$$

Hence, one density estimator apaproximating $dF(x_0)/dx$ is

$$(1/(Nh^k)) \sum_i (1/2) \cdot 1[x_0 - h < x_i < x_0 + h]$$

$$= (1/(Nh^k)) \sum_i (1/2) \cdot 1[-h < x_i - x_0 < h]. \tag{2.1}$$

For this approximation to work, h should be small. If h is too small, however, there will be no observation satisfying $-h < x_i - x_0 < h$. Thus, we can let $h \to +0$ only as $N \to \infty$.

Viewing the role of the indicator function in (2.1) as a weighting function giving the weight 1 if x_i is within h-distance from x_0 and 0 otherwise, we can generalize the estimator (2.1) with a smooth weighting function K (Rosenblatt, 1956):

$$\hat{f}(x_0) \equiv (1/(Nh^k)) \sum_i K((x_i - x_0)/h), \tag{2.2}$$

where K is called a *kernel*. Choosing $K(z) = 1[-1 < z < 1]/2$ in (2.2) yields (2.1). If we want to estimate $\partial f(x_0)/\partial x$, assuming that K is differentiable, differentiate $\hat{f}(x_0)$ wrt x_0:

$$\partial \hat{f}(x_0)/\partial x \equiv -(1/(Nh^{k+1})) \sum_i K'((x_i - x_0)/h). \tag{2.3}$$

The choice of the kernel function is up to the user, but usually functions with the following properties are used:

(i) $K(z)$ is symmetric around 0 and continuous.

(ii) $\int K(z)dz = 1$, $\int K(z)zdz = 0_k$, and $\int |K(z)|dz < \infty$.   (2.4)

(iii) (a) $K(z) = 0$ if $|z| \geq z_0$ for some z_0, or
     (b) $|z| \cdot K(z) \to 0$ as $|z| \to \infty$.

(iv) $K(z) = \prod_{j=1}^{k} L(z_j)$, where L satisfies (i) to (iii), and $\int t^2 L(t)dt = \kappa$.

Obviously, the condition (iii)(a) implies (iii)(b). In addition to (2.4), we often require $K(z) \geq 0$. But the so-called "high-order kernels" to be later discussed can take negative values. We can impose restrictions other than (2.4) on K. So long as we can find a kernel satisfying the restrictions, imposing them should not matter. Not all the conditions in (2.4) are necessary for our purpose; almost all kernels used in practice satisfy (i)–(iii).
     Examples of K satisfying (2.4) for $k = 1$ are

(i) $1[|z| < 1]/2$: uniform kernel;

(ii) $N(0,1)$ density $\phi(z)$;   (2.5)

(iii) $(3/4) \cdot (1 - z^2) \cdot 1[|z| < 1]$: (trimmed) quadratic kernel;

(iv) $(15/16) \cdot (1 - z^2)^2 1[|z| < 1]$: quartic or biweight kernel.

The uniform kernel is not smooth and so rarely used (using it renders a histogram), while the other three are used frequently. The normal kernel has an unbounded support and is continuously differentiable up to any order. The trimmed quadratic kernel has a bounded support and continuously differentiable up to the second order except at $z = \pm 1$. The quartic kernel has a bounded support and continuously differentiable once; except at $z = \pm 1$, it is continuously differentiable up to the fourth. For $k > 1$, products of one-dimensional kernels such as $\prod_{j=1}^{k} \phi(z_j)$ can be used, but multivariate kernels such as a multivariate normal density with the variance Ω may be used as well:

$$K(z) = (2\pi)^{-k/2} \cdot |\Omega|^{-1/2} \cdot \exp\{-(1/2) \cdot z'\Omega^{-1}z\}.$$ (2.6)

     Later we will examine how to choose h in detail; h is called a *smoothing parameter* or *bandwidth* or *window size*. From (2.1), if h is small, only a few observations are used in calculating $\hat{f}(x_0)$. Then $\hat{f}(x_0)$ may be too jagged as x_0 varies. If h is too large, $\hat{f}(x_0)$ will show little change as x_0 varies; at the extreme, $\hat{f}(x_0)$ can become a constant not changing as x_0 changes. Thus, a "good" h should be found between these two extremes.

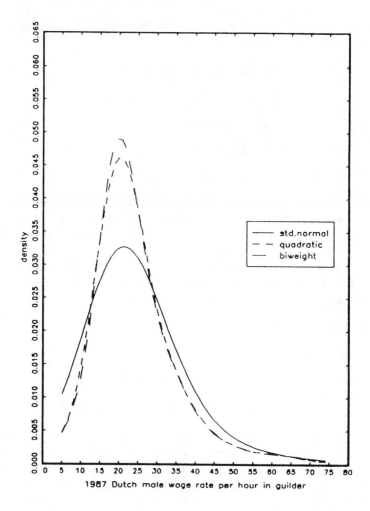

FIGURE 7.1. Over-smoothing $[h = 0.5 \times SD(x), n = 2421]$.

In Figures 7.1 and 7.2, we compute $\hat{f}(x)$ for hourly male wage with data ($N = 2421$) drawn from the 1987 "wave" of the Dutch Socio Economic Panel. In Figure 7.1, three lines with $N(0,1)$, quadratic and biweight kernels are shown where h is 0.5 times $SD(x)$; in Figure 7.2, h is 10 times smaller. Small peaks and troughs in Figure 7.2 indicate that the lines are slightly under-smoothed (too jagged to be true); note that the lines over-

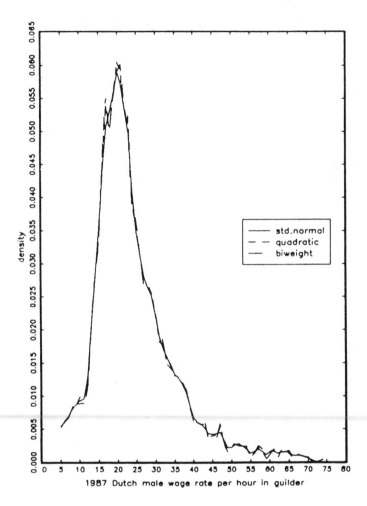

FIGURE 7.2. Under-smoothing $[h = 0.05 \times SD(x),\ N = 2421]$.

lap almost perfectly despite different kernels. In Figure 7.1, lines are over-smoothed (relative to the under-smoothed Figure 7.2), and the one with $N(0,1)$ kernel is noticeably different from the other two. In both Figures 7.1 and 7.2, all lines have a long right tail due to a small proportion of people making a lot of money (the maximum x is around 500); the lines shows the mode $\cong 20$, while the sample mean is 25.1.

7.3   Bias of Kernel Density Estimator

Calculating $E\hat{f}(x_0)$,

$$E\hat{f}(x_0) = \int h^{-k} \cdot K((x - x_0)/h)f(x)dx = \int K(z)f(x_0 + hz)dz, \quad (3.1)$$

where $z = (x - x_0)/h$; h^{-k} disappears due to the Jacobian of transformation. For unbiasedness, we want $E\hat{f}(x_0) = f(x_0)$. But this is impossible in general. Instead we will make the bias shrink as $N \to \infty$.

Assuming f has a $k \times 1$ bounded continuous first derivative vector $\partial f/\partial x$, for some $x^* \in (x_0, x_0 + hz)$, we get

$$E\hat{f}(x_0) = \int K(z)\{f(x_0) + hz'(\partial f(x^*)/\partial x)\}dz$$

$$= \int K(z)f(x_0)dz + \int hK(z)z'(\partial f(x^*)/\partial x)dz$$

$$= f(x_0) + h \cdot \int K(z)z'(\partial f(x^*)/\partial x)dz. \quad (3.2)$$

Hence the bias is $O(h)$. If f has a $k \times k$ bounded continuous second derivative matrix $\partial^2 f/\partial x \partial x'$, then using $\int K(z)zdz = 0$,

$$E\hat{f}(x_0) = \int K(z)\{f(x_0) + hz'(\partial f(x)/\partial x)$$

$$+ (h^2/2)z'(\partial^2 f(x^*)/\partial x \partial x')z\}dz = f(x_0) + O(h^2). \quad (3.3)$$

If we further assume that f has bounded and continuous partial derivatives up to an order $m \geq 3$, then we get ($\sum_{j=.}^{J} = 0$ when $J < j$)

$$E\hat{f}(x_0) = f(x_0) + (h^2/2) \int z'\{\partial^2 f(x_0)/\partial x \partial x'\}zK(z)dz$$

$$+ \sum_{q=3}^{m-1}(h^q/q!) \int \sum_{i_1=1}^{k} \cdots \sum_{i_q=1}^{k}\left\{(\partial^q f(x_0)/\partial x_{i_1}\cdots\partial x_{i_q}) \cdot \prod_{j=1}^{q}z_{i_j}\right\}K(z)dz$$

$$+ (h^m/m!) \int \sum_{i_1=1}^{k} \cdots \sum_{i_m=1}^{k}\left\{(\partial^m f(x^*)/\partial x_{i_1}\cdots\partial x_{i_m})\right.$$

$$\left. \cdot \prod_{j=1}^{m}z_{i_j}\right\}K(z)dz. \quad (3.4)$$

In general, the second order term does not disappear and so the bias is still $O(h^2)$. Hereafter assume that $\partial^2 f/\partial x \partial x'$ is continuous and bounded.

If we choose $K(z)$ such that the terms of higher order than the first in (3.4) disappear, we can make the bias smaller than $O(h^2)$. Such a kernel is called a *high-order kernel*, which, however, has the following problem. For a scalar z, the second-order term becomes

$$(h^2/2) \cdot \{\partial^2 f(x_0)/\partial x^2\} \cdot \int K(z)z^2 dz.$$

To have $\int K(z)z^2 dz = 0$, $K(z)$ should be negative for some z, which backs away from the notion of weights given to $K(z)$. For instance, consider

$$K(z) = a_0\phi(z) + a_1 z^2 \phi(z)$$

where a_0 and a_1 are chosen such that

$\int K(z)dz = 1 \Leftrightarrow a_0 + a_1 = 1,$

$\int zK(z)dz = 0$ is satisfied for any a_0 and $a_1,$

$\int z^2 K(z)dz = 0 \Leftrightarrow a_0 + a_1 \cdot \int z^4 \phi(z)dz = 0;$

one of a_1 and a_2 should be negative in view of the last line.

Two examples of ϕ-based high-order kernels are:

$$(3/2) \cdot \phi(z) - (1/2) \cdot z^2 \phi(z) : \quad \int z^2 K(z)dz = 0;$$

$$(3/2) \cdot \phi(z) - (1/2) \cdot z^2 \phi(z) + (1/35) \cdot z^4 \phi(z) : \quad \int z^4 K(z)dz = 0. \quad (3.5)$$

See Bierens (1987, p. 112) for multivariate versions of these. A polynomial kernel with $\int z^2 K(z)dz = 0$ is

$$(15/32) \cdot (3 - 10z^2 + 7z^4) \cdot 1[|z| < 1],$$

an extension of the quadratic kernel. A table in Müller (1988, p. 68) lists univariate polynomial kernels indexed by three parameters ν, k, and μ; set $\nu = 0$ and choose k and μ which mean, respectively, $\int z^j K(z) = 0$ for $0 \leq j < k$ and $\mu - 1$ times continuous differentiability of $K(z)$ on R.

In general, a k-variate $K(z)$ is called a *kernel of order m* if

$$\int z_1^{j_1} \ldots z_k^{j_k} K(z)dz = 0 \quad \text{for } 0 < j_1 + \cdots + j_k \leq m - 1,$$

$$\int |z_1^{j_1} \ldots z_k^{j_k} K(z)|dz < \infty \quad \text{for } 0 < j_1 + \cdots + j_k = m. \quad (3.6)$$

Using an order m kernel makes the bias $O(h^m)$ due to the first line. In a small sample, high-order kernels may make $\hat{f}(x_0)$ negative. In this chapter, unless otherwise noted, we will use kernels with $K(z) \geq 0$. We put the above discussion as a theorem.

Theorem 3.7. *Suppose f is twice continuously differentiable with bounded derivatives, and $K(z)$ satisfies (2.4)(i) and (ii). Then the bias is*

$$E\hat{f}(x_0) - f(x_0) = h^2(1/2) \int z'\{\partial^2 f(*)/\partial x \partial x'\} z K(z) dz = O(h^2).$$

If f is m times continuously differentiable with bounded derivatives, and the kernel is of order $m \geq 2$, then the bias is $O(h^m)$.

7.4 Variance and Consistency of Kernel Estimator

In this section, first we will show that, for a random sequence $\{y_n\}$, if $V(y_n) \to 0$ and $\lim_n Ey_n \to \mu$, then $y_n =^p \mu$. Using this, we will prove the (weak) consistency of $\hat{f}(x_0)$ by showing that $V\hat{f}(x_0) \to 0$ and $E\hat{f}(x_0) \to f(x_0)$ as $N \to \infty$.

Suppose $\lim_n Ey_n \to \mu$ and $V(y_n) \to 0$. Invoking the triangle inequality,

$$|y_n - \mu| \leq |y_n - Ey_n| + |Ey_n - \mu|. \tag{4.1}$$

Then the event $|y_n - \mu| > 2\varepsilon$ for any $\varepsilon > 0$ implies that the right-hand side (rhs) of (4.1) is greater than 2ε. Hence,

$$P(|y_n - \mu| \geq 2\varepsilon) \leq P(|y_n - Ey_n| + |Ey_n - \mu| \geq 2\varepsilon). \tag{4.2}$$

Since $\lim_n Ey_n = \mu$, there exists n_0 such that for all $n \geq n_0$ $|Ey_n - \mu| \leq \varepsilon$. Hence, for all $n \geq n_0$, the rhs of (4.2) should be less than $P(|y_n - Ey_n| \geq \varepsilon)$. By Chebyshev's inequality,

$$P(|y_n - Ey_n| \geq \varepsilon) \leq V(y_n)/\varepsilon^2. \tag{4.3}$$

Then as $V(y_n) \to 0$, $P(|y_n - \mu| > 2\varepsilon) \to 0 \Leftrightarrow y_n =^p \mu$. We state this as a theorem.

Theorem 4.4. *For a random variable sequence $\{y_n\}$, if $Ey_n \to \mu$ and $V(y_n) \to 0$ as $n \to \infty$, then $y_n =^p \mu$.*

Deriving $V(\hat{f}(x_0))$,

$$V(\hat{f}(x_0)) = (1/N) \cdot E\{h^{-k} K((x-x_0)/h)\}^2 - (1/N) \cdot E^2\{h^{-k} K((x-x_0)/h)\}. \tag{4.5}$$

Similarly to (3.2), we get

$$(1/N) \cdot E\{h^{-k}K((x-x_0)/h)\}^2 = (1/N) \cdot h^{-2k} \int K^2((x-x_0)/h)f(x)dx$$

$$= (1/(Nh^k)) \cdot f(x_0) \int K^2(z)dz$$

$$+ o(1/(Nh^k)).$$

Therefore, from 3.7 and (3.5),

$$V(\hat{f}(x_0)) = (1/(Nh^k)) \cdot f(x_0) \int K^2(z)dz + o(1/(Nh^k))$$

$$- (1/N)\{f(x_0) + O(h^2)\}^2.$$

Ignore the last term, for $1/N$ goes to 0 faster than $1/(Nh^k)$. Therefore, we have the following theorem:

Theorem 4.6. *If f is twice continuously differentiable with bounded derivatives, and K satisfies (2.4)(i) and (ii), then*

$$V(\hat{f}(x_0)) = (1/(Nh^k)) \cdot f(x_0) \int K^2(z)dz + o(1/(Nh^k)).$$

For $\hat{f}(x_0) =^p f(x_0)$, it is sufficient to have $V\hat{f}(x_0) \to 0$ and $E\hat{f}(x_0) \to f(x_0)$ as $N \to \infty$. Hence, we need $Nh^k \to \infty$ for the variance to go to 0, and $h \to 0$ for the bias to go to 0. The smoothing parameter h should be kept small for the bias, but not too small for the variance. If h is too small, the small number of data points around x_0 will result in a high variance for $\hat{f}(x_0)$. If h is too large, the large number of data far away from x_0 will result in a high bias for $\hat{f}(x_0)$.

Theorem 4.7. *Suppose f is twice continuously differentiable with bounded derivatives, and K satisfies (2.4)(i) and (ii). If $Nh^k \to \infty$ and $h \to 0$ as $N \to \infty$, then $\hat{f}(x_0) =^p f(x_0)$.*

7.5   Uniform Consistency of Kernel Estimator

Although $\hat{f}(x_0) =^p f(x_0)$, it would be better to establish

$$\sup_{x_0} |\hat{f}(x_0) - f(x_0)| = o_p(1), \tag{5.1}$$

which is called *uniform consistency*. It means that the maximum deviation of $\hat{f}(x_0)$ from $f(x_0)$ converges to 0. The reason uniform consistency is desirable is as follows. What we wish to have is the "graph" $f(x)$. But what $\lim_N \hat{f}(x_0)$ offers is the "graph of the limits" [imagine connecting $\hat{f}(x_0), \hat{f}(x_1), \ldots, \hat{f}(x_M)$]. Under the consistency result in the preceding section, there is no guarantee that the graph of limits is close to $f(x)$. Uniform consistency gives the "limit of graph" rather than the "graph of limits," for the maximum difference between the estimated graph and the true graph goes to 0.

In order to appreciate the difference between the (usual) pointwise convergence and the uniform convergence better, recall that (*pointwise*) *convergence* of $\{g_n(x)\}$ to $g(x)$ at x: for any $\varepsilon > 0$, there exists a $n_0(\varepsilon, x)$ such that

$$|g_n(x) - g(x)| < \varepsilon, \quad \text{for all } n > n_0(\varepsilon, x).$$

Uniform convergence is: for any $\varepsilon > 0$, there exists a $n_0(\varepsilon)$ such that

$$\sup_x |g_n(x) - g(x)| < \varepsilon, \quad \text{for all } n > n_0(\varepsilon).$$

Note the difference between $n_0(\varepsilon, x)$ and $n_0(\varepsilon)$, the latter not depending on x. For instance, consider $g_n(x) \equiv x/n$, where $|x| \leq A$. Then for a given x, $g_n(x)$ converges to $g(x) \equiv 0$ as $n \to \infty$. For pointwise convergence, suppose we choose $n_0(x, \varepsilon) = |x|/\varepsilon$. Then

$$|g_n(x) - g(x)| = |x/n| < \varepsilon \quad \text{for all } n > n_0(x, \varepsilon).$$

For uniform convergence, define $n_0(\varepsilon) = A/\varepsilon$. Then

$$\sup_x |g_n(x) - g(x)| = \sup_x |x/n| \leq A/n < \varepsilon \quad \text{for all } n > n_0(\varepsilon).$$

Note that $n_0(\varepsilon) \geq n_0(x, \varepsilon)$; it takes a larger n for uniform convergence.

Uniform convergence guarantees that if n is larger than a certain number, the approximation of $g(x)$ by $g_n(x)$ is at least as good as a certain tolerance level no matter what x is. Pointwise convergence only ensures such a thing at a given point x_0; if we change x_0 into x_1, we may need a larger n to get as good an approximation as at x_0. For semi-nonparametric methods to be introduced in later chapters, we need not only (5.1) but also its specific convergence rate.

The following is a uniform consistency theorem in Prakasa Rao (1983, p. 185); the conditions are not the weakest available.

Theorem 5.2. *Suppose f is twice continuously differentiable with bounded derivatives, K satisfies (2.4)(i)–(iii) and K is Lipschitz-continuous*

$$|K(z_1) - K(z_0)| \leq C|z_1 - z_0| \quad \text{for a constant } C > 0.$$

Also $\int |x|^\gamma f(x)dx < \infty$ for some $\gamma > 0$. If $Nh^k/\ln(N) \to \infty$ and $h \to 0$ as $N \to \infty$, then $\sup_x |\hat{f}(x) - f(x)| =^{as} 0$. A fortiori, $\hat{f}(x) =^{as} f(x)$.

The rate $Nh^k/\ln(N) \to \infty$ requires h to be larger than h of $Nh^k \to \infty$ in the pointwise convergence. A smaller h makes $\hat{f}(x_0)$ more jagged, which goes against the uniformity we desire in (5.1). The rate $Nh^k/\ln(N)$ with $k = 1$ was also shown by Silverman (1978).

7.6   Choosing Smoothing Parameter and Kernel

In this section, we discuss the basics of how to choose h and K; more will be discussed in the following chapter. There are many ways to choose h and K, but we will use the mean squared error $E\{\hat{f}(x_0) - f(x_0)\}^2$ and its relatives. Assume (2.4) holds.

Mean squared error (MSE) is Variance + Bias^2. From 3.7 and 4.6,

$$MSE(\hat{f}(x_0)) = E\{\hat{f}(x_0) - f(x_0)\}^2 \cong (Nh^k)^{-1}f(x_0) \int K^2(z)dz$$

$$+ (h^4/4)\left[\int z'\{\partial^2 f(x_0)/\partial x \partial x'\}zK(z)dz\right]^2.$$

From (2.4)(iv), $\int K(z)zz' = \kappa I_k$ for a positive scalar κ. The term inside $[\cdot]$ is

$$\int \text{trace}(z'\{\partial^2 f(x_0)/\partial x \partial x'\}z) \cdot K(z)dz$$

$$= \int \text{trace}(zz'\{\partial^2 f(x_0)/\partial x \partial x'\}) \cdot K(z)dz = \kappa \cdot \sum_{j=1}^{k} \partial^2 f(x_0)/\partial x_j^2.$$

Therefore MSE is

$$\text{MSE}(\hat{f}(x_0)) \cong (Nh^k)^{-1}f(x_0)\int K^2(z)dz + (h^4/4)\left\{\kappa \cdot \sum_j \partial^2 f(x_0)/\partial x_j^2\right\}^2.$$
(6.1)

MSE measures the local error of estimation around x_0. The global error can be measured by *mean integrated squared error* (MISE), which removes x_0 in (6.1) by integration:

$$\text{MISE}(\hat{f}(x_0)) = E\int\{\hat{f}(x_0) - f(x_0)\}^2 dx_0 = \int E\{\hat{f}(x_0) - f(x_0)\}^2 dx_0$$

$$\cong (1/(Nh^k))\int K^2(z)dz + (h^4/4)\cdot\kappa^2\cdot\int\left\{\sum_j \partial^2 f(s)/\partial x_j^2\right\}^2 ds$$

$$\equiv (1/(Nh^k))\cdot A + (h^4/4)\cdot B,$$
(6.2)

where

$$A \equiv \int K^2(z)dz, \quad B \equiv \kappa^2 \cdot \int \left\{ \sum_j \partial^2 f(s)/\partial x_j^2 \right\}^2 ds. \qquad (6.3)$$

MISE can be controlled by the choice of h and K. The choice of K is much more difficult to address than that of h, for K is a function. But it is known in the literature that K is not as crucial as choosing h is, $\hat{f}(x_0)$ varying not much as K varies. Hence we first choose h by minimizing MISE. Then we discuss how to choose K in the simple case $k = 1$.

Differentiating MISE wrt h, the optimal h, h_0, is

$$h_0 = (kA/B)^{1/(k+4)} \cdot N^{-1/(k+4)}; \qquad (6.4)$$

$$N \cdot h_0^k = (kA/B)^{k/(k+4)} \cdot N^{1-k/(k+4)} = (kA/B)^{k/(k+4)} \cdot N^{4/(k+4)}. \qquad (6.5)$$

The result is more illuminating for $k = 1$:

$$\text{MISE} = (Nh)^{-1} \int K^2(z)dz + (h^4/4)\kappa^2 \int f''(s)^2 ds, \qquad (6.6)$$

$$h_0 = \left\{ \int K^2(z)dz \right\}^{1/5} \left\{ \kappa^2 \int f''(s)^2 ds \right\}^{-1/5} N^{-1/5}. \qquad (6.7)$$

Here $\int f''(s)^2 ds$ measures the variation in f. So, if $f(x)$ is highly variable, then h_0 is small. If $K = \phi$ and f is also a normal density, then $h_0 \cong N^{-1/5} SD(x)$, where $SD(x) = \{V(x)\}^{1/2}$. This simple h_0 often works well even if K and f are not normal: in Figure 7.3, $h_0 = 0.21 \cdot SD(x)$ and the lines look good when compared with Figures 7.1 and 7.2 despite that $f(x)$ is clearly not normal. Extending the rule of thumb to $k > 1$ yields $h = N^{-1/(k+4)}$ when all variables are standardized [see Scott (1992, p. 152)].

With $h = O(N^{-1/(k+4)})$, the bias^2 is

$$\text{bias}^2 = O(h^4) = O(N^{-4/(k+4)}),$$

and the variance is $O((Nh^k)^{-1}) = O(N^{-4/(k+4)})$ as shown in (6.5). That is, both variance and bias^2 in MISE converge to 0 at the same rate if $h = h_0$. Decreasing one faster than the other increases MISE. Recall that for a sample mean $\bar{x}_N = (1/N)\sum_i x_i$, where $V(x) = \sigma^2$, its MSE goes to zero at $O(N^{-1})$, for $\text{MSE}(\bar{x}_N) = V(\bar{x}_N) = \sigma^2/N$. In kernel density estimation, MISE with h_0 converges to 0 at $O(N^{-4/(k+4)})$, slower than $O(N^{-1})$. MISE converges to 0 more slowly as k increases.

Turning to choosing K, set $k = 1$ and $\kappa = 1$ to simplify exposition. Substituting h_0 of (6.7) into MISE of (6.6), MISE is proportional to

$$\left\{ \int K^2(z)dz \right\}^{4/5} N^{-4/5}. \qquad (6.8)$$

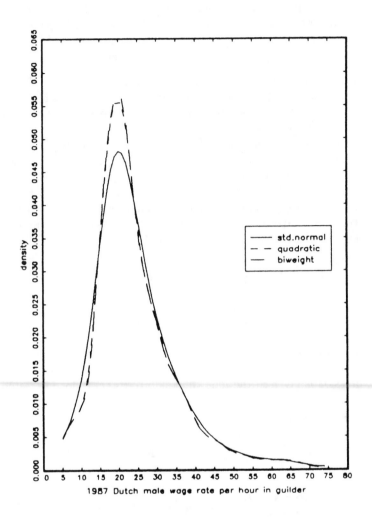

FIGURE 7.3. Rule-of-thumb $[h = N^{-1/5} \times SD(x), \; N = 2421]$.

The problem of choosing K becomes minimizing $\int K^2(z)dz$ subject to $K(z) = K(-z)$, $\int K(z)dz = 1$ and $\int K(z)z^2dz = 1$. This can be solved by a dynamic optimization technique to yield the so-called *Epanechinikov kernel*

$$K(z) = \{3/(4\sqrt{5})\} \cdot (1 - 0.2z^2) \cdot 1[|z| \le \sqrt{5}], \qquad (6.9)$$

which is equivalent to the trimmed quadratic kernel in (2.5). Although this kernel minimizes MISE, the crude uniform kernel in (2.5) has 93% efficiency compared with the optimal kernel [see Silverman (1986)]. This indicates that the choice of K is not crucial.

Going back to the choice of h, h_0 depends on unknown f''. Plugging $\hat{f}''$, an estimate for f'', into f'', we can estimate h_0, which is called a "plug-in method." But estimating f'' to estimate f appears unattractive, for f'' is more difficult to estimate than f [see, for instance, Scott (1992, p. 131–132)]. There are many data-driven alternative ideas to select h. Among them, one favorite is the *least squares cross validation* minimizing an estimate for the *integrated squared error* (ISE):

$$\int \{\hat{f}(x_0) - f(x_0)\}^2 dx_0 = \int \hat{f}(x_0)^2 dx_0$$

$$- 2\int \hat{f}(x_0)f(x_0)dx_0 + \int f(x_0)^2 dx_0 \qquad (6.10)$$

wrt h. Drop the last term independent of h. To minimize ISE, we should rewrite the first and the second term into a function of h and the data. One such expression is given in the following; skip to (6.17) if not interested in the details.

The first term in (6.10) is

$$\int N^{-2}h^{-2k} \sum_i \sum_j K((x_i - x)/h) \cdot K((x_j - x)/h)dx$$

$$= N^{-2}h^{-2k} \sum_i \sum_j \int K((x_i - x)/h) \cdot K((x_j - x)/h)dx. \qquad (6.11)$$

Let $-z = (x_j - x)/h \Leftrightarrow x/h = x_j/h + z$ to get

$$\int K(x_i/h - x/h) \cdot K((x_j - x)/h)dx = h^k \cdot \int K(x_i/h - x_j/h - z) \cdot K(z)dz$$

$$= h^k \cdot \int K((x_i - x_j)/h - z) \cdot K(z)dz = h^k K^{(2)}((x_i - x_j)/h), \qquad (6.12)$$

where $K^{(2)}(a) \equiv \int K(a - m)K(m)dm$, the density of the sum of two independent random variables from the distribution whose density is K; that is, $K^{(2)}$ is the convolution of K with itself. For instance, if $K(z) = \phi(z)$, then $K^{(2)}(z)$ is $(1/\sqrt{2})\phi(z/\sqrt{2})$, the density of $N(0,2)$. Substituting (6.12) into (6.11), the first term in (6.10) becomes

$$(N^2 h^k)^{-1} \sum_i \sum_j K^{(2)}((x_i - x_j)/h). \qquad (6.13)$$

As for the second term in (6.10),

$$\int \hat{f}(x_0) f(x_0) dx_0 = (Nh^k)^{-1} \sum_i \int K((x_i - x)/h) f(x) dx. \qquad (6.14)$$

From this,

$$E\left\{ \int \hat{f}(x_0) f(x_0) dx_0 \right\} = h^{-k} \int \int K((y - x)/h) f(x) f(y) dx dy$$

$$= E\{h^{-k} K((x - y)/h)\} = E\left[\{N(N-1)h^k\}^{-1} \sum_{i \neq j} K((x_i - x_j)/h) \right],$$
$$(6.15)$$

where $\sum_{i \neq j}$ is $\sum_i \sum_{j, i \neq j}$. Thus, an unbiased estimate for the second term in (6.10) is

$$-2\{N(N-1)h^k\}^{-1} \sum_{i \neq j} K((x_i - x_j)/h)$$

$$\cong -2(N^2 h^k)^{-1} \sum_{i \neq j} K((x_i - x_j)/h). \qquad (6.16)$$

In short, the cross-validation minimizes

$$(N^2 h^k)^{-1} \sum_i \sum_j K^{(2)}((x_i - x_j)/h) - 2(N^2 h^k)^{-1} \sum_{i \neq j} K((x_i - x_j)/h).$$
$$(6.17)$$

In the second double sum, we have $i \neq j$, which means that x_j is not used in estimating $\hat{f}(x_j)$. This "leave-one-out" scheme yields the name "cross-validation." One drawback of the cross-validation idea is that the minimand tends to have several local minima practicularly in h-small areas (Marron, 1988, p. 196); that is, there is a good chance for undersmoothing. Also, since the minimand is ISE, not MISE, the optimal choice here is only good for the particular data set, not for any sample of the same size; another data set from the same population may require a rather different h. Nevertheless, Hall (1983) and Stone (1984) prove that the least squares cross validation choice h_{cv} minimizing (6.17) is optimal in the sense that (6.17) evaluated at h_{cv} converges to $\min_h \int \{\hat{f}(x, h) - f(x)\}^2 dx - \int f(x)^2 dx$ a.s.; recall (6.10). See Marron (1988) and Park and Marron (1990) for more on choosing h and data-driven methods.

To discuss the practical aspects of choosing h and K, recall Figures 7.1 and 7.2, where h is chosen by "eye-balling." Despite the above theoretical discussion, when k is 1 or 2 (so when we can draw two- or three-dimensional graphs), eye-balling or trial-and-error seems to be the best way of choosing h. Data-driven methods will be helpful, however, if $k \geq 3$. Note that despite the big difference between the h's in Figures 7.1 and 7.2, all the lines show more or less the same shape of the density.

In the over-smoothed Figure 7.1, the line with $N(0,1)$ kernel is different from the other two. This is due to the fact that while the other two have bounded supports, $N(0,1)$ kernel has the unbounded support so that observations far away from x_0 affect $\hat{f}(x_0)$ when h is large. Thus, although it seems to be agreed upon in the literature that choosing K does not matter, it may not hold when h is not chosen well as in Figure 7.1; neither may it hold when k is large. Using K with a bounded support is preferable for two reasons. One is already shown in Figure 7.1. The other is the so-called "boundary problem": near the boundary of x, fewer observations are used for $\hat{f}$ and the weights given by the kernel become asymmetric. For a kernel with a bounded support, this boundary effect is limited, while this is not the case for kernels with unbounded support.

7.7 Asymptotic Distribution

Since

$$V(\hat{f}(x_0)) = (Nh^k)^{-1} \cdot f(x_0) \int K^2(z)dz + o(1/(Nh^k)),$$

multiplying $\hat{f}(x_0) - f(x_0)$ by $(Nh^k)^{0.5}$ will give a nondegenerate asymptotic variance with the asymptotic law being normal. That is, the result we desire to establish in this section is

$$(Nh^k)^{0.5}\{\hat{f}(x_0) - f(x_0)\} =^d N\left(0, f(x_0)\int K^2(z)dz\right). \qquad (7.1)$$

Observe that

$$(Nh^k)^{0.5}\hat{f}(x_0) = (Nh^k)^{-0.5}\sum_i K((x_i - x_0)/h)$$

$$= \sum_i (Nh^k)^{-0.5}K((x_i - x_0)/h). \qquad (7.2)$$

We will be applying the Lindeberg CLT rewritten for "triangular arrays" or "double arrays," because the summands in the sum are changing with N. In an ordinary CLT, we deal with independent arrays: for $N = 10$, we draw 10 observations, and for $N = 11$, we throw away those 10 and redraw 11 new observations to have each array (the observations with a given N) independent of the other arrays. In (7.2), h_N (now we use h_N to make the dependence of h on N explicit) gets smaller as $N \to \infty$; for instance, $h_{11} < h_{10}$. So an "observation" $K((x_i - x_0)/h_{11})$ with $N = 11$ tends to be smaller than an observation $K((x_i - x_0)/h_{10})$ with $N = 10$. Hence, an array with a given N is not independent of another array with a different N. A CLT good for a triangular array allows dependence across arrays, so long as the terms in a given array are independent. The expression "triangular" is due to that putting each array horizontally and stacking the arrays vertically yields a triangular shape; with N being the row number, each row has one more term than the preceding row. Such a CLT is also needed for power comparison of tests under local alternatives. The following is the Lindeberg CLT for triangular arrays.

7.3. Lindeberg CLT for triangular arrays: For columnwise independent triangular arrays $\{z_{Ni}\}$, let

$$S_N \equiv \sum_{i=1}^{N} z_{Ni}, \quad E z_{Ni} = 0, \quad \sigma_N^2 \equiv \sum_{i=1}^{N} E z_{Ni}^2.$$

If, for any given $\varepsilon > 0$, the Lindeberg condition

$$\sum_{i=1}^{N} E\{(z_{Ni}/\sigma_N)^2 \cdot 1[|z_{Ni}/\sigma_N| \geq \varepsilon]\} \to 0 \quad \text{as } N \to 0$$

is satisfied, then $S_N/\sigma_N =^d N(0,1)$.

An example of the Lindeberg CLT can be seen in $S_N = (1/\sqrt{N}) \sum_i z_i = \sum_i (1/\sqrt{N}) z_i$ where z_i's are iid with $E(z_i) = 0$ and $V(z_i) = 1$. Set $z_{Ni} = (1/\sqrt{N}) z_i$, then $\sigma_N^2 = 1$ and the Lindeberg condition is

$$(1/N) \sum_i E(z_i^2 1[|z_i| \geq \varepsilon \sqrt{N}]) = E(z_i^2 1[|z_i| \geq \varepsilon \sqrt{N}]). \quad (7.4)$$

The term $E(z_i^2 1[\cdot])$ is dominated by $E(z_i^2) = 1$. Since $1[|z_i| \geq \varepsilon \sqrt{N}] =^p 0$, $E(z_i^2 1[|z_i| \geq \varepsilon \sqrt{N}])$ goes to 0 due to the dominated convergence theorem.

Instead of (7.2), consider its centered version to have $E z_{Ni} = 0$ before we apply the CLT:

$$(Nh^k)^{0.5} \hat{f}(x_0) - (Nh^k)^{0.5} E\hat{f}(x) = (Nh^k)^{0.5}\{\hat{f}(x_0) - E\hat{f}(x)\}$$

$$= \sum_i (Nh^k)^{-0.5}\{K((x_i - x_0)/h) - EK((x_i - x_0)/h)\}. \quad (7.5)$$

Let

$$z_{Ni} = (Nh^k)^{-0.5}\{K((x_i - x_0)/h) - EK((x_i - x_0)/h)\},$$

and note

$$\sigma_N^2 = Nh^k \cdot V(\hat{f}(x_0)) = f(x_0) \int K(z)^2 dz + o(1).$$

The Lindeberg condition holds by doing analogously to (7.4); the only role $\sqrt{N}$ plays in (7.4) is going to ∞ which is now played by $(Nh^k)^{0.5}$. Therefore we have

$$(Nh^k)^{0.5}\{\hat{f}(x_0) - E\hat{f}(x_0)\} =^d N\left(0, f(x_0) \int K^2(z)dz\right). \tag{7.6}$$

To go to (7.1) from (7.6), we need

$$(Nh^k)^{0.5}\{E\hat{f}(x_0) - \hat{f}(x_0)\} = o(1). \tag{7.7}$$

Since we know $E\hat{f}(x_0) - f(x_0) = O(h^2)$, (7.7) becomes

$$O\{(Nh^k)^{0.5} \cdot h^2\} = O\{(Nh^{k+4})^{0.5}\}.$$

By choosing h such that $Nh^{k+4} \to 0$, we can have (7.7) and thus (7.1). If (7.7) does not hold but $(Nh^k)^{0.5}\{E\hat{f}(x_0) - \hat{f}(x_0)\}$ converges to a constant, then the constant is called the "asymptotic bias." One problem with $Nh^{k+4} \to 0$ is that the optimal h in (6.4) does not satisfy $Nh^{k+4} \to 0$: the optimal rate was $h = O(N^{-1/(k+4)})$ such that Nh^{k+4} is a nonzero constant. The optimal rate decreases both variance and bias^2 in MSE at the same rate. For (7.1), we should remove bias^2 faster than variance with $Nh^{k+4} \to 0$. We put the discussion up to this point into a theorem:

Theorem 7.8. *Suppose $f(x)$ is twice continuously differentiable with bounded derivatives, and (2.4)(i) and (ii) hold. Let $h \to 0$ and $Nh^k \to \infty$ as $N \to \infty$. Then*

$$(Nh^k)^{1/2}\{\hat{f}(x_0) - f(x_0)\} - \text{Asym. Bias} =^d N\left(0, f(x_0) \int K^2(z)dz\right),$$

where Asym. Bias (asymptotic bias) is $O((Nh^{k+4})^{0.5})$. If $Nh^{k+4} \to 0$, then

$$(Nh^k)^{1/2}(\hat{f}(x_0) - f(x_0)) =^d N\left(0, f(x_0) \int K^2(z)dz\right).$$

The theorem renders the pointwise convergence rate $(Nh^k)^{0.5}$; or to put it differently, $\hat{f}(x_0)$ is "$(Nh^k)^{0.5}$-consistent" for $f(x_0)$. It can be shown that the asymptotic covariance between $(Nh^k)^{1/2}\{\hat{f}(x_1) - f(x_1)\}$ and $(Nh^k)^{1/2}\{\hat{f}(x_2) - f(x_2)\}$ for $x_1 \neq x_2$ is zero, so that the vector version of the theorem is straightforward with the diagonal covariance matrix. Under

$Nh^{k+4} \to 0$, the theorem gives a "pointwise" confidence interval; a 95% confidence interval for $f(x_0)$ is

$$\hat{f}(x_0) \pm 1.96 \left\{ \hat{f}(x_0) \int K^2(z)dz/(Nh^k) \right\}^{0.5}. \qquad (7.9)$$

Getting $(1-\alpha)100\%$ confidence intervals at $x_1 \ldots x_m$ and connecting them together does not give $(1-\alpha)100\%$ confidence "region," for the coverage probability becomes $(1-\alpha)^m$.

Summarizing the conditions on convergence rates for h so far,

(i) for (weak) consistency, $Nh^k \to \infty$;

(ii) for a.e. uniform consistency, $Nh^k/\ln(N) \to \infty$;

(iii) for asymptotic unbiasedness in distribution, $Nh^{k+4} \to 0$.

7.8   Other Density Estimation Ideas

In this section, we introduce three more nonparametric density estimation ideas with $k = 1$, drawing mainly upon Silverman (1986). While the third is not related to the kernel method, the first two are related to the kernel method by weakening boundary effects.

(1) Nearest-Neighbor Method

On average, $N \cdot 2hf(x_0)$ observations in a data set will fall in the interval $[x_0 - h, x_0 + h]$, since $2hf(x_0) \cong P(x_0 - h \le x_i \le x_0 + h)$. Rearrange the data in the increasing order of distance from x_0 and denote the distance of the ith nearest datum by $d_i(x_0)$. Then in the interval of $[x_0 - d_s(x_0), x_0 + d_s(x_0)]$, there are s observations. Approximately, s should be equal to $N \cdot 2d_s(x_0)\hat{f}(x_0)$. Solving this for $\hat{f}(x_0)$,

$$\hat{f}(x_0) = s/\{N2d_s(x_0)\}, \qquad (8.1)$$

which is called the sth nearest-neighbor estimator. In (8.1), s is the smoothing parameter as h is in the kernel method, and s should be chosen such that $s/N \to 0$ as $N \to \infty$; for instance, $s = \sqrt{N}$. The advantage of nearest-neighbor method over the kernel estimator is that the same number of observations are used for estimating $f(x_1)$ as well as $f(x_0)$. The disadvantage is that the curve estimate is not smooth and it is not a density $[\int \hat{f}(x_0)dx_0 \ne 1]$.

(2) Adaptive Kernel Estimation

Adaptive kernel estimation proceeds as follows:

(i) First, get an initial estimate $\hat{f}(x_0)$ as x_0 varies.

(ii) Define a local smoothing parameter λ_i for each x_i such that

$$\lambda_i = \{\hat{f}(x_i)/g\}^{-\alpha},$$

where $0 \leq \alpha \leq 1$ and g is the geometric mean of the $\hat{f}(x_i)$:

$$\ln(g) = (1/N) \sum_i \ln\{\hat{f}(x_i)\}.$$

(iii) Then, the adaptive kernel estimator is defined by

$$(1/N) \cdot \sum_i \{1/(h\lambda_i)\} \cdot K\{(x_i - x_0)/(h\lambda_i)\}. \tag{8.2}$$

The bandwidth $h\lambda_i$ is stretched when f is small, that is, when only a few observations are available near x_0. Setting $\alpha = 0$ gives the usual kernel estimator.

(3) Maximum Penalized Likelihood Estimator

Extending the idea of MLE, we may consider maximizing the likelihood over the space of the possible probability density functions for x. However, there is a major difference between this idea and the usual MLE: in the latter we have only a finite number of parameters, while in the former we have an infinite number of parameters. One modification to make this idea work is penalizing the "overfit." The maximum penalized likelihood estimator is defined by

$$\max_{g \in G} \sum_i \log(g(x_i)) - \alpha \int g''(x)^2 dx, \tag{8.3}$$

where G is the set of the probability densities satisfying certain properties. Here, α determines the degree of smoothing. With $\alpha = 0$, the resulting estimate will be too jagged. Note that we use g'', not g', to measure the variation of g; a straight line with a slope has $g' \neq 0$ but it is just a transposition of a flat line. With a high α, the total variation in g is highly penalized. So in this method, α plays the role of h in the kernel method. This idea is related to the "spline smoothing" to be seen later in relation to nonparametric regression.

8
Nonparametric Regression

8.1 Introduction

In the previous chapter, we discussed nonparametric density estimation. In this chapter, we discuss nonparametric estimation of the regression function $E(y \mid x) \equiv r(x)$, where $y = r(x) + u$ and $E(u \mid x) = 0$. More generally, we can consider functionals of the conditional distribution $F_{y|x}$, such as $\partial r(x)/\partial x$ and $V(y \mid x)$. But usually, estimation methods for the functionals can be inferred from those for $E(y \mid x)$.

Since $E(y \mid x)$ is $\int \{f(y,x)/f(x)\}y \, dy$, we can estimate $E(y \mid x)$ as long as we can estimate the densities. As in the previous chapter, kernel estimation will be our main method, although some alternatives to the kernel method will also be discussed briefly in the last section. The problems of choosing the smoothing parameter and the kernel are similar to those in density estimation. As in the previous chapter, the choice of kernel will not make much difference. So we will discuss only how to choose the smoothing parameter in detail. See Prakasa Rao (1983), Bierens (1987), Müller (1988), Härdle (1990), and Härdle and Linton (1994) for more on (kernel) nonparametric estimation; references for applied studies in econometrics can be found in Härdle and Linton (1994).

The rest of the chapter is organized as follows. In Section 2, a kernel regression estimator is introduced. In Sections 3 and 4, consistency and uniform consistency are discussed, respectively. In Section 5, asymptotic distribution is examined. In Section 6, we discuss how to choose smoothing parameter and kernel. In Section 7, three (unrelated) topics are discussed:

nonparametrics when some regressors are continuous and some are discrete, estimating derivatives of $E(y \mid x)$, and estimating $E(y \mid x)$ in MLE-based models. In Section 8, nonparametric methods other than the kernel method are examined.

8.2 Kernel Nonparametric Regression

If many y_i's are available for a given x_0, we can estimate $E(y \mid x_0)$ by the sample mean of those observations. In a nonexperimental setting with a continuous random variable (rv) x, however, this is impossible, for we get all different x_i's by random sampling. Assuming $E(y \mid x) = r(x)$ is continuous in x, y_i's with its x_i close to x_0 may be treated as (y_i, x_0'). This gives multiple observations for a given x_0 so that $E(y \mid x_0)$ can be estimated by the sample mean. More generally, $E(y \mid x_0)$ is estimated by a weighted mean $\sum_i w_i y_i$, where w_i is large if x_i is close to x_0, small otherwise, and $\sum_i w_i = 1$.

Typically, popular choices of w_i's in estimating $E(y \mid x_0)$ are various densities (or kernels). With K denoting a kernel, one potential estimate for $r(x_0)$ is

$$\sum_i K((x_i - x_0)/h)y_i. \tag{2.1}$$

The role of h is the same as the group interval length in a histogram. To ensure for the weights to add up to 1, normalize (2.1) to get a kernel estimator

$$r_N(x_0) = \sum_i K((x_i - x_0)/h)y_i / \sum_i K((x_i - x_0)/h)$$

$$= \sum_i \left\{ K((x_i - x_0)/h) / \sum_i K((x_i - x_0)/h) \right\} y_i = \sum_i w_i y_i$$

$$= (Nh^k)^{-1} \sum_i K((x_i - x_0)/h)y_i / (Nh^k)^{-1} \sum_i K((x_i - x_0)/h)$$

$$= g_N(x_0)/f_N(x_0), \tag{2.2}$$

where

$$g_N(x_0) \equiv (Nh^k)^{-1} \sum_i K((x_i - x_0)/h)y_i$$

and

$$f_N(x_0) \equiv (Nh^k)^{-1} \sum_i K((x_i - x_0)/h).$$

The estimator (2.2) was suggested by Nadaraya (1964) and Watson (1964), and is called a Nadaraya–Watson kernel estimator; there are also other kernel estimators suitable for experimental data with nonrandom x.

Since $f_N(x_0) =^p f(x_0)$, for $r_N(x_0) =^p r(x_0)$ to hold, we should have

$$g_N(x_0) =^p g(x_0) \equiv r(x_0) \cdot f(x_0). \tag{2.3}$$

This can be better understood in the following discrete case. Suppose x is discrete. Then we can estimate $E(y \mid x_0)$ by the sample average of y_i's with $x_i = x_0$:

$$\sum_i 1[x_i = x_0]y_i / \sum_i 1[x_i = x_0]$$

$$= (1/N) \sum_i 1[x_i = x_0]y_i / (1/N) \sum_i 1[x_i = x_0]$$

$$=^p E(1[x = x_0] \cdot y)/P(x = x_0) = E(y \mid x = x_0).$$

Thus it must be that

$$(1/N) \sum_i 1[x_i = x_0]y_i = E(1[x = x_0] \cdot y) = E(y \mid x = x_0) \cdot P(x = x_0);$$

(2.3) is an analog of this for a continuous x.

Since the norming factor in g_N and f_N is Nh^k not N, the asymptotic distribution of $r_N(x_0)$ is obtained by multiplying $r_N(x_0)$ by $(Nh^k)^{0.5}$, not by $\sqrt{N}$. Since

$$(Nh^k)^{0.5}g_N(x_0)/f_N(x_0) - (Nh^k)^{0.5}g_N(x_0)/f(x_0)$$

$$= (Nh^k)^{0.5}g_N(x_0)\{f_N(x_0)^{-1} - f(x_0)^{-1}\} = o_p(1)$$

due to $f_N(x_0) =^p f(x_0)$, the asymptotic distribution of $r_N(x_0)$ can be obtained from $(Nh^k)^{0.5}g_N(x_0)/f(x_0)$. Keeping $f_N(x_0)$ in the denominator would make the derivation complicated.

The kernel nonparametric method can be used for other functional estimations. For instance, suppose we want to estimate $V(y \mid x)$. Since $V(y \mid x) = E(y^2 \mid x) - \{E(y \mid x)\}^2$, having already estimated $E(y \mid x)$, we only need to estimate $E(y^2 \mid x)$ in addition. This is done by replacing y by y^2 in $r_N(x_0)$. Also $\partial E(y \mid x_0)/\partial x = \partial r(x_0)/\partial x$ can be estimated by differentiating $r_N(x_0)$ wrt x_0. If $r(x_0) = x_0'\beta$, then $\partial r(x_0)/\partial x = \beta$. Hence, in nonparametric regression, an estimate for $\partial r(x_0)/\partial x$ reflects the marginal effect of x on $r(\cdot)$ at $x = x_0$. In the chapter for semi-nonparametrics, we will see more of this idea.

One problem in kernel regression estimator $r_N(x)$ (in fact, in most nonparametric methods based on a "local weighted average" idea) is that $r_N(x)$ may have a large bias near the boundary of the range of x. Suppose $k = 1$ and that $r(x)$ is increasing. If we estimate $r(x)$ at $\max_{1 \le i \le N} x_i$, then $r_N(\max_i x_i)$ has a downward bias, since $r(x)$ is increasing but the data only come from the left-hand side of $\max_i x_i$. The opposite upward bias will occur near $\min_{1 \le i \le N} x_i$. There are ways to correct for the bias, but so long as $r(x)$ at the boundary is not of interest, we just have to take caution in interpreting the estimation result near the boundaries.

For ease of reference, we show the assumptions for kernels, which were already given in the preceding chapter except for a part of the last assumption:

   (i) $K(z)$ is symmetric around 0 and continuous.

  (ii) $\int K(z)dz = 1$, $\int K(z)z\,dz = 0_k$, and $\int |K(z)|dz < \infty$.

 (iii) (a) $K(z) = 0$ if $|z| \ge z_0$ for some z_0, or $\hspace{2cm}$ (2.4)
       (b) $|z| \cdot K(z) \to 0$ as $|z| \to \infty$.

 (iv) $K(z) = \prod_{j=1}^{k} L(z_j)$, where L satisfies (i) to (iii), and $\int t^2 L(t)dt = \kappa$.

8.3   Consistency of Kernel Estimator

Since we already know that $f_N(x_0) =^p f(x_0)$, we just have to show $g_N(x_0) =^p g(x_0)$, the intuition of which was already provided in the preceding section. The reader not interested in the proof may simply read (3.3), (3.7), and 3.8, skipping the rest of this section.

Substitute $y_i = r(x_i) + u_i$ into $g_N(x_0)$ to get

$$g_N(x_0) = (Nh^k)^{-1} \sum_i K((x_i - x_0)/h)\{r(x_i) + u_i\}. \hspace{1cm} (3.1)$$

Recall $E(u \mid x) = 0$. Deriving $Eg_N(x_0)$,

$$E[h^{-k}K((x - x_0)/h)\{r(x) + u\}] = E[h^{-k}K((x - x_0)/h)r(x)]$$

$$= \int K(z)r(x_0 + hz)f(x_0 + hz)dz = \int K(z)g(x_0 + hz)dz. \hspace{1cm} (3.2)$$

Assuming that $r(x)$ and $f(x)$ [thus $g(x)$] are twice continuously differentiable with bounded derivatives, (3.2) becomes

$$Eg_N(x_0) = r(x_0)f(x_0) + O(h^2). \hspace{1cm} (3.3)$$

To derive $V\{g_N(x_0)\}$, define v:

$$v \equiv h^{-k}K((x - x_0)/h)y. \hspace{1cm} (3.4)$$

From (3.3), we already know that $E(v) = r(x_0)f(x_0) + O(h^2)$. Since $V\{g_N(x_0)\} = (1/N)V(v) = (1/N)\{E(v^2) - E^2(v)\}$, we need to derive $E(v^2)$:

$$E(v^2) = E_x E_{u|x} h^{-2k}K^2((x - x_0)/h)\{r(x) + u\}^2. \hspace{1cm} (3.5)$$

Define $s(x) \equiv V(u \mid x) = E(u^2 \mid x)$. Then,

$$E(v^2) = E_x h^{-2k} K^2((x - x_0)/h)\{r^2(x) + s(x)\}$$

$$= \int h^{-k} K^2(z)\{r^2(x_0 + hz) + s(x_0 + hz)\} f(x_0 + hz) dz. \qquad (3.6)$$

Expanding $r^2 + s$ and f around x_0, we have, respectively,

$$r^2(x_0) + s(x_0) + h\{2r(*)r_x(*) + s_x(*)\}'z$$

and

$$f(x_0) + h\{\partial f(**)/\partial x\}'z,$$

where $r_x \equiv \partial r/\partial x$ and $s_x \equiv \partial s/\partial x$. Assume that s_x is continuous and bounded. Substituting the expansion into (3.6), (3.6) is $A(h^{-k}) + o(h^{-k})$ where

$$A \equiv \{r^2(x_0) + s(x_0)\} f(x_0) \int K^2(z) dz.$$

Thus,

$$V(v) = E(v^2) - E^2(v) = A \cdot h^{-k} + o(h^{-k}).$$

Ignore the smaller-order term and focus on the leading term to get

$$V\{g_N(x_0)\} = (1/N) \cdot V(v) = O(1/(Nh^k)). \qquad (3.7)$$

Hence, we require $Nh^k \to \infty$ and $h \to 0$ and invoke Theorem 4.4 of the preceding chapter to get $g_N(x_0) =^p g(x_0)$.

Theorem 3.8. *Assume the following: $K(z)$ satisfies (2.4)(i) and (ii), $f(x) > 0$ for all x, $r(x)$ and $f(x)$ have bounded and continuous second derivatives, and $s(x) \equiv V(u^2 \mid x)$ is continuously differentiable with bounded first derivatives. If $Nh^k \to \infty$ and $h \to 0$ as $N \to \infty$, then $r_N(x_0) =^p r(x_0)$.*

8.4   Uniform Consistency

Although $r_N(x) =^p r(x)$, it is desirable to have the following uniform consistency

$$\sup_x |r_N(x) - r(x)| = o_p(1), \qquad (4.1)$$

which means that the maximum deviation of $r_N(x)$ from $r(x)$ goes to 0. The following is a slight adaptation of uniform convergence theorems in Györfi et al. (1989, pp. 24–32), who in turn adapt a theorem in Collomb and Härdle (1986, p. 82); see also Mack and Silverman (1982) and Härdle, Janssen and Serfling (1988) for the case $k = 1$.

Theorem 4.2. *Suppose that $r(x)$ and $f(x)$ are $k+1$ times continuously differentiable with bounded derivatives on R^k, that $f(x)$ is bounded from below by a positive scalar η on a compact set $G \subset R^k$, that $|Y| < M$ for a positive constant M, and that $V(y \mid x)$ is continuous on G. Assume that K satisfies (2.4) and is of order $k+1$, and that for some $L > 0$ and $\alpha > 0$, $K(z)$ satisfies a Lipschitz-continuity condition*

$$|K(z_1) - K(z_0)| \le L|z_1 - z_0|^\alpha.$$

Assuming $h \to 0$, $N^{1-2\nu}h^{2k} \to \infty$, and $Nh^{2k+2} \to 0$ for some $\nu > 0$, the following holds: for any $\varepsilon > 0$,

$$\sup_{x \in G} |r_N(x) - r(x)| =^{as} O(1/(N^{1-\varepsilon}h^k)^{0.5}).$$

The rate $(N^{1-\varepsilon}h^k)^{0.5}$ in Theorem 4.2 is the best known in the literature. Requiring $N^{1-\varepsilon}h^k \to 0$, is for any $\varepsilon > 0$ equivalent to $Nh^k/\ln(N) \to 0$, which appeared in the preceding chapter for the uniform consistency of the kernel density estimate. Theorem 4.2 will play an important role in establishing $\sqrt{N}$-convergence rate for various two-stage methods with a nonparametric first stage in later chapters. In 4.2, if $|Y| < M$ is replaced by $E|Y|^\alpha$ for some $\alpha > 2$, then the rate $\frac{Nh^k}{\ln(N)}$ is replaced by $\frac{Nh^k}{N^\lambda \ln(N)}$ where $\lambda \in \left(\frac{4}{\alpha+2}, 1\right)$; i.e., $N^{1-\varepsilon}h^k$ is replaced by $N^{1-\lambda-\varepsilon}h^k$.

The convergence rate $(N^{1-\varepsilon}h^k)^{0.5}$ is a tiny bit slower than $(Nh^k)^{0.5}$ for the pointwise convergence, which is in turn slower than $N^{0.5}$. Note that these rates are the ones at which the variances shrink; the bias of $r_N(x)$ was eliminated by the assumption of high-order kernel and $Nh^{2k+2} \to 0$. In density estimation, the optimal h minimizing the mean-square-error (MSE) reduces the variance and bias at the same rate. In the above theorem, we deliberately choose a smaller than the optimal h (under-smoothing) with $Nh^{2k+2} \to 0$ to remove the bias.

8.5   Asymptotic Distribution of Kernel Estimator

In this section, we derive the asymptotic distribution of $(Nh^k)^{0.5}\{r_N(x_0) - r(x_0)\}$ drawing upon Bierens (1987, p. 106–108); it is possible to use the "linearization" approach, but the Bierens' approach seems easier; both approaches amount to replacing f_N in r_N with f.

Observe that

$$r_N(x_0) - r(x_0) = (Nh^k)^{-1} \sum_i K((x_i - x_0)/h)\{y_i - r(x_0)\}/f_N(x_0). \quad (5.1)$$

Thus,

$$(Nh^k)^{0.5}\{r_N(x_0) - r(x_0)\}f_N(x_0)$$

$$= (Nh^k)^{-0.5} \sum_i K((x_i - x_0)/h)\{r(x_i) - r(x_0) + u_i\}, \qquad (5.2)$$

which will be shown to converge to a normal distribution, say $N(0, \zeta(x)^2)$. Then, since (due to the Slutsky lemma),

$$(Nh^k)^{0.5}\{r_N(x_0) - r(x_0)\}f_N(x_0) =^p (Nh^k)^{0.5}\{r_N(x_0) - r(x_0)\}f(x_0),$$

the desired result will be

$$(Nh^k)^{0.5}\{r_N(x_0) - r(x_0)\} =^d N(0, \zeta(x_0)^2/f(x_0)^2). \qquad (5.3)$$

So the main task is showing the convergence of (5.2) and finding $\zeta(x_0)^2$. The end result is given in Theorem 5.14 to which the reader may jump.

Rewrite (5.2) as

$$(Nh^k)^{-0.5} \sum_i K((x_i-x_0)/h)u_i + (Nh^k)^{-0.5} \sum_i K((x_i-x_0)/h)\{r(x_i)-r(x_0)\}.$$

Subtract and add the expected value of the second term to get

$$(Nh^k)^{-0.5} \sum_i K((x_i - x_0)/h) \cdot u_i$$

$$+ (Nh^k)^{-0.5} \sum_i [K((x_i - x_0)/h)(r(x_i) - r(x_0))$$

$$- E\{K((x - x_0)/h)(r(x) - r(x_0))\}]$$

$$+ (Nh^k)^{-0.5} \sum_i E\{K((x - x_0)/h)(r(x) - r(x_0))\}. \qquad (5.4)$$

The following can be shown to hold for the three terms:

(i) First $=^d N(0, V(u \mid x_0)f(x_0) \cdot \int K(z)^2 dz)$.

(ii) $E(\text{second}^2) \to 0$ as $N \to \infty$, making the second term $o_p(1)$. \qquad (5.5)

(iii) Third $\to 0$ while yielding a bias term as $N \to \infty$.

From this, we will get

$$(Nh^k)^{0.5}\{r_N(x_0) - r(x_0)\}f_N(x_0) =^d N\left(0, V(u \mid x_0)f(x_0)\int K(z)^2 dz\right),$$
$$(5.6)$$

and the variance in (5.6) is $\zeta(x_0)^2$ in (5.3). Thus, the desired variance in (5.3) will become $V(u \mid x_0)\int K(z)^2 dz/f(x_0)$.

Since $V[(Nh^k)^{0.5}(f_N(x_0) - f(x_0))] = f(x_0)\int K(z)^2 dz$, the only difference between this variance and the one in (5.6) is $V(u \mid x_0)$. This is natural, for the first term in (5.4) becomes $f_N(x_0)$ if u_i is replaced by 1. In the remainder of this section, we prove (iii) first and then (i); (ii) is omitted [see Bierens (1987, (2.2.12))].

In showing (iii), first consider $E\{(r(x)-r(x_0))\cdot K((x-x_0)/h)\}$. Defining $z \equiv (x-x_0)/h$, we get

$$h^k \cdot \int \{r(x_0 + hz) - r(x_0)\} \cdot f(x_0 + hz) \cdot K(z)dz.$$

Apply a Taylor's expansion to $r(\cdot)$ and $f(\cdot)$:

$$r(x_0 + hz) - r(x_0) = hz'\partial r(x_0)/\partial x + (1/2)h^2 z'\{\partial^2 r(x^*)/\partial x \partial x'\}z,$$
$$f(x_0 + hz) = f(x_0) + \{\partial f(x^{**})/\partial x'\}hz.$$

Substituting the product of these into the integral, we get four terms. Among these, the term $f(x_0) \cdot \partial r(x_0)/\partial x \cdot hz$ disappears, since

$$\int f(x_0) \cdot \partial r(x_0)/\partial x \cdot hz \cdot K(z)dz = h \cdot f(x_0) \cdot \partial r(x_0)/\partial x \cdot \int zK(z)dz = 0,$$

leaving two $O(h^{k+2})$ terms and one $O(h^{k+3})$ term; the $O(h^{k+2})$ terms are

$$h^{k+2} \int z'\{\partial r(x_0)/\partial x \cdot \partial f(x^{**})/\partial x'\}z \cdot K(z)dz$$

$$+ h^{k+2}\{f(x_0)/2\} \cdot \int z'\{\partial^2 r(x^*)/\partial x \partial x'\}z \cdot K(z)dz.$$

Taking the trace on this and using $\int zz'K(z) = \kappa I_k$, this becomes h^{k+2} times

$$\kappa \sum_{j=1}^{k} [\{\partial r(x_0)/\partial x_j\}\{\partial f(x_0)/\partial x_j\} + (1/2)f(x_0) \cdot \partial^2 r(x_0)/\partial x_j^2]. \quad (5.7)$$

Thus, the third term of (5.4) is $O(Nh^{k+2})/(Nh^k)^{0.5} = O((Nh^{k+4})^{0.5})$. If $Nh^{k+4} \to 0$, the bias disappears as in the density estimation case. The asymptotic bias for $(Nh^k)^{0.5}\{r_N(x_0) - r(x_0)\}$ can be obtained by dividing (5.7) by $f(x_0)$:

$$(Nh^{k+4})^{0.5} \cdot \kappa \sum_{j=1}^{k} [\{\partial r(x_0)/\partial x_j\}\{\partial f(x_0)/\partial x_j\}/f(x_0)$$

$$+ (1/2)\partial^2 r(x_0)/\partial x_j^2]. \quad (5.8)$$

For $k = 1$, this becomes (now we use prime to denote the derivatives)

$$(Nh^5)^{0.5} \cdot \kappa[r'(x_0)f'(x_0)/f(x_0) + (1/2)r''(x_0)]. \quad (5.9)$$

As in the density estimation, the second derivative r'' reflecting the variation of $r(x)$ appears; but it is somewhat troubling to have r', for it means
a bias even when $r(x) = x'\beta$. Later we will introduce an estimator without
r' in its bias.

To prove (5.5)(i), define a generic term in the first term of (5.4) as v_i:

$$v_i \equiv u_i \cdot K((x_i - x_0)/h)/(Nh^k)^{0.5}. \tag{5.10}$$

Clearly, $E(v_i) = 0$. The second moment of v_i is

$$E\{u_i^2 K^2((x_i - x_0)/h)/Nh^k\} = (1/N) \int V(u \mid x_0 + hz) f(x_0 + hz) \cdot K^2(z) dz$$

$$\to (1/N) V(u \mid x_0) f(x_0) \cdot \int K^2(z) dz, \quad \text{as } N \to \infty. \tag{5.11}$$

To apply the Lindeberg CLT, observe that

$$\sigma_N^2 \equiv \sum_{i=1}^{N} E(v_i^2) = V(u \mid x_0) f(x_0) \cdot \int K^2(z) dz. \tag{5.12}$$

The Lindeberg condition is

$$\sum_{i=1}^{N} E\{(v_i/\sigma_N)^2 \cdot 1[|u_i K(\cdot)| \geq \varepsilon \sigma_N (Nh^k)^{0.5}]\} \tag{5.13}$$

which is dominated by $\sum_i E(v_i^2)/\sigma_N^2 = 1$. So applying the dominated convergence theorem, the Lindeberg condition is satisfied and the CLT holds.
We put our discussion up to this point as a theorem:

Theorem 5.14. *Assume the following: $K(z)$ satisfies (2.4), $f(x) > 0$ for
all x, $r(x)$ and $f(x)$ have bounded and continuous second derivatives, and
$s(x) \equiv V(u^2 \mid x)$ is continuously differentiable with bounded first derivatives. Then*

$$(Nh^k)^{0.5}\{r_N(x_0) - r(x_0)\} =^d N(\text{Asym.Bias}, V(u \mid x_0) \int K(z)^2 dz/f(x_0) \tag{5.15}$$

*where Asym.Bias = (5.8). If $Nh^{k+4} \to 0$, then the bias goes to 0. Under
homoskedasticity $[V(u \mid x_0) = \sigma^2]$, we have*

$$(Nh^k)^{0.5}\{r_N(x_0) - r(x_0)\} =^d N\left(\text{Asym.Bias}, \sigma^2 \int K(z)^2 dx/f(x_0)\right).$$

In the variance, if $f(x_0)$ is higher, then the variance becomes lower, since
more data are available for $r_N(x_0)$. Asymptotically, $(Nh^k)^{0.5}\{r_N(x_0) -
r(x_0)\}$ and $(Nh^k)^{0.5}\{r_N(x_1) - r(x_1)\}$ have zero covariance if $x_0 \neq x_1$.

Therefore, a vector version of 5.14 is straightforward with a diagonal covariance matrix.

If we want a (pointwise) confidence interval, we need an estimate for $V(u \mid x_0)$ in (5.15) and for σ^2 for the homoskedastic case. Estimating σ^2 is easy: take the squared residuals $\hat{u}_i^2 \equiv \{y_i - r_N(x_i)\}^2$ and get their sample mean. Note that, to visualize $r_N(x)$ over x, we can pick some values of x, say $x_{(1)} \ldots x_{(m)}$, and then connect $r_N(x_{(1)}) \ldots r_N(x_{(m)})$; that is, we need only m-many estimates. But getting $\hat{u}_i$ requires $r_N(x_i)$, $i = 1 \ldots N$, N-many estimates. Under heteroskedasticity, estimating $V(u \mid x_0) = E(u^2 \mid x_0)$ requires another kernel method:

$$E_N(u^2 \mid x_0) \equiv \sum_i K((x_i - x_0)/hu) \cdot \hat{u}_i^2 / \sum_i K((x_i - x_0)/hu); \quad (5.16)$$

hu here should be chosen as well. Perhaps a better alternative is using the fact $V(u \mid x_0) = V(y \mid x_0)$ and estimating $E(u^2 \mid x_0)$ with

$$E_N(y^2 \mid x_0) - \{E_N(y \mid x_0)\}^2. \quad (5.17)$$

As mentioned in the preceding chapter, connecting 95% pointwise confidence intervals over $x_1 \ldots x_m$ does not give the 95% confidence region. If we want the 95% coverage probability at all m points, then we need to solve $(1 - \alpha)^m = 0.95$ for α to obtain $(1 - \alpha)100\%$ pointwise intervals at all m points. For instance, $m = 20$ (40) gives $\alpha = 0.0026$ (0.0013), and the $\alpha/2$-quantile from $N(0,1)$ is -3.02 (-3.22). Alternatively, one can use uniform confidence interval for all x; see Härdle and Linton (1994, p. 2317).

One problem in estimating the variance in (5.15) is $\int K(z)^2 dz$. This can be evaluated analytically. But suppose we use the standard normal density for $K(z)$. Then

$$\int K(z)^2 dz = \int K(z)K(z)dz = E\{K(z)\}, \quad (5.18)$$

which is the expected value of the function $K(z)$ of z with $z =^d N(0,1)$. Using a LLN, $E\{K(z)\} =^{as} (1/T) \sum_{i=1}^{T} K(z_i)$, where $z_i =^d N(0,1)$. Hence, $\int K(z)^2 dz$ can be estimated by the mean of a simulated rv's $\{z_i\}$. This way of approximating an integral is called *Monte Carlo integration*, the idea already employed in the method of simulated moments.

8.6   Choosing Smoothing Parameter and Kernel

In the preceding chapter, we discussed choosing h by minimizing MISE or by the least squares cross validation. For $r_N(x)$, however, MISE is difficult to evaluate, although one may use the bias and variance in (5.15) as an approximation for MSE, which then can be integrated over x to yield MISE.

From this, the MISE-minimizing h can be found, and a plug-in method can be implemented if $\partial f/\partial x$, $\partial r/\partial x$, and $\partial^2 r/\partial x \partial x'$ are estimated. In practice, however, data-driven automatic methods shown below look more attractive.

Define the "leave-one-out" kernel estimator for $r(x_j)$:

$$r_{Nj}(x_j) \equiv \sum_{i \neq j} K((x_i - x_j)/h)y_i \Big/ \sum_{i \neq j} K((x_i - x_j)/h). \qquad (6.1)$$

Then we can choose h minimizing the *cross-validation* (CV) criterion

$$(1/N) \sum_{j=1}^{N} \{y_j - r_{Nj}(x_j)\}^2 \cdot w(x_j), \qquad (6.2)$$

where $w(x_j)$ is a weighting function to downgrade the "prediction errors" when x_j falls near the boundary of its range. Choice of $w(x_j)$ is up to the researcher; obviously the simplest is $w(x_j) = 1$ for all j. The $\hat{h}$ which minimizes CV has been shown to be optimal by Härdle and Marron (1985), which is discussed in the following.

Introduce distances between r_N and r: with SE $\equiv$ squared error,

Average SE: $d_A(r_N, r) \equiv (1/N) \sum_j \{r_N(x_j) - r(x_j)\}^2 w(x_j)$,

Integrated SE: $d_I(r_N, r) \equiv \int \{r_N(x) - r(x)\}^2 w(x)f(x)dx$, $\qquad (6.3)$

Conditional Mean SE: $d_C(r_N, r) \equiv E\{d_I(r_N, r) \mid x_1 \ldots x_N\}$.

For notational simplicity, let $w(x) = 1$ from now on. Consider d_I which is analogous to ISE in density estimation:

$$d_I(r_N, r) = \int r_N(x)^2 f(x)dx - 2 \int r_N(x)r(x)f(x)dx + \int r(x)^2 f(x)dx. \qquad (6.4)$$

Ignore the last term without h and observe that, owing to $y = r(x) + u$,

$$\int r_N(x)r(x)f(x)dx = \int r_N(x)yf(x)dx \cong (1/N) \sum_j r_N(x_j)y_j. \qquad (6.5)$$

Likewise, the first term in (6.4) can be estimated by

$$(1/N) \sum_j r_N(x_j)^2. \qquad (6.6)$$

Substituting (6.5) and (6.6) into (6.4), we get (6.2). Regard a bandwidth selection rule $\hat{h}$ as a function from $\{(x_i, y_i)\}_{i=1}^{N}$ to H_N (say, $H_N = [N^{-a}, N^{-b}]$). Then $\hat{h}$ is said to be "optimal wrt d," if

$$\lim_{N \to \infty} \left[d\{r_N(x; \hat{h}), r\} \Big/ \inf_{h \in H_N} d\{r_N(x; h), r)\} \right] =^p 1. \qquad (6.7)$$

Härdle and Marron (1985) show that the CV-minimizing $\hat{h}$ is optimal wrt d_A, d_I, and d_C under some conditions.

Consider a generic nonparametric estimator

$$m_N(x_j; h) = \sum_{i=1}^{N} w_N(x_1 \ldots x_N; h) y_i \quad j = 1 \ldots N,$$

(6.8)

$$\equiv W_N(h) \cdot Y,$$

where $Y \equiv (y_1 \ldots y_N)'$ and $W_N(h)$ is a $N \times N$ matrix. For r_N,

$W_N(h) \equiv$

$$\begin{bmatrix} K((x_1 - x_1)/h)/\sum_i K((x_i - x_1)/h) \ldots K((x_N - x_1)/h)/\sum_i K((x_i - x_1)/h) \\ K((x_1 - x_N)/h)/\sum_i K((x_i - x_N)/h) \ldots K((x_N - x_N)/h)/\sum_i K((x_i - x_N)/h). \end{bmatrix}$$

(6.9)

Then, the *generalized cross validation* (GCV) is another popular bandwidth selection rule minimizing

$$(1/N) \sum_{j=1}^{N} \{y_j - m_N(x_j; h)\}^2 \cdot [1 - \text{tr}\{W_N(h)\}/N]^{-2}.$$

(6.10)

For r_N, this becomes

$$(1/N) \sum_{j=1}^{N} \{y_j - r_N(x_j; h)\}^2$$

$$\cdot \left[1 - \{K(0)/N\} \cdot \sum_j \left\{ \sum_i K((x_i - x_j)/h) \right\}^{-1} \right]^{-2}.$$

(6.11)

Compared with CV, GCV is slightly more convenient due to no need for leave-one-out. See, however, Andrews (1991a) who shows that CV is better under heteroskedasticity; see also Härdle, Hall, and Marron (1988).

So far we used only one h even when $k > 1$. In practice, one may use k different bandwidths, say $h_1 \ldots h_k$, because the regressors have different scales. Then we would have $\prod_{j=1}^{k} h_j$ instead of h^k. Although using different bandwidths should be more advantageous in principle, this makes choosing bandwidths too involved a task. An easier alternative is to standardize the regressors and use one bandwidth h. This is equivalent to $h_j = h \cdot SD(x_j)$, $j = 1 \ldots k$, for nonstandardized data; as in density estimation, one rule of thumb for h is $N^{-1/(k+4)}$. Let w denote the standardized x, say $w = Sx$ for a $k \times k$ invertible matrix S. Then $r(x) = q(Sx)$, where $E(y \mid w) \equiv q(w)$,

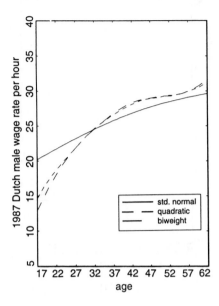

FIGURE 8.1.

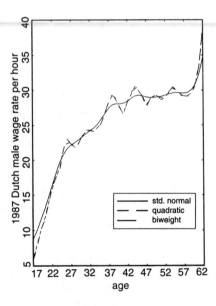

FIGURE 8.2.

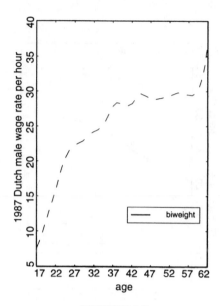

FIGURE 8.3.

because w and x are one-to-one. This, however, should not be taken as $r(x) = q(x)$.

If the dimension of x is small, then the best way to choose h seems to be trial and error. In Figures 8.1–8.3, we show three graphs for wage rate vs. age; the same data and the same kernels were used in the preceding chapter except that now we have the regressor age ranging from 16 to 63. Figure 8.1 is slightly over-smoothed with $h = SD(x)$, and Figure 8.2 is slightly under-smoothed with $h = 0.2 \cdot SD(x)$. In Figure 8.1, we can see a boundary effect in the line with $N(0,1)$ kernel; namely, at age $= 17$, it is above the other two lines, since the observations used for $r_N(17)$ comes mostly from the right of age $= 17$ and all of those have higher incomes than at age $= 17$. If there were observations to the left of age $= 17$, they would have lower incomes than at age $= 17$ and so $r_N(17)$ would not be so upward biased. This problem should be also present in the other two lines, but to a lesser degree, since the other kernels have bounded supports. In Figure 8.3, we show one under-smoothed line with biweight kernel and $h = 0.38 \cdot SD(x)$ chosen by CV; GCV is slightly worse with $h = 0.32 \cdot SD(x)$.

Choosing kernels in nonparametric regression is similar to that in density estimation; namely, one can set up an optimum criterion and choose an optimal kernel [see Müller (1988)]. However, there seems to be a consensus that choosing kernels does not matter much, being of secondary importance compared with choosing bandwidth; Figures 8.1 and 8.2 show little difference between using the quadratic kernel and the biweight kernel. But kernels with bounded support are preferable in general to kernels with

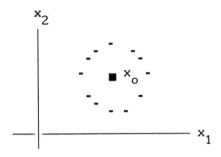

FIGURE 8.4.

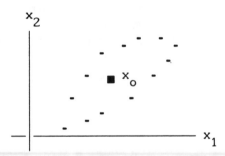

FIGURE 8.5.

unbounded support; r_N using the former looks at least less sensitive to the choice of h, in addition to a lesser boundary effect mentioned already.

Another point deserving discussion is whether to use a product kernel or a multivariate kernel. To simplify the discussion, consider $r(x_0)$ with $k = 2$ and kernels with bounded support. If $\text{CORR}(x_1, x_2) = 0$, then the data plot of $\{(x_{1i}, x_{2i})\}_{i=1}^{N}$ will show no pattern. If $\text{CORR}(x_1, x_2) > 0$, there will be an upward sloping pattern. Using a product kernel means that we pick observations around x_0 using a circle centered at x_0 (Figure 8.4) and take a weighted average of y_i's of those selected. With a bivariate kernel, we pick observations around x_0 using an ellipse (Figure 8.5) and take a weighted average of y_i's of those selected. The question of which is better depends on how close the distributions of $y \mid x$ of those selected are to the distribution of $y \mid x_0$. If u in $y = r(x) + u$ is independent of x, then the distribution of $y \mid x$ should have no bearing with x. This gives a support for product kernels in addition to their theoretical convenience.

8.7   Discrete Regressors, Estimating Derivatives, and MLE-Based Models

In this section, we discuss three unrelated topics. The first topic, nonparametrics with discrete regressors, has practical importance because there are almost always discrete regressors in practice. The second topic is estimating derivatives of $r(x)$. The third topic, combining MLE-based models with nonparametrics, deals with the case of an unknown regression function and a known error term distribution.

So far we dealt with the case where x has a continuous density function $f(x)$. Suppose we have a mixed case where $x = (x_c', x_d')'$, x_c is a $k_c \times 1$ continuous regressor, and x_d is a $k_d \times 1$ finitely discrete regressor. Then we can form a number of cells based on the values that x_d can take. For instance, if x_d has two variables d_1 and d_2, where d_1 takes 0,1,2,3, and d_2 takes $1, \ldots, J$, then there will be $4 \times J$ cells. Within a given cell, we can apply the results derived for the kernel estimator up to this point. The only change needed is to replace k by k_c and $f(x)$ by $f(x_c \mid x_d)$.

As an example, suppose that $k_c = 3$ and x_d has the above two variables d_1 and d_2. Let $d_1 = 0$ and $d_2 = 1$ in the first cell, and we estimate

$$r(x_0, 0, 1) = E(y \mid x_c = x_0, \ d_1 = 0, \ d_2 = 1) \tag{7.1}$$

with

$$r_N(x_0, 0, 1) \equiv \sum_i K((x_{ci} - x_0)/h) \cdot 1[d_{1i} = 0, d_{2i} = 1] \cdot y_i \Big/$$

$$\sum_i K((x_{ci} - x_0)/h) \cdot 1[d_{1i} = 0, d_{2i} = 1], \tag{7.2}$$

where $K(\cdot)$ is a three- ($= k_c$) dimensional kernel. $r_N(x_0, 0, 1)$ is consistent for $r(x_0, 0, 1)$, and, ignoring the asymptotic bias,

$$(Nh^3)^{0.5}\{r_N(x_0, 0, 1) - r(x_0, 0, 1)\} =^d N\Big(0, V(u \mid x_0, 0, 1)$$

$$\cdot \int K(z)^2 dz / f(x_0 \mid d_1 = 0, d_2 = 1)\Big). \tag{7.3}$$

$V(u \mid x_0, 0, 1)$ and $f(x_0 \mid d_1 = 0, d_2 = 1)$ can be estimated, respectively, by

$$E_N(u^2 \mid x_0, 0, 1) = \sum_i K((x_{ci} - x_0)/h) \cdot 1[d_{1i} = 0, d_{2i} = 1] \cdot \hat{u}_i^2 \Big/$$

$$\sum_i K((x_{ci} - x_0)/h) \cdot 1[d_{1i} = 0, d_{2i} = 1];$$

$$f_N(x_0 \mid d_1 = 0, d_2 = 1) = (Nh^3)^{-1} \sum_i K((x_{ci} - x_0)/h) \cdot 1[d_{1i} = 0, d_{2i} = 1] \Big/$$

$$(1/N) \sum_i 1[d_{1i} = 0, d_{2i} = 1], \qquad (7.4)$$

where the smoothing parameter h is not necessarily the same as the one in $r_N(x_0, 0, 1)$. The numerator in $f_N(x_0 \mid d_1 = 0, d_2 = 1)$ estimates

$$f(x_0 \mid d_1 = 0, d_2 = 1) \cdot P(d_1 = 0, d_2 = 1); \qquad (7.5)$$

so we need to divide the numerator by the estimate for $P(d_1 = 0, d_2 = 1)$ to get the conditional density. The expression (7.5) may look somewhat strange at the first glance. If all involved variables are continuous, we get $f(x_c, d_1, d_2)$ instead of (7.5); if all are discrete, we get $P(x_c, d_1, d_2)$. When they are continuous and discrete, we get (7.5).

The above method of applying nonparametrics on each cell can be cumbersome when there are too many cells. When all regressors are discrete, Bierens (1987, p. 115–117) shows that applying the kernel method yields $r_N(x_0) =^p r(x_0)$, and

$$\sqrt{N}\{r_N(x_0) - r(x_0)\} =^d N(0, V(u \mid x_0)/P(x_0)); \qquad (7.6)$$

compare this to (5.15). The convergence rate $\sqrt{N}$ and no kernel shows up in the asymptotic distribution. If we take the sample average of the observations with $x = x_0$, then we would get the same result. Based on this, Bierens (1987, p. 117–118) states that, when the regressors are mixed (continuous and discrete), if we apply the kernel method as if all regressors are continuous, then we still get consistent estimates and (7.3), with the only change being $\int K(z)^2 dz$ replaced by $\int K(z_1, 0)^2 dz_1$, where $K(z_1, z_2)$ is the kernel with z_1 and z_2 corresponding to continuous and discrete regressors, respectively. This result means that whether the regressors are continuous or discrete, we can always apply the kernel method; the only thing we should be careful about is the convergence rate and the asymptotic distribution. Still, we may learn more by forming cells and doing nonparametrics on each cell in the mixed case.

Define the gradient of $r(x)$ at x_0:

$$\nabla r(x_0) \equiv \partial r(x_0)/\partial x \equiv (\partial r(x_0)/\partial x_1 \dots \partial r(x_0)/\partial x_k)'. \qquad (7.7)$$

The gradient is often interesting. For instance, if x is the factors of a production function, then the gradient reflects the marginal contributions of x to the output y. $\nabla r(x_0)$ can be estimated by

$$
\begin{aligned}
\nabla r_N(x_0) &= \partial\{g_N(x_0)/f_N(x_0)\} \\
&= f_N(x_0)^{-2}\{f_N(x_0)\nabla g_N(x_0) - g_N(x_0)\nabla f_N(x_0)\} \qquad (7.8) \\
&= \nabla g_N(x_0)/f_N(x_0) - r_N(x_0) \cdot \nabla f_N(x_0)/f_N(x_0).
\end{aligned}
$$

Recall that CV-like methods were based on $y_i - r_N(x_i)$ in choosing the bandwidth. But there is no analogous expression in estimating $\nabla r(x)$, making the choice of a bandwidth difficult. Müller (1988) shows some suggestions; note that an optimal h for r_N is not optimal for ∇r_N in general. Second-order derivatives are estimated by differentiating ∇r_N again, but the number of terms goes up by twice due to the denominator of $r(x_0)$; the problem gets worse for higher-order derivatives.

Under certain conditions and the assumption (2.4), Vinod and Ullah (1988) show that $\nabla r_N(x_0) =^p \nabla r(x_0)$, and denoting the jth component of $\nabla r_N(x_0)$ ($\nabla r(x_0)$) as $\nabla r_{Nj}(x_0)$ ($\nabla r_j(x_0)$),

$$(Nh^{k+2})^{0.5}\{\nabla r_{Nj}(x_0) - \nabla r_j(x_0)\}$$

$$=^d N\left(0, V(u \mid x_0) \int \{\partial K(z)/\partial z_j\}^2 dz/f(x_0)\right). \qquad (7.9)$$

Compared to (5.15), there are two differences. One is the convergence rate slower by the factor h, and the other is $\int \{\partial K(z)/\partial z_j\}^2 dz$ instead of $\int K(z)^2 dz$; naturally, $\partial K(z)/\partial z_j$ is due to the differentiation. If one uses h_j for x_j, $j = 1 \ldots k$, Nh^{k+2} should be replaced by $N \cdot h_j^2 \prod_{j=1}^{k} h_j$.

To avoid the problem of having too many terms in high order derivatives, Mack and Müller (1989) suggest an estimate for $r^{(\nu)}(x_0) \equiv \partial^\nu r(x_0)/\partial x^\nu$ when $k = 1$:

$$r_N^{(\nu)}(x_0) \equiv (Nh^{1+\nu})^{-1} \sum_i K^{(\nu)}((x_i - x_0)/h) \cdot \{y_i/f_N(x_i)\}, \qquad (7.10)$$

where (ν) stands for the νth derivative. The main idea is to regard the regression function as $E\{y/f(x) \mid x\} \cdot f(x)$, where $y/f(x)$ is taken as the response variable, which explains $y_i/f_N(x_i)$ in (7.9). They show

$$(Nh^{1+2\nu})^{0.5}\{r_N^{(\nu)}(x_0) - r^{(\nu)}(x_0)\}$$

$$=^d N\left(0, V(u \mid x_0) \int \{K^{(\nu)}(z)\}^2 dz/f(x_0)\right). \qquad (7.11)$$

Compare this to (7.9). For (7.11), the bandwidth for $f_N(x_i)$ is required to converge to 0 more slowly than the h in the numerator of (7.10).

It is possible to combine nonparametrics with MLE specifying the error term distribution; one example is a Weibull duration model:

$$f(t \mid x) = \alpha\theta t^{\alpha-1}\exp(-\theta t^\alpha), \quad \theta = e^{r(x)}, \qquad (7.12)$$

where $f(t \mid x)$ is the density of a duration t given x. Observe that the kernel estimator $r_N(x_0)$ can be obtained by minimizing

$$\sum_i K((x_i - x_0)/h)(y_i - \lambda_0)^2 \qquad (7.13)$$

wrt λ_0. Analogously, if $\sum_i \ln f(y_i, x_i, \lambda(x_i))$ is the log-likelihood of (x_i', y_i), one can think of maximizing

$$\sum_i K((x_i - x_0)/h) \cdot \ln\{f(y_i, x_0, \lambda_0)\} \qquad (7.14)$$

wrt λ_0 to estimate $\lambda(x_0)$; $\lambda_0 = r(x_0)$ in (7.12). Staniswalis (1989) shows $\lambda_N(x_0) =^p \lambda(x_0)$, where $\lambda_N(x_0)$ is the maximizer of (7.14), and

$$(Nh^k)^{0.5}\{\lambda_N(x_0) - \lambda(x_0)\}$$

$$=^d N\left(0, \int K(z)^2 dz/E[\{\partial \ln f(y, x_0, \lambda_0)/\partial \lambda_0\}^2 \mid x_0]\right). \qquad (7.15)$$

Compare this to (5.15) ignoring the asymptotic bias.

8.8   Other Nonparametric Regression Methods

(1) Local Linear Regression

The expression (7.13) shows that the kernel method predicts y_i's whose x_i is close to x_0 by an "intercept" λ_0. But it may be better to use a line (intercept and slope) centered at x_0. Using this idea, *local linear regression* (LLR) minimizes

$$\sum_i \{y_i - a - b'(x_i - x_0)\}^2 \cdot K((x_i - x_0)/h) \qquad (8.1)$$

with respect to a and b. The estimate $a_N(x_0)$ for a is the LLR estimate for $r(x_0)$, while the estimate for b is the LLR estimate for $\partial r(x_0)/\partial x$. To be specific, the LLR estimator is

$$a_N(x_0) = (1, 0 \ldots 0) \cdot \{X(x_0)'W(x_0)X(x_0)\}^{-1} \cdot \{X(x_0)'W(x_0)Y\}, \quad (8.2)$$

where $Y \equiv (y_1 \ldots y_N)'$, $W(x_0) \equiv \text{diag}\{K((x_1 - x_0)/h) \ldots K((x_N - x_0)/h)\}$, and

$$X \equiv \begin{bmatrix} 1 \ (x_1 - x_0)' \\ \ldots\ldots\ldots \\ 1 \ (x_N - x_0)' \end{bmatrix}.$$

The asymptotic distribution, other than the bias, is the same as that of the usual (local constant) kernel method.

LLR has some advantages over the kernel method. First, the asymptotic bias of the LLR does not depend on $\partial r/\partial x$ as in the kernel method; the kernel method has a bias even when $r(x) = x'\beta$. Second, LLR works better at the boundaries due to the local linear fitting; see Hastie and Loader (1993) for a graphic illustration. Third, LLR has certain efficiency property

(Fan, 1992, 1993). Fourth, as a by-product, LLR provides an estimate for $\partial r(x_0)/\partial x$ as well.

LLR can be generalized to local quadratic (and higher polynomial) regression; for more, see Hastie and Loader (1993) and Rupper and Wand (1994), which is the most relevant for multivariate applications. Also there are some other (and earlier) versions of LLR using different weighting schemes; see Cleveland et al. (1988).

(2) Nearest-Neighbor Estimator

Generalizing the nearest-neighbor method for density estimation, we can estimate $r(x_0)$ with an (weighted) average of y_i's whose x_i's fall within the sth nearest neighbor (NN) of x_0. One such estimate is

$$r_N(x_0) = (1/s) \sum_i 1[x_i \in \text{sth NN of } x_0] \cdot y_i$$

$$= \sum_i \{1[x_i \in \text{sth NN of } x_0]/s\} y_i = \sum_i w_i(x_0, s) \cdot y_i \qquad (8.3)$$

with $w_i(x_0, s) \equiv 1[x_i \in \text{sth NN of } x_0]/s$. Note that $\sum_i w_i(x_0, s) = 1$ for all x_0 and s, so long as we include x_0 in the sth NN observations if $x_0 = x_i$ for some i. One advantage of (8.3) over kernel methods is that the same number of observations are used for each $r_N(x_0)$ as x_0 varies, which makes estimates in data-scarce areas more reliable. One disadvantage is the nonsmoothness in $r_N(x_0)$, which makes the asymptotic analysis difficult; making the weighting function smooth when it is not zero [subject to $\sum_i w_i(x_0, s) = 1$] yields a smoother estimate.

For the case $k = 1$, with $s/N \to 0$ as $N \to \infty$, Härdle (1990, p. 43) shows

$$Er_N(x_0) - r(x_0) \cong \{24f(x_0)^2\}^{-1}\{r''(x_0) + 2r'(x_0)f'(x_0)/f'(x_0)\}$$

$$\cdot (s/N)^2, \quad V(r_N(x_0)) \cong V(u \mid x_0)/s. \qquad (8.4)$$

Regarding s/N as h in the kernel method, the bias is of order h^2 while the variance is of order $(Nh)^{-1}$, which is the same as in the kernel method. Minimizing the MSE, we get $s = O(N^{4/5})$, which makes MSE converge to 0 at $O(N^{-4/5})$, the same rate in the kernel method. The variance does not have $f(x_0)^{-1}$, which the kernel method has; that is, even if $f(x_0)$ is small, the variance is not affected. The bias however is larger than that of the kernel method by the factor $f(x_0)^{-2}$.

(3) Spline Smoothing

Suppose x is a scalar over $[0, 1]$, and $r(x)$ has a continuous $(m - 1)$th derivative and a square-integrable mth derivative $r^{(m)}$ ($\int \{r^{(m)}(x)\}^2 dx <$

∞). Imagine (x_i, y_i) scattered over the xy plane and we want to fit a line that has a good fit as well as smoothness. These two contradicting goals can be achieved by minimizing the following wrt q for a $\lambda > 0$ over a function space to which $r(x)$ belongs:

$$(1/N)\sum_i \{y_i - q(x_i)\}^2 + \lambda \int_0^1 \{q^{(m)}(x)\}^2 dx. \qquad (8.5)$$

Here λ penalizes overfit, so it is a smoothing parameter. The solution to (8.5) is a piecewise polynomial. Choosing the optimal $q(x)$ this way is a *spline smoothing* (Wahba, 1990). If we replace the first term in (8.5) by $(1/N)\sum_i -\ln f(z_i)$ and the second term by a measure of likelihood variation, we have "likelihood spline smoothing." The smoothing spline for (8.5) is attractive not so much for its practicality (at least for $k > 1$) but rather for its relation to our prior information on $r(x)$. We will show this point below, drawing upon Eubank (1989).

Apply Taylor expansion with an integral remainder to $r(x)$ around 0:

$$r(x) = \sum_{i=0}^{m-1} \beta_i x^i + (1/(m-1)!)\int_0^1 r^{(m)}(z) \cdot (1-z)^{m-1} dz. \qquad (8.6)$$

If we want approximate $r(x)$ with a polynomial in x with $m-1$ degree, the result depends on the extent that the remainder term is negligible. Using the Schwartz inequality,

$$\left[\int_0^1 r^{(m)}(z) \cdot (1-z)^{m-1} dz\right]^2 \leq \left[\int_0^1 \{r^{(m)}(z)\}^2 dz\right]$$

$$\cdot \int_0^1 (1-z)^{2m-2} dz = \left[\int_0^1 \{r^{(m)}(z)\}^2 dz\right] \cdot (1-2m)^{-1}(1-z)^{2m-1}\Big|_0^1$$

$$= \left[\int_0^1 \{r^{(m)}(z)\}^2 dz\right] \cdot (2m-1)^{-1}. \qquad (8.7)$$

With this, we get

$$|\text{remainder in (8.6)}| \leq (2m-1)^{-0.5} \cdot \left[\int_0^1 \{r^{(m)}(z)\}^2 dz\right]^{0.5}$$

$$\equiv (2m-1)^{-0.5} \cdot J_m(r)^{0.5}. \qquad (8.8)$$

An assumption such as $J_m(r) \leq \rho$ reflects our prior belief on how the model deviates from the polynomial regression. It can be shown that, if we minimize $(1/N)\sum_i\{y_i - q(x_i)\}^2$ over q subject to the condition that q has $(m-1)$ continuous derivatives and $J_m(q) \leq \rho$, then there is a $\lambda > 0$ such that the same q minimizes (8.5). Choosing $\lambda = 0$ implies a polynomial regression where y is regressed on 1, x, $x^2, \ldots, x^{m-1}$. Hence, smoothing spline is an extension of polynomial regressions with guarding against the departure from the assumption that the regression function is polynomial.

Implementing smoothing spline in practice is somewhat complicated. Instead we mention a result in Silverman (1984): for a smoothing spline, there exists an equivalent kernel regression with its local bandwidth proportional to $f(x)^{-1/4}$. Therefore, we will not lose much by using adaptive kernel-type methods in most cases. Also Jennen–Steinmetz and Gasser (1988) present a further generalization where various nonparametric methods are shown to be equivalent to the kernel method with the local bandwidth proportional to $f^{-\alpha}$, $0 \leq \alpha \leq 1$. They also suggest selecting α adaptively instead of fixing it in advance. These findings show that we will not miss much in practice by focusing on kernel nonparametric regression method or its variations.

(4) Series Approximation

If x is in R^k, then we can write x as $x = \sum_{j=1}^{k} x_j e_j$ where e_j's are the orthonormal basis vectors for R^k $[e_i' e_j = \delta_{ij}$ (Kronecker delta)]. A regression function $r(x)$ at m different x points, $r(x_{(1)}) \ldots r(x_{(m)})$, may be viewed as a point in R^m; more generally, $r(x)$ can be regarded as a point in an infinite-dimensional space. Suppose x belongs to a compact set in R^k. If $\int r(x)^2 dx < \infty$, then $r(x)$ can be written as

$$r(x) = \sum_{j=1}^{\infty} \beta_j \psi_j(x), \tag{8.9}$$

where $\int \psi_i(x)\psi_j(x)dx = \delta_{ij}$. Here $r(x)$ is decomposed into orthogonal components and $\psi_j(x)$ may be regarded as an "axis" or "direction." The rhs is an "orthogonal series." The idea of orthogonal *series estimator* for $r(x)$ is to estimate the coefficient β_j by an estimator $b_{Nj}(\psi_j(x)$ are known functions), so that $r(x)$ can be estimated by

$$r_N(x) = \sum_{j=1}^{\infty} b_{Nj}\psi_j(x). \tag{8.10}$$

While a kernel method uses a local approximation idea, a series estimator uses a global approximation, because the same basis functions are used for all x and all observations contribute to $r_N(x_0)$.

Recalling $y = r(x) + u$, (8.9) implies

$$y_i = \sum_{j=1}^{\infty} \beta_j \psi_j(x_i) + u_i. \tag{8.11}$$

With a given data set, we can only estimate a finite number of β_j's. So we need to trim the series at a number, say a positive integer h, to get

$$y_i = \sum_{j=1}^{h} \beta_j \psi_j(x_i) + \left\{ u_i - \sum_{j=h+1}^{\infty} \beta_j \psi_j(x_i) \right\}. \qquad (8.12)$$

This can be estimated with LSE, where $\{\cdot\}$ is the error term. Although $\int \psi_i(x)\psi_j(x)dx = 0$ if $i \neq j$, $\int \psi_i(x)\psi_j(x)f(x)dx = E\{\psi_i(x)\psi_j(x)\} \neq 0$ in general. Thus, we will incur an omitted variable basis $[\sum_{j=h+1}^{\infty} \beta_j \psi_j(x_i)$ is omitted]. This is natural as other nonparametric estimators have biases. In (8.12), h plays the role of a smoothing parameter.

There are many sets of orthonormal functions for $\{\psi_j\}$. One example for $x \in [-1, 1]$ is the *Legendre polynomial*, which is obtained by applying the so-called "Gram–Schmidt procedure" to $1, x, x^2, x^3, \ldots$. The general form of Legendre polynomial is

$$\{(2n+1)/2\}^{0.5} \cdot \{(-1)^n/(2^n n!)\}$$
$$\cdot d^n\{(1-x^2)^n\}/dx^n, \quad n = 0, 1, 2, \ldots . \qquad (8.13)$$

Specifically, substituting $n = 0, 1, 2, 3, \ldots$, we get

$$1/\sqrt{2}, \; x/(2/3)^{0.5}, \; 0.5(3x^2 - 1)/(2/5)^{0.5},$$
$$0.5(5x^3 - 3x)/(2/7)^{0.5}, \ldots . \qquad (8.14)$$

Suppose we want to apply MLE under the assumption of normality but are unsure of the validity of the assumption. One way to test the assumption (or to estimate the nonnormal likelihood itself) is to apply a series expansion to the likelihood. Let $f(x)$ be the true density function of x admitting the following series expansion:

$$f(x) = \sum_{j=1}^{\infty} \beta_j \psi_j(x), \qquad (8.15)$$

where $\{\psi_j(x)\}$ are not necessarily orthogonal. Orthogonality makes estimating β_j "less wasteful" but the overall fit of the series remains the same for any basis of the same dimension; also nonorthogonal basis can be convenient as in the following. If $f(x)$ can be written as $f(x) = \phi(x) + \beta \cdot \psi(x)$, where $\phi(x)$ is the normal density, then an estimate of β will shed light on the validity of the normality assumption; $\psi(x)$ is not orthogonal to $\phi(x)$, if $\psi(x)$ is continuous.

To be specific on (8.15), Gallant and Nychka (1987) propose a variation of "Hermite polynomial":

$$H(x) = \left\{ \sum_{|\alpha|=0}^{K} \beta_\alpha x^\alpha \right\}^2 \cdot \phi(x)/C, \qquad (8.16)$$

where $\alpha \equiv (\alpha_1, \ldots, \alpha_k)'$, α_j's are nonnegative integers,

$$|\alpha| \equiv \sum_{j=1}^{k} \alpha_j, \quad x^\alpha \equiv \prod_{j=1}^{k} x_j^{\alpha_j}, \quad C = \int \left\{ \sum_{|\alpha|=0}^{K} \beta_\alpha x^\alpha \right\}^2 \cdot \phi(x) dx;$$

C is a normalizing constant. For instance, with $K = 2$, (8.16) is

$$\left\{ \beta_0 + \sum_{j=1}^{k} \beta_j x_j + \sum_{j=1}^{k} \sum_{m=1}^{k} \beta_{jm} x_j x_m \right\}^2 \cdot \phi(x)/C. \qquad (8.17)$$

This becomes $\phi(x)$ when all β_j's and β_{jm}'s but β_0 are zero.

9

Semiparametrics

9.1 Introduction

In the previous chapters, we examined parametric and nonparametric methods. In parametrics, we specify the regression function and the conditional distribution of $u \mid x$, $F_{u|x}$. In nonparametrics, neither is specified. So semantically speaking, anything between the two extremes can be called "semiparametric" or "semi-nonparametric."

We will call an estimator "semiparametric" if getting the estimator does not require a nonparametric smoothing technique, although the asymptotic variance matrix may do. Typically such estimators specify the regression function, but not $F_{u|x}$. Under this definition, LSE and many method-of-moments estimators (MME) are semiparametric. In the following chapter, under the heading of semi-nonparametrics, we will discuss methods that require (first-stage) nonparametric techniques to obtain the estimates. Sometimes such estimators do not specify the regression function fully. These definitions of semiparametrics and semi-nonparametrics are helpful at least in classifying the literature.

The semiparametric methods to be discussed in this chapter are mostly model-specific, so we will treat each paper (topic) more or less separately rather than trying to unify the literature. We will present motivations and asymptotic distributions (if available) for various semiparametric methods. The reader will find that, for most parametric methods we have discussed so far, there exist competing semiparametric methods for the same model.

See Robinson (1988b) and Powell (1994) for a survey on semiparametric methods.

In Section 2, a median regression estimator and its extensions for binary and multinomial responses are introduced. In Section 3, median regression for ordered discrete response is studied. In Section 4, mode regression that is good for either contaminated data or truncated data is examined. In Section 5, median regression for censored regression is introduced. In Section 6, versions of LSE for censored and truncated regressions are studied. In Section 7, robust trimmed LSE's for truncated and censored regressions are introduced. In Section 8, we examine single-index models. In Section 9, a rank correlation estimator, which is the only $\sqrt{N}$-consistent "smoothing parameter free" estimator for binary response models, is examined. In Section 10, two estimators for binary response and censored response panel data are discussed. In Section 11, we introduce some specification tests using semiparametric estimators.

9.2   Median Regression Estimators for Multinomial Responses

Manski (1975, 1985) proposed the maximum score estimation (MSC) for binary and multinomial choice models. Suppose $y^* = x'\beta + u$ and only $y = 1[y^* \geq 0]$ is observed along with x. Assume

$$\text{Med}(y^* \mid x) = x'\beta \Leftrightarrow \text{Med}(u \mid x) = 0,$$

where Med stands for median; recall that the median of a distribution F is $\inf\{\mu: F(\mu) \geq 0.5\}$. Then

$$\text{Med}(y \mid x) = 0 \quad \text{if } x'\beta < 0, \quad \text{Med}(y \mid x) = 1 \quad \text{if } x'\beta \geq 0,$$

$$\Leftrightarrow \text{Med}(y \mid x) = 1[x'\beta \geq 0],$$

(2.1)

where $\text{Med}(y \mid x)$ is either 0 or 1 for y is binary; in the following we discuss this key point further.

Perhaps the best way to understand (2.1) is the fact that, for a random variable z with $\text{Med}(z) = m$, $\text{Med}(\tau(z)) = \tau(m)$ for any nondecreasing function $\tau(\cdot)$; $\tau(\cdot) = 1[\cdot \geq 0]$ in (2.1). Graphically, examine Figure 9.1 with $x'\beta < 0$, where $P(y = 1 \mid x) = P(y^* \geq 0 \mid x) < 0.5$ while $P(y = 0 \mid x) \geq 0.5$; thus $\text{Med}(y \mid x) = 0$ in Figure 9.1. In Figure 9.2 with $x'\beta > 0$, $P(y = 1 \mid x) = P(y^* \geq 0 \mid x) \geq 0.5$, so $\text{Med}(y \mid x) = 1$. Combining the two cases, we get (2.1).

Since the least absolute deviation loss function is minimized at the median, Manski suggests to minimize

$$(1/N) \sum_i |y_i - 1[x_i'b \geq 0]|$$

(2.2)

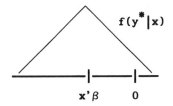

FIGURE 9.1.

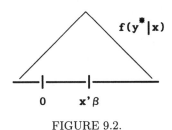

FIGURE 9.2.

wrt b. Since $|y_i - 1[x_i'b \geq 0]|$ is either 0 or 1, square it to get

$$y_i - 2y_i \cdot 1[x_i'b \geq 0] + 1[x_i'b \geq 0] = y_i - (2y_i - 1) \cdot 1[x_i'b \geq 0]. \qquad (2.3)$$

Dropping y_i, minimizing (2.2) is equivalent to maximizing

$$(1/N) \sum_i (2y_i - 1) \cdot 1[x_i'b \geq 0]. \qquad (2.4)$$

Further observing $\text{sgn}(x_i'b) = 2 \cdot 1[x_i'b \geq 0] - 1$, where $\text{sgn}(A) = 1$ if $A > 0$, 0 if $A = 0$ and -1 if $A < 0$, maximizing (2.4) is equivalent to maximizing $(1/N) \sum_i (2y_i - 1) \cdot \text{sgn}(x_i'b)$.

The minimand (2.2) may be viewed as a nonlinear regression. But the difficulty is that the derivative of $1[x'b \geq 0]$ is 0 except at $x'b = 0$ so that the usual method of deriving asymptotic distributions is not applicable. One critical assumption for MSE to overcome the lack of smoothness in (2.2) is that there is a regressor, say x_k, with a nonzero coefficient such that $x_k \mid (x_1 \ldots x_{k-1})$ has a nonzero density on all of R^1 for any value of $(x_1 \ldots x_{k-1})$. Also the support of the distribution of $u \mid x$ should be R^1 for all x.

More points are in order in comparing MSC with MLE such as probit. First, algorithm-wise, we need one that uses no gradient, for the indicator function has no informative derivative. Second, while MLE does not produce a direct predictor for y, MSC gives a natural predictor $1[x'b_N \geq 0]$. Third, while MLE does not allow heteroskedasticity of unknown form, MSC does. Fourth, the main drawback of MSC is that its asymptotic distribution is not practical with $N^{1/3}$-consistency (Kim and Pollard, 1990). Finally, suppose $E(y \mid x) = G(x'\beta)$ but MLE misspecifies $G(x'\beta)$ as $F(x'\beta)$. Then the performance of MLE depends on the difference between $G(x'\beta)$ and $F(x'\beta)$. If $k = 1$ and x is a dummy variable, MLE requires G to agree with F only at two points; if x takes more values, it will become harder to make $G(x'\beta) = F(x'\beta)$. On the contrary, MSC requires at least one continuous regressor as mentioned already. Hence, the continuity of x works for MSC but against MLE when the likelihood is misspecified. Since continuity is "realized" when N is large, MLE may perform better in a small sample where $G(x'\beta) = F(x'\beta)$ has a better chance to hold, while MSC may work better in a large sample.

As MLE, MSC cannot identify β fully, since if b_N minimizes (2.2), λb_N with $\lambda > 0$ also minimizes (2.2). If the threshold for y^* is not zero but unknown, say γ, it will be absorbed into the intercept estimate. Denote β_1 by α and the regressors without one by x. Then for any $\lambda > 0$,

$$\alpha + x'\beta + u \geq \gamma \Leftrightarrow \alpha - \gamma + x'\beta + u \geq 0 \Leftrightarrow \alpha^* + x'\beta + u \geq 0$$

$$\Leftrightarrow \lambda(a^* + x'\beta + u) \geq 0, \tag{2.5}$$

where $\alpha^* \equiv \alpha - \gamma$. That is, we can estimate α^* and β only up to a positive scale constant. In other words, in the binary model, only the ratios of the estimates make sense and the ratios involving the intercept cannot be interpreted.

In probit, we set $\lambda = 1/SD(u)$ to use $N(0,1)$, but $\lambda = 1/|\beta_k|$ is better for MSC where x_k is the continuous regressor with unbounded support. That is, in (2.5), we estimate $\alpha^*|\beta_k|$, $\beta_2/|\beta_k| \ldots \beta_{k-1}/|\beta_k|$, $\text{sgn}(\beta_k)$. To implement this normalization in practice, first set $\beta_k = 1$ and estimate β_{kc} where $\beta = (\beta'_{kc}, 1)'$. Then set $\beta_k = -1$ and again estimate β_{kc}. By comparing the optimands, choose one estimate out of the two estimates for the cases $\beta_k = 1$ and $\beta_k = -1$. If we had a prior information $\beta_k > 0$, then the step with $\beta_k = -1$ would be unnecessary.

For a multinomial model $y^*_{ij} = x'_{ij}\beta + u_{ij}$ where j indexes the choice set $\{1, \ldots, J\}$, and $y_{ij} = 1$ if i chooses j and 0 otherwise, MSC maximizes

$$(1/N) \sum_{i=1}^{N} \sum_{j=1}^{J} y_{ij} \cdot 1[x'_{ij}b \geq x'_{i1}b \ldots x'_{ij}b \geq x'_{iJ}b], \tag{2.6}$$

since $\sum_{j=1}^{J} y_{ij} \cdot 1[\cdot] = 1$ if the prediction is right and 0 otherwise. The population version is

$$E\left\{ \sum_{j=1}^{J} y_j \cdot 1[x'_j b \geq x'_1 b \ldots x'_j b \geq x'_J b] \right\}$$

$$= E_x\left\{ \sum_{j=1}^{J} P(y_j = 1 \mid x) \cdot 1[x'_j b \geq x'_1 b \ldots x'_j b \geq x'_J b] \right\}, \tag{2.7}$$

which is maximized by matching $1[\cdot]$ with the highest choice probability. Suppose $P(y_1 = 1 \mid x)$ is the highest. Then, in order to identify β, it is necessary to have

$$P(y_1 = 1 \mid x) \geq P(y_2 = 1 \mid x) \ \ldots \ P(y_1 = 1 \mid x) \geq P(y_J = 1 \mid x)$$

$$\text{iff} \ \ x_1'\beta \geq x_2'\beta \ \ldots \ x_1'\beta \geq x_J'\beta. \tag{2.8}$$

Manski (1975) imposed two assumptions for (2.8): all regressors are continuous, and $u_{i1} \ldots u_{iJ}$ are uncorrelated with one another.

Horowitz (1992) improves on MSC by replacing the indicator function in MSC with a smooth function that is an integral of a kernel. This version of MSC has a faster convergence rate which, however, still falls short of the usual $\sqrt{N}$-rate. Specifically, Horowitz (1992) proposes to maximize [compare to (2.4)]

$$(1/N) \sum_i (2y_i - 1) \cdot J(x_i'b/h)] \tag{2.9}$$

wrt b where h is a smoothing parameter, $J(-\infty) = 0$ and $J(\infty) = 1$. Thus as $h \to 0$, J takes either 0 or 1 because $x_i'b/h$ goes to $\pm\infty$ depending on $\text{sgn}(x_i'b)$. See also Horowitz (1993a, p. 66) for a quick overview.

Matzkin (1992) proposes estimating both the regression function and the error term distribution for the binary response model. Matzkin (1991) suggests estimating the regression function for the multinomial choice model when the error term distribution is parametric, and then relaxes the latter restriction in Matzkin (1993). Das (1991) applies MSC to a decision on whether to idle a cement kiln. Horowitz (1993b) applies MSC and its smoothed version to a binary trip mode choice problem. Horowitz (1993a) reviews the literature on semiparametric methods for binary response models.

9.3 Median Regression for Ordered Discrete Response (MEO)

Generalizing MSC "horizontally" (MSC with a multinomial choice model may be called a "vertical" generalization of MSC), suppose

$$y^* = x'\beta + u, \ \ y = \sum_{r=1}^{R-1} 1[y^* \geq \gamma_r], \ \ \gamma_0 = -\infty, \ \ \text{Med}(y^* \mid x) = x'\beta; \tag{3.1}$$

that is,

$$y = r - 1 \ \text{if} \ \gamma_{r-1} \leq y^* < y_r, \ \ \gamma_0 = -\infty, \ \ \gamma_R = \infty, \ \ r = 1 \ldots R. \tag{3.2}$$

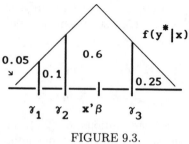

FIGURE 9.3.

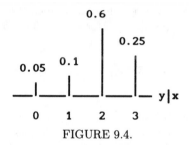

FIGURE 9.4.

One example is income surveys: instead of continuous income y^*, a group representative value $y = r$ is reported if y^* belongs to an interval $[\gamma_r, \gamma_{r+1})$. Here, y takes an integer between 0 to $R - 1$. Another example is y^* being a continuous demand for cars. Since only positive integers can be realized, the observed y is an (integer) transformation of y^*.

Owing to $\text{Med}(y^* \mid x) = x'\beta$ and the transformation $\tau(\cdot) = \sum_{r=1}^{R-1} 1[\cdot \geq \gamma_r]$, $\text{Med}(y \mid x) = \sum_{r=1}^{R-1} 1[x'\beta \geq \gamma_r]$, the representative value of the group where $x'\beta$ belongs; recall the paragraph following (2.1). To see this graphically, examine Figures 9.3 and 9.4: the density of $y^* \mid x$ is chopped into four pieces at γ_r's in Figure 9.3 to yield Figure 9.4. Piling up the probability masses at γ_r's gives the probability distribution of $y \mid x$. Since $\text{Med}(y^* \mid x) = x'\beta$ implies that the probability sum up to $x'\beta$ is equal to 0.5 in Figure 9.3, the probability sum up to the category where $x'\beta$ belongs in Figure 9.4 should be equal to or greater than 0.5. That is, $x'\beta$ is in the category $y = 2$, $\sum_{j=0}^{1} P(y = j \mid x) < 0.5$ and $\sum_{j=0}^{2} P(y = j \mid x) > 0.5$, which imply $\text{Med}(y \mid x) = 2$. Based upon this idea, Lee (1992a) suggests minimizing the following wrt b and $c \equiv (c_1, \ldots, c_{R-1})'$:

$$(1/N) \sum_i \left| y_i - \sum_{r=1}^{R-1} 1[x'b \geq c_r] \right|. \tag{3.3}$$

Call the estimator MEO (median regression estimator for ordered discrete response).

In the income survey example, $\gamma \equiv (\gamma_1, \ldots, \gamma_{R-1})'$ is known, while only some bounds on γ can be known in the car demand example. Also there exist cases where γ is completely unknown. Depending on the case under consideration, the identifiability of the positive scale factor [λ in (2.5)] varies. If γ is completely unknown, then the scale factor cannot be identified. Suppose y^* is a continuous demand for cassettes and

$$y = r - 1 \quad \text{if } \gamma_{r-1} \leq y^* < \gamma_r, r = 1 \ldots R$$

where $r - 1 < \gamma_r \leq r$, $\gamma_0 = -\infty$, $\gamma_R = \infty$, $r = 1 \ldots R - 1;$ \hfill (3.4)

that is, if the realized demand y is 2, then y^* must be between 1 and 3. Using the bounds on γ in (3.4), Lee (1992a) draws the following bounds on the scale factor λ good for any estimation methods including MLE:

for $R = 3$, $0 < \lambda < \infty$; for $R > 3$, $(R-3)/(R-1) < \lambda \leq (R-2)/(R-3)$.
\hfill (3.5)

For instance, $1/2 < \lambda \leq 3/2$ with $R = 5$. As $R \to \infty$, λ gets pinned down.

MEO is likely to work better when γ_r is known. Particularly, suppose that γ_r's are equally spaced, say $\gamma_r - \gamma_{r-1} = s$, which is stronger than (3.4). Then by subtracting γ_1 from $\gamma_{r-1} \leq y^* < \gamma_r$, $r = 2 \ldots R^{-1}$, and dividing this by s, we get $y = [y^*/s]$, where $[y^*]$ is the integer part of y^*, except when y^*/s is in the tails. Thus for all y^*, we have

$$y = \max\{0, \min(R - 1, [y^*/s])\}. \hfill (3.6)$$

In this case, the sample minimand (3.3) becomes ($a \equiv b/s$)

$$(1/N) \sum_i |y_i - \max\{0, \min(R - 1, [x_i'a])\}|. \hfill (3.7)$$

The several points mentioned for MSC in comparison to MLE also apply to MEO. A computation algorithm has been suggested in Pinske (1993, appendix). Melenberg and Van Soest (1995b) smooth (3.3) following the idea of Horowitz (1992), and estimate household "equivalence scale" based upon ordered data.

9.4   Mode Regression

Mean and median are not the only measures of location. Another well-known location measure is the mode. Suppose we have a sample with its dependent variable truncated from below at c:

$$y^* = x'\beta + u, \ (x', y^*)' \text{ is observed only when } y^* = x'\beta + u > c. \hfill (4.1)$$

Denoting the truncated dependent variable by y, $E(y \mid x) \neq x'\beta$ even if $E(y^* \mid x) = x'\beta$. Consider maximizing

$$E\{1[|y^* - q(x)| < w]\} = E_x[F\{q(x) + w\} - F\{q(x) - w\}]$$

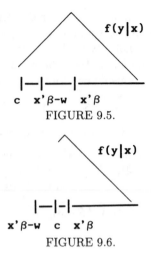

FIGURE 9.5.

FIGURE 9.6.

wrt $q(x)$, where $w > 0$ and F is the distribution function of $y^* \mid x$. The optimal choice $q^*(x)$ is the location measure of $y^* \mid x$ such that the interval $[q^*(x) - w, \; q^*(x) + w]$ captures the most probability mass under $f(z) \equiv dF(z)/dz$. Assuming that f is strictly unimodal and symmetric around the mode up to $\pm w$, the mode maximizes $E\{1[|y^* - q(x)| < w]\}$. That is, the optimal predictor $q^*(x)$ is $\text{Mode}(y^* \mid x)$.

Suppose $\text{Mode}(y^* \mid x) = x'\beta$. The maximizer $\hat{q}(x)$ of $E\{1[|y - q(x)| < w]\}$ is not necessarily $x'\beta$. Consider Figures 9.5 and 9.6. In Figure 9.5, truncation is done in the interval $(-\infty, x'\beta - w)$ and we still capture the most probability mass under $f(y \mid x)$ with the interval $x'\beta \pm w$; $\hat{q}(x)$ is $x'\beta = \text{Mode}(y \mid x) = \text{Mode}(y^* \mid x)$. In Figure 9.6, the truncation takes place somewhere in $(x'\beta - w, \infty)$, then $\hat{q}(x) = c + w$; that is, we capture the most probability mass under $f(y \mid x)$ with the interval $[c, c + 2w]$. Combining the two cases, the optimal predictor $\hat{q}(x)$ in the truncated case is

$$\max\{\text{Mode}(y^* \mid x), c + w\} = \max\{x'\beta, c + w\}. \tag{4.2}$$

Based on the above observation, Lee (1989) proposes the mode regression estimator for the truncated regression by maximizing

$$(1/N) \sum_i 1[|y_i - \max(x_i'b, c + w)| < w]. \tag{4.3}$$

Letting $c \to -\infty$, we get the usual case of no truncation where we maximize

$$(1/N) \sum_i 1[|y_i - x_i'b| < w]. \tag{4.4}$$

The mode is arguably the most attractive location parameter when the distribution is asymmetric. Also as shown below, the mode is robust to outliers (or "high-influence" observations).

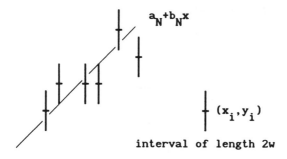

interval of length 2w

FIGURE 9.7.

Consider a simple case for (4.4) with $x'b = a + bx$ where x is a scalar. In Figure 9.7, each datum is given a vertical interval with the length $2w$. Then a_N and b_N are obtained by piercing the most number of intervals. Several points are in order:

(i) Owing to the nondifferentiability of the indicator function, the asymptotic distribution of the mode estimator is not known. It can be proven that the mode estimator is $N^{1/3}$-consistent with a nonpractical form of asymptotic distribution. Also the computation requires an algorithm not using gradients; see Pinske (1993, appendix) for instance.

(ii) For a random variable z, the optimal predictor maximizing $E\{1[|z - q| < w]\}$ may not be exactly the same as $\text{Mode}(z)$, which is clear if we assume f_z is asymmetric. The optimal predictor is the middle value of the optimal interval of size $2w$ which captures the most probability under f_z.

(iii) Although we assumed the symmetry up to $\pm w$, it is not necessary under independence of u from x. Without symmetry, only the slope coefficients are identified.

(iv) The choice of w is open. Analogously to the kernel density estimation with the uniform kernel, a small w may reduce the bias of the estimator and a large w may make the estimator more efficient.

(v) In (4.3), each datum is given an equal weight of 1 or 0, no matter how large the error $y - \max(x'b, c + w)$ may be. Hence an outlier cannot influence the estimator by more than its share $1/N$. Therefore, the mode regression is robust. It can be shown that, depending on data configuration, the mode regression for (4.4) can resist up to 50% data contamination, the best one can hope for.

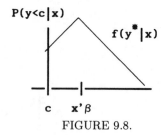

FIGURE 9.8.

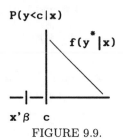

FIGURE 9.9.

9.5 Censored Least Absolute Deviations Estimator (CLAD)

Consider a median regression model with y censored from below at c:

$$y = \max(y^*, c) = \max(x'\beta + u, c) \quad \text{and} \quad \text{Med}(y^* \mid x) = x'\beta; \qquad (5.1)$$

see the next section for how to transform the data when censoring is from above. Since y^* is transformed into y by a nondecreasing function $\tau(\cdot) = \max(\cdot, c)$,

$$\text{Med}(y \mid x)\{= \text{Med}(\tau(y^*) \mid x) = \tau(\text{Med}(y^* \mid x))$$

$$= \tau(x'\beta, c) = \max(x'\beta, c). \qquad (5.2)$$

To understand this graphically, imagine $f^*_{y|x}$ centered at $x'\beta$. Censoring cuts $f^*_{y|x}$ at c and piles up $P(y^* \leq c|x)$ at c. In Figure 9.8, $c < x'\beta$ and the censoring takes place at the lower tail left of $x'\beta$, which implies $P(y \leq x'\beta \mid x) = 1/2 \Leftrightarrow \text{Med}(y \mid x) = x'\beta$. In Figure 9.9, $c \geq x'\beta$ and c is the smallest value such that $P(y \leq c \mid x) \geq 1/2$, thus $\text{Med}(y \mid x) = c$ in Figure 9.9. Combine these two cases to get (5.2). Since an absolute value loss function is minimized at the median, Powell (1984) proposed the censored least absolute deviation estimator (CLAD) minimizing

$$(1/N) \sum_i |y_i - \max(x_i'b, c)|. \qquad (5.3)$$

Compared with LSE and the mode regression, the optimand is one degree less smoother than that of LSE and one degree more smoother than that of the mode regression. Recall that the asymptotic distribution of an estimator b_N minimizing $(1/N) \sum_i q(b)$ is

$$\sqrt{N}(b_N - \beta) =^d N(0, E^{-1}(\partial^2 q/\partial b \partial b') \cdot E\{(\partial q/\partial b)(\partial q/\partial b')\}$$

$$\cdot E^{-1}(\partial^2 q/\partial b \partial b')). \tag{5.4}$$

Although (5.3) is not differentiable, still the asymptotic distribution of CLAD can be obtained analogously to yield

$$\sqrt{N}(b_{\text{CLAD}} - \beta) =^d N(0, (1/4) \cdot A^{-1} \cdot E\{xx'1[x'\beta > c]\} \cdot A^{-1}),$$

$$A \equiv E\{f_{u|x}(0) \cdot xx'1[x'\beta > c]\}. \tag{5.5}$$

Comparing (5.4) and (5.5), the matrix A takes the place of the second derivative of $E|y - \max(x'b, c)|$ at $b = \beta$. Suppose $x'b > c$ to get $E|y - x'b|$. The first derivative wrt b is $-E\{\text{sgn}(y - x'b) \cdot x\}$. Although $\text{sgn}(u)$ is not differentiable, it changes by 2 (from -1 to 1) at $u = 0$. Then taking $2\delta_0(u)$ as the derivative of $\text{sgn}(u)$, where $\delta_0(u) = 1$ if $u = 0$ and 0 otherwise, the second derivative of $E|y - x'b|$ wrt b at $b = \beta$ is

$$E\{2\delta_0(u) \cdot xx'\} = 2 \cdot E\{f_{u|x}(0) \cdot xx'\},$$

which explains A. Under independence of u from x, the variance becomes

$$(1/4) \cdot \{f_u(0)\}^{-2} \cdot E^{-1}\{xx'1[x'\beta \geq c]\}. \tag{5.6}$$

Denoting the residuals by r_i, a sample estimate for A in (5.5) is

$$(1/N) \sum_i \{1[0 < r_i < h]/h\} \cdot x_i x_i' \cdot 1[x_i b_N > c], \tag{5.7}$$

where h is a smoothing parameter; $f_u(0)$ is estimated nonparametrically.

If the censoring percentage is high, then a αth quantile with $\alpha > 0.5$ may be more meaningful than the median. From $y^* = x'\beta + u$,

$$Q_\alpha(y^* \mid x) = x'\beta + Q_\alpha(u \mid x), \tag{5.8}$$

where $Q_\alpha(\cdot \mid \cdot)$ denotes the αth conditional quantile. If some components of x appear in $Q_\alpha(u \mid x)$, then the corresponding parameter β_j's are not identified. If $Q_\alpha(u \mid x) = 0$ for some α, say 0.75, then $Q_{0.75}(y^* \mid x) = x'\beta$. Suppose the first component x_1 of x is 1 as usual. If u is independent of x, then $Q_\alpha(u \mid x) = Q_\alpha(u)$, an unknown function of α. In this case,

$$Q_\alpha(y^* \mid x) = 1 \cdot \{\beta_1 + Q_\alpha(u)\} + x_2\beta_2 + \cdots + x_k\beta_k; \tag{5.9}$$

the slope parameters are identified for any α. In short, $Q_\alpha(y^* \mid x) = x'\beta$ is valid for any α under the independence of u from x, while it is valid only for the α satisfying $Q_\alpha(u \mid x) = 0$ if the independence does not hold.

Using (5.8) for the censored model $y = \max(x'\beta + u, c)$, we get

$$Q_\alpha(y \mid x) = \max(x'\beta, c). \tag{5.10}$$

Then the "censored αth quantile estimator" (Powell, 1986b) minimizes

$$(1/N) \sum_i \{y_i - \max(x_i'b, c)\} \cdot \{\alpha - 1[y_i - \max(x_i'b, c) < 0]\}. \tag{5.11}$$

The asymptotic variance matrix is

$$\{\alpha(1-\alpha)\}^{-1} \cdot E^{-1}\{f_{u|x}(Q_\alpha(u \mid x)) \cdot xx'1[x'\beta > c]\} \cdot E\{xx'1[x'\beta > c]\}$$

$$\cdot E^{-1}\{f_{u|x}(Q_\alpha(u \mid x)) \cdot xx'1[x'\beta > c]\}. \tag{5.12}$$

Under the independence of u from x, this matrix becomes

$$\{\alpha(1-\alpha)\}^{-1} \cdot \{f_u(Q_\alpha(u))\}^{-2} \cdot E^{-1}\{xx'1[x'\beta > c]\}; \tag{5.13}$$

(5.11) and (5.12) include, respectively, (5.5) and (5.6) as a special case when $\alpha = 1/2$ and $Q_\alpha(u \mid x) = 0$.

A more efficient two stage version of CLAD was proposed by Newey and Powell (1990). Hall and Horowitz (1990) discuss how to choose h in CLAD and other related estimators. CLAD has been applied by Horowitz and Neumann (1987, 1989) to duration data, by Lee (1995a) to female labor supply data and by Melenberg and Van Soest (1995a) to vacation expenditure data. The censored quantile estimator is applied to a returns-to-schooling problem by Buchinski (1994).

9.6   Symmetry-Based Estimators (STLS, SCLS)

In the linear model $y^* = x'\beta + u$, LSE satisfies the orthogonality condition $E(ux) = 0$, which is implied by $E(u \mid x) = 0$. Consider a truncation in $u \mid x$, which ruins the orthogonality condition. Assuming that $f_{u|x}$ is symmetric and unimodal around 0, Powell (1986a) suggests one way to restore the orthogonality condition by "trimming" the error density. Suppose $x'\beta > c$ and $y^* \mid x$ is truncated from below at c. Then $u \mid x$ is truncated from below at $-x'\beta + c$. Let y denote the truncated version of y^*. If we trim $u \mid x$ from above at $x'\beta - c$, then $u \cdot 1[|u| < x'\beta - c]$ will be symmetric; see Figure 9.10 (it may be easier to see this point with $c = 0$). Therefore

$$E_x[x \cdot \{E_{u|x} u \cdot 1[|u| < x'\beta - c]\}] = E\{xu \cdot 1[|u| < x'\beta - c]\} = 0. \tag{6.1}$$

Trimming $u \mid x$ at $\pm(x'\beta - c)$ is equivalent to trimming $y \mid x$ at c and $2x'\beta - c$; see Figure 9.11.

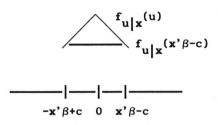

FIGURE 9.10.

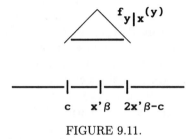

FIGURE 9.11.

One minimand yielding an unique minimizer satisfying (6.1) as the asymptotic first-order condition is

$$(1/N) \sum_i \{y_i - \max(0.5y_i + 0.5c, x_i b)\}^2. \tag{6.2}$$

When $b = \beta$, the terms with $0.5y_i + 0.5c < x_i' \beta$ yield $\{y_i - x_i \beta\}^2$. Powell names this estimator *symmetrically trimmed least squares estimator* (STLS). The asymptotic distribution of STLS is

$$\sqrt{N}(b_{\text{STLS}} - \beta) =^d N(0, (W - V)^{-1} Z (W - V)^{-1}), \tag{6.3}$$

where

$$W = E\{1[|u| < x'\beta - c]xx'\},$$
$$V = E[1[x'\beta > c]2(x'\beta - c)\{f_{u|x}(x'\beta - c)/F_{u|x}(x'\beta - c\}xx'],$$
$$Z = E\{1[|u| < x'\beta - c]u^2 xx'\}.$$

An examination of $W - V$ will reveal that $W - V$ is

$$E[1[x'\beta > c]xx' \cdot \{\text{area under } f_{u|x} \text{ between}$$

$$\pm(x'\beta - c) \text{ above } f_{u|x}(x'\beta - c)\}],$$

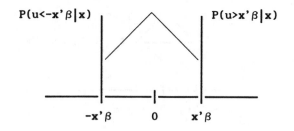

FIGURE 9.12.

where $\{\cdots\}$ is the triangular area in Figure 9.10; $F_{u|x}(x'\beta - c)$ is a normalizing constant for the truncation. For the nontruncated case, take $c = -\infty$. Then the covariance matrix becomes that of LSE. Estimating V requires a nonparametric technique as in (5.7) due to $f_{u|x}$.

Following STLS, Powell (1986a) proposes another estimator for y^* censored from below at c; let $c = 0$ for simplicity. If $c \neq 0$ or y^* is censored from above, we can transform the data such that $y = \max(x'\beta+u, 0)$ holds. For instance, suppose y^* is censored from above at γ: $y = \min(x'\beta + u, \gamma)$. Then subtract γ from both sides and multiply both sides by -1 to get

$$\gamma - y = -\min(x'\beta + u - \gamma, 0) = \max(-x'\beta - u + \gamma, 0).$$

Absorb γ into the intercept in $x'\beta$, and defining $v \equiv -u$,

$$\gamma - y = \max((-x)'\beta + v, 0). \tag{6.4}$$

Transforming y into $\gamma - y \equiv \tilde{y}$ and x into $-x \equiv \tilde{x}$, $\tilde{y} = \max(\tilde{x}'\beta + v, 0)$ holds, where $\imath$ has the same distribution function as u due to the symmetry. After the estimates for this transformed model are obtained, simply add γ to the intecept estimate to get the estimates for the original model.

With $x'\beta > 0$, censoring $f_{u|x}$ at $-x'\beta$ in the lower tail results in a probability mass piling at $-x'\beta$, while truncation simply cuts off $f_{u|x}$ at $-x'\beta$. The idea is to censor the upper tail of $f_{u|x}$ at $x'\beta$ to restore the symmetry; see Figure 9.12 and compare it to Figure 9.10 with $c = 0$. Then with the assumption of symmetry, the following orthogonality condition holds:

$$E_x[x \cdot 1[x'\beta > 0]E_{u|x}\{u \cdot 1[|u| < x'\beta]$$
$$+ x'\beta(1[u > x'\beta] - 1[u < -x'\beta])\}] = 0, \tag{6.5}$$

where u in $E(xu)$ is replaced by $x'\beta \cdot \text{sgn}(u)$ if $|u| > x'\beta$. The symmetry of $f_{u|x}$ is sufficient for (6.5).

One minimand yielding a unique minimum with (6.5) as its asymptotic first order condition is

$$(1/N)\sum\{y_i - \max(0.5y_i, x_i'b)\}^2 + (1/N)\sum 1[y_i > 2x_ib]$$

$$\cdot [(0.5 y_i)^2 - \{\max(0, x_i' b)\}^2]. \tag{6.6}$$

Minimizing this gives the *symmetrically censored least squares estimator* (SCLS). Observe that the first term is the minimand for STLS when $c = 0$. For censored cases, there is more information than truncated cases, and the second term makes use of the extra information. The asymptotic distribution of SCLS is attractive in that it does not require density estimation as STLS and CLAD do. The asymptotic distribution is

$$\sqrt{N}(b_{\mathrm{SCLS}} - \beta) =^d N(0, H^{-1} \cdot E\{1[x'\beta > 0] \cdot \min(u^2, (x'\beta)^2) \cdot xx'\} \cdot H^{-1}),$$

$$H \equiv E\{1[|u| < x'\beta] \cdot xx'\}. \tag{6.7}$$

The middle term in the variance matrix again indicates that u^2 is replaced by $(x'\beta)^2$ whenever $|u| > x'\beta$.

In Chapter 4, we discussed a duration model under a Weibull specification where the duration y has the Weibull distribution

$$F(y) = 1 - \exp(-\theta y^\alpha). \tag{6.8}$$

When $(x', y)'$ is fully observed, assuming $\theta = \exp(x'\beta)$, we showed

$$\ln(y) = x'\beta/(-\alpha) + u \tag{6.9}$$

to which LSE is applicable. When y is censored from above, the LSE is not valid but the Weibull MLE is. Usually duration has a density with a long right tail. The $\ln(y)$ in (6.9) can be viewed as a way to transform y so that $\ln(y)$ given x may become symmetric. If this is the case, then we can use SCLS for (6.9) with y censored from above; the log-linear model does not necessarily have to be motivated by Weibull.

Powell (1986a) suggests an iterative scheme to get b_{SCLS}. Start with an initial estimate b_0, say LSE, and then iterate the following, updating b_0 until $|b_1 - b_0|$ gets small:

$$b_1 = \left\{ \sum_i 1[x_i' b_0 > 0] \cdot x_i x_i' \right\}^{-1} \sum_i \{1[x_i b_0 > 0] \cdot \min(y_i, 2x_i b) \cdot x_i\}. \tag{6.10}$$

This does not guarantee global convergence. Also the matrix to be inverted may not be invertible; however, from our limited experience, (6.10) seems to work well. Honoré (1992) extends STLS and SCLS to panel data with truncated or censored dependent variables. Newey (1991) proposes semiparametrically efficient versions of STLS and SCLS by taking the full advantage of the symmetry assumption. Lee (1995a) applies SCLS to female labor supply data.

9.7 Partial-Symmetry-Based Estimators (QME, WME)

Generalizing the mode regression further, Lee (1993a) suggests smoothing the mode regression estimator to obtain a $\sqrt{N}$-consistent estimator good for $y^* = x'\beta + u$ truncated from below at c. Assume

(x', y^*) is observed only when $y^* = x'\beta + u > c$, $\text{Mode}(y^* \mid x) = x'\beta$,

$\ln(f_{u|x})$ is strictly concave, and $f_{u|x}$ is symmetric over $\pm w$;   (7.1)

the concavity of $\ln(f_{u|x})$ is called the "strong unimodality" of $f_{u|x}$ (Dharmadhikari and Joag-dev, 1988). Denoting the truncated version of y^* as y, the estimator named *quadratic mode regression estimator* (QME) maximizes

$$(1/N) \sum_i [w^2 - \{y_i - \max(x_i'b, c + w)\}^2] \cdot 1[|y_i - \max(x_i'b, c + w)| < w]$$

$$= (1/N) \sum_i \max[w^2 - \{y_i - \max(x_i'b, c + w)\}^2, 0],   (7.2)$$

where w is to be chosen by the researcher such that $w > c$. The idea of QME is similar to that of STLS: $y^* < c \Leftrightarrow u > c - x'\beta$, which implies $u > c - w$ if $x'\beta > w$. Thus, QME trims $f_{u|x}$ at $c - w$ and $w - c$ when $x'\beta > w$, while STLS trims $f_{u|x}$ at $c - x'\beta$ and $x'\beta - c$ when $x'\beta > c$. The trimming points $\pm(c - w)$ in QME are constants, while the trimming points $\pm(c - x'\beta)$ in STLS vary as x varies.

The asymptotic distribution of QME is analogous to that of STLS:

$$\sqrt{N}(b_{\text{QME}} - \beta) =^d N(0, (W - V)^{-1} Z (W - V)^{-1}),   (7.3)$$

where

$$Z = E\{1[x'\beta > c + w, |u| < w] \cdot xx'u^2\},$$
$$W = E\{1[x'\beta > c + w, |u| < w] \cdot xx'\},$$
$$V = E[1[x'\beta > c + w] \cdot \{2w f_{u|x}(w)/(1 - F_{u|x}(c - x'\beta))\} \cdot xx'],$$

where $W - V$ is xx' times a triangle area analogous to that in Figure 9.10. Letting $c = -\infty$ for no truncation, (7.2) becomes

$$(1/N) \sum_i \{w^2 - (y_i - x_i'b)^2\} \cdot 1[|y_i - x_i'b| < w].   (7.4)$$

So QME is similar to the mode regression but it imposes the quadratic weight on the data with $|y_i - x_i'b| < w$. Maximizing (7.4) is equivalent to minimizing

$$(1/N) \sum_i \{(y_i - x_i'b)^2 1[|y_i - x_i'b| < w] + w^2 1[|y_i - x_i'b| \geq w]\},   (7.5)$$

which is the regression version of Huber's "trimmed (or skipped) mean" (Huber, 1981). This was suggested to robustify LSE, limiting the effect of outliers by trimming the residuals at $\pm w$.

While STLS becomes LSE with no truncation, QME becomes trimmed LSE with no truncation. Hence, QME has two advantages over STLS. One is the weaker symmetry assumption on $f_{u|x}$ and the other is the robustness inherited from the trimmed LSE. On the other hand, STLS has two advantages over QME. One is not needing to choose w as in QME, and the other is that STLS appears to be more efficient in simulation studies.

In (7.5), we saw a way of robustifying LSE. One trade-off of doing it is the loss of efficiency by not using the quantitative information in the data with $|y - x'b| \geq w$. Another robust estimator is obtained by combining the mean and median regressions. Consider minimizing

$$(1/N) \sum_i [(1/2)(y_i - x_i'b)^2 1[|y_i - x_i'b| < w]$$

$$+ (w|y_i - x_i'b| - w^2/2) \cdot 1[|y_i - x_i'b| \geq w]]. \tag{7.6}$$

The information in $|y - x'b| \geq w$ is not thrown away; instead its impact is downgraded by taking the absolute value, not the squared value. If we divide (7.6) by w and let $w \to 0$, then (7.6) becomes the minimand of LAD $(1/N) \sum_i |y_i - x_i'b|$. Recall that CLAD is applicable to censored regression by modifying LAD. So it seems possible to modify (7.6) for censored regression. Indeed, that is the case as explained in Lee (1992b).

For $y = \max(x'\beta + u, 0)$ where $f_{u|x}$ is symmetric around 0 up to $\pm w$, Lee (1992b) suggests to minimize

$$(1/N) \sum_i [(1/2)\{y_i - \max(x_i'b, w)\}^2 1[|y_i - \max(x_i'b, w)| < w]$$

$$+ \{w|y_i - \max(x_i'b, w)| - w/2\} \cdot 1[|y_i - \max(x_i', b, w)| \geq w]]; \tag{7.7}$$

the first term in (7.7) is (7.2) for QME. The estimator is called the *winsorized mean estimator* (WME). The asymptotic first-order condition is

$$E_x[1[x'\beta > w] \cdot x \cdot E_{u|x}\{1[|u| < w]u + (1[u > w] - 1[u < -w])w\}] = 0; \tag{7.8}$$

given $x'\beta > w$, we replace u by $-w$ if $u < -w$ and by w if $u > w$. This is analogous to SCLS, where u is replaced by $-x'\beta$ if $u < -x'\beta$ and by $x'\beta$ if $u > x'\beta$ given $x'\beta > 0$; recall Figure 9.12. While SCLS needs the symmetry of $f_{u|x}$ up to $\pm x'\beta$, WME needs it only up to $\pm w$. WME is consistent for β and

$$\sqrt{N}(b_{WME} - \beta) =^d N(0, C^{-1} \cdot E\{1[x'\beta > w] \cdot \min(u^2, w^2) \cdot xx'\} \cdot C^{-1}),$$

$$C \equiv E\{1[x'\beta > w, |u| < w] \cdot xx'\}. \tag{7.9}$$

Lee (1995a) applies WME to female labor supply data.

Comparing WME to CLAD, WME includes CLAD as a limiting case when $w \to 0$. But differently from CLAD, (7.9) has no $f_{u|x}$, so WME does not require a nonparametric technique for its variance matrix estimation. Later in this chapter, we will show how to use w to test the symmetry and heteroskedasticity assumptions in the censored and truncated models. So the fact that WME (and QME) is consistent for various w's is not necessarily a disadvantage of WME (and QME).

9.8   Single-Index Models

In a wide sense, a "single-index model" is a regression model where x affects y only through $x'\beta$; several functions of the conditional distribution of $y \mid x$ may depend on x only through $x'\beta$. In a narrower sense, the definition is restricted to models with $E(y \mid x) = E(y \mid x'\beta)$, which may be called "mean index sufficiency;" $V(y \mid x) = V(u \mid x)$ may be unrestricted, or restricted to $V(u \mid x) = V(u \mid x'\beta)$ or even to independence between u and x depending on the estimator at hand. We will use the latter definition in this section. Consider a single index model

$$y = G(x'\beta) + u, \quad E(u \mid x) = 0, \tag{8.1}$$

and $G(\cdot)$ is an uknown bounded function. Several examples have been given already in this chapter. Since any nonzero rescaling of β can be absorbed into the unknown G, we can identify β only up to a scale.

Stoker (1986) proposes an estimation idea for (8.1). Suppose x has a continuous density $f(x)$, which is zero on the boundary of x that is denoted as ∂X. Then

$$E_x\{\partial G(x'\beta)/\partial x\} = E[\{-\partial \ln(f(x))/\partial x\} \cdot y], \tag{8.2}$$

the proof of which is as follows. Using integration by parts, the lhs is

$$\int \partial G(x'\beta)/\partial x f(x)dx = G(x'\beta)f(x)\big|_{\partial X} - \int G(x'\beta)\partial f(x)/\partial x dx; \tag{8.3}$$

the first term on the rhs disappears. Then, (8.2) becomes

$$\int \{\partial G(x'\beta)/\partial x \cdot f(x)\}dx = -\int [G(x'\beta)$$

$$\cdot\{(\partial f(x)/\partial x)/f(x)\}] \cdot f(x)dx \tag{8.4}$$

$$= -\int [\{G(x'\beta) + u\} \cdot \{(\partial f(x)/\partial x)/f(x)\}] \cdot f(x)dx \quad [\text{for } E(u \mid x) = 0]$$

$$= \int [y \cdot \{(\partial f(x)/\partial x)/f(x)\}] \cdot f(x)dx. \tag{8.5}$$

Since

$$E_x\{\partial G(x'\beta)/\partial x\} = E_x\{dG(x'\beta)/d(x'\beta)\} \cdot \beta \equiv \gamma\beta \qquad (8.6)$$

(note that γ depends on β), combining this with (8.2), we get

$$\gamma\beta = E[\{-\partial \ln(f(x))/\partial x\} \cdot y]. \qquad (8.7)$$

If $f(x)$ is known, then $\gamma\beta$ is estimated by

$$(\gamma\beta)_N = (1/N)\sum_i \{-\partial \ln(f(x_i))/\partial x\} \cdot y_i. \qquad (8.8)$$

This equation links β to the marginal distribution of x, which is some-what strange at first sight. In terms of parametric assumptions, (8.8) is not any better than specifying G, for we have to specify f. Therefore, (8.8) as such was not promising for application. However, it opened a way to semi-nonparametric methods for single-index models; the obvious thing to do is replacing $-\partial \ln(f(x_i))/\partial x$ in (8.8) by a nonparametric estimate. More estimators for (8.1) will appear in the next chapter.

Generalizing the single-index model is a *double-index model* where $E(y \mid x, z) = G(x'\beta, z'\lambda)$. For instance, recall the selection model where $y^* = x'\beta + u$ is observed only when $z'\gamma + v > 0$. Then denoting the joint density of u and v as $f(u, v)$, where (u, v) is independent of (x, z),

$$E(y \mid z, v > -z'\gamma) = x'\beta + \int_{-\infty}^{\infty}\int_{-z'\gamma}^{\infty} uf(u,v)dvdu/A, \qquad (8.9)$$

where $A \equiv \int_{-\infty}^{\infty}\int_{-z'\gamma}^{\infty} f(u,v) \cdot dvdu = P(v > -z'\gamma \mid z)$. Hence, (8.9) is a double-index model. Note that if $f(u, v) = f(u)f(v)$ and $E(u) = 0$, then $E(y \mid x, z, v > -z'\gamma) = x'\beta$ despite the selection process.

A difficulty with double- or multiple-index models, in general, is that the parameters are identified only under exclusion restrictions. For instance, consider $E(y \mid x, z) = G(x'\beta, z'\lambda)$ but with $x = z$ and all nonzero coefficients (no exclusion restriction), which then leads to $E(y \mid x) = G(x'\beta, x'\lambda)$. In this case,

$$\partial E(y \mid x)/\partial x = G_1(x)\beta + G_2(x)\gamma \text{ where } G_j(x) \equiv \partial G(w_1, w_2)/\partial w_j, j = 1, 2$$

$$\Rightarrow E\{\partial E(y \mid x)/\partial x\} = E\{G_1(x)\}\beta + E\{G_2(x)\}\gamma; \qquad (8.10)$$

here β and γ are not separated. If the $(k-1)$th and kth element of x are in $x'\beta$ but not in $z'\lambda$ (an exclusion restriction), then $E\{\partial r(x)/\partial x_j\} = E\{G_1(x)\} \cdot \beta_j$, $j = k-1, k$, and β_k/β_{k-1} is identified. Interested readers may want to see Ichimura and Lee (1991).

9.9 Rank Correlation Estimator (RCE) for Binary Response

So far we have introduced two consistent semiparametric estimators for binary response models: maximum score estimator (MSC) and the single-index estimator. In this section, we examine a $\sqrt{N}$-consistent estimator for the binary response model $y = 1[x'\beta + u \geq 0]$.

Han (1987) proposed a "rank correlation estimator" (RCE) maximizing

$$Q_N(b) \equiv \{1/(N(N-1))\} \cdot \sum_{i \neq j} 1[y_i > y_j, x_i'b > x_j'b]. \tag{9.1}$$

Since multiplying $x'b$ by a positive constant c does not change the ordering $x_i'b > x_j'b$, the parameters for RCE are identified only up to a positive constant. For this reason, we will discuss RCE only for the binary model, although RCE is also applicable to other models of limited dependent variable models (LDV).

The main assumptions in RCE are the independence between x and u and the existence of at least one continuous regressor with unbounded support and an almost everywhere positive density conditional on almost every value of the other regressors. The latter also appeared for MSC, which makes ignoring the ties in (9.1) harmless. The idea of RCE is similar to MSC: if $x_i'\beta > x_j'\beta$, then it is more likely that $y_i > y_j$ than $y_i < y_j$. That is,

$$P(y_i > y_j \mid x_i, x_j) > P(y_i < y_j \mid x_i, x_j) \text{ whenever } x_i'\beta > x_j'\beta. \tag{9.2}$$

So we can estimate β by maximizing the pairwise prediction score.

To understand the idea (9.2) better, rewrite (9.1) as

$$\{(N(N-1))\}^{-1} \cdot \sum_{i<j} \left\{ 1[x_i'b > x_j'b, y_i > y_j] + 1[x_i'b < x_j'b, y_i < y_j] \right\}; \tag{9.3}$$

note that $\sum$ is over $i < j$, not over $i \neq j$. Examine the population version of (9.3) conditional on x_i and x_j:

$$P(y_i = 1, y_j = 0 \mid x_i, x_j) \cdot 1[x_i'b > x_j'b]$$

$$+ P(y_i = 0, y_j = 1 \mid x_i, x_j) \cdot 1[x_i'b < x_j'b]$$

$$P(y_i = 1 \mid x_i)P(y_j = 0 \mid x_j) \cdot 1[x_i'b > x_j'b]$$

$$+ P(y_i = 0 \mid x_i)P(y_j = 1 \mid x_j) \cdot 1[x_i'b < x_j'b]$$

$$= S(-x_i'\beta) \cdot F(-x_j'\beta) \cdot 1[x_i'b > x_j'b] + F(-x_i'\beta) \cdot S(-x_j'\beta) \cdot 1[x_i'b < x_j'b];$$

where $F(t) = P(u \leq t \mid x)$ and $S(t) = 1 - F(t)$. Suppose $x_i'\beta > x_j'\beta$, given x_i and x_j. Since $F(-t)$ is nonincreasing in t and $S(-t)$ is nondecreasing in t,

$$S(-x_i'\beta) \cdot F(-x_j'\beta) > S(-x_i'\beta) \cdot F(-x_i'\beta) > S(-x_j'\beta) \cdot F(-x_i'\beta). \quad (9.4)$$

Thus, when $x_i'\beta > x_j'\beta$, the conditional maximand is maximized by choosing $1[x_i'b > x_j'b]$, not $1[x_i'b < x_j'b]$; the opposite holds analogously when $x_i'\beta < x_j'\beta$. That is, maximizing (9.1) means matching the sign of $x_i'b - x_j'b$ with that of $x_i'\beta - x_j'\beta$; recall that we try to match $x'b$ with $x'\beta$ given x in LSE. In the following, we briefly examine "U-statistics" drawing upon Serfling (1980) and Lehman (1975), and then study the distribution of RCE in Sherman (1993).

Consider an iid sample x_i, $i = 1, \dots, N$, where x_i is a scalar, $E(x) = \mu$ and $V(x) = \sigma^2$. Suppose we want to estimate a parameter θ in the distribution $F(x)$ of x, and θ can be represented as

$$\theta = E\{h(x_1, x_2)\}, \quad (9.5)$$

where $h(x_1, x_2)$ is called a "kernel," which is symmetric around $x_1 = x_2$ line: $h(x_1, x_2) = h(x_2, x_1)$. The name "kernel" is not the same one as we used for kernel nonparametrics. We can estimate θ by averaging the kernel h over the $\binom{N}{2}$ combinations:

$$U_N \equiv \left\{ 1 \Big/ \binom{N}{2} \right\} \sum_{i<j} h(x_i, x_j) = \{2/(N(N-1))\} \cdot \sum_{i<j} h(x_i, x_j) \quad (9.6)$$

which is called a U-statistic. Then it can be shown that,

$$\sqrt{N}(U_N - \theta) =^{as} (2/\sqrt{N}) \sum_i [E\{h(x_i, x_j) \mid x_i\} - \theta] =^d N(0, 4 \cdot \lambda^2),$$

$$\lambda^2 \equiv E[E\{h(x_i, x_j) \mid x_i\} - \theta]^2, \quad (9.7)$$

where the number 2 is due to the two variables in $h(x_i, x_j)$ ("order 2 kernel"). The term $E\{h(x_i, x_j) \mid x_i\}$ is called the projection of $h(x_i, x_j)$ on x_i. If λ is zero, then the limiting distribution to U_N is not normal. It is possible to extend the second-order kernel $h(x_i, x_j)$ to the third order $h(x_i, x_j, x_k)$ or higher; see Serfling (1980) for more.

If $h(x_i, x_j)$ is not symmetric and we have

$$A_N \equiv \{N(N-1)\}^{-1} \cdot \sum_{i \neq j} h(x_i, x_j) = \{N(N-1)\}^{-1} \cdot \sum_{i \neq j} h(x_j, x_i) \quad (9.8)$$

in (9.6), then we can "symmetrize" A_N:

$$A_N = (A_N + A_N)/2 = \{N(N-1)\}^{-1} \cdot \sum_{i \neq j} \{h(x_i, x_j) + h(x_j, x_i)\}/2$$

$$\equiv \{N(N-1)\}^{-1} \cdot \sum_{i\neq j} g(x_i, x_j) = \{2/(N(N-1))\}^{-1} \cdot \sum_{i<j} g(x_i, x_j), \quad (9.9)$$

where $g(x_i, x_j) \equiv \{h(x_i, x_j) + h(x_j, x_i)\}/2$ is symmetric.

As an example, consider estimating $E^2(x) \equiv \mu^2$. One way to estimate μ^2 is $\bar{x}^2$; with the δ-method, $\sqrt{N}(\bar{x}^2 - \mu^2) =^d N(0, (2\mu)^2 \sigma^2)$. Another way is

$$Z_N \equiv \{2/(N(N-1))\} \cdot \sum_{i<j} x_i x_j; \quad (9.10)$$

$x_i x_j$ is symmetric. Since $E(Z_N) = \mu^2$, Z_N is unbiased for μ^2. Observe that

$$E(x_i x_j \mid x_i) = x_i E(x_j) = x_i \mu.$$

The expression corresponding to λ^2 in (9.7) is

$$E(x_i \mu - \mu^2)^2 = \mu^2 E(x_i - \mu)^2 = \mu^2 \sigma^2.$$

Thus,

$$\sqrt{N}(Z_N - \mu^2) =^d N(0, 4 \cdot \mu^2 \sigma^2)$$

which is the same as the asymptotic distribution of $\sqrt{N}(\bar{x}^2 - \mu^2)$.

It is clear that $Q_N(b)$ in (9.1) is of the form of a U-statistic with the second-order kernel, which is not symmetric. $Q_N(b)$, however, is different from U_N in that $Q_N(b)$ involves b. So if we view $Q_N(b)$ as a process indexed by b, then we can call $Q_N(b)$ a "U-process" [Nolan and Pollard (1987, 1988) and Sherman (1994)]. $Q_N(b)$ has an indicator function not differentiable as in MSC and the mode regression. But $Q_N(b)$ has double sums, and one sum smooths the indicator function to yield $\sqrt{N} \cdot Q_N(b) \cong (1/\sqrt{N}) \sum_i q(b)$, while the other provides a CLT to $(1/\sqrt{N}) \sum_i q(b)$. This is the key idea for the $\sqrt{N}$-consistency and the distribution of RCE.

Let $z_i \equiv (x_i', y_i)'$. Sherman (1993) rewrites $Q_N(b)$ into

$$Q_N(b) = Q(b) + N^{-1} \sum_i [E\{1[y_i > y_j, x_i'b > x_j'b] \mid z_i\} - Q(b)]$$

$$+ N^{-1} \sum_j [E\{1[y_i > y_j, x_i'b > x_j'b] \mid z_j\} - Q(b)] \quad (9.11)$$

$$+ \{N(N-1)\}^{-1} \sum_{i\neq j} r(z_i, z_j, b),$$

where $Q(b) \equiv E\{1[y_i > y_j, x_i'b > x_j'b]\}$ and

$$r(z_i, z_j, b) \equiv 1[y_i > y_j, x_i'b > x_j'b] - [E\{1[y_i > y_j, x_i'b > x_j'b] \mid z_i\} - Q(b)]$$

$$- [E\{1[y_i > y_j, x_i'b > x_j'b] \mid z_j\} - Q(b)] + Q(b).$$

In (9.11), the first term $Q(b)$ yields the second-order matrix as in (5.4), the middle two terms yield an asymptotic distribution, and that the last term with $r(\cdot)$ is $o_p(1/N)$ and negligible.

Set $\beta_k = 1$ without loss of generality, and define

$$x_{kc} \equiv (x_1 \ldots x_{k-1})', \quad \beta_{kc} \equiv (\beta_1 \ldots \beta_{k-1})';$$

$\text{sgn}(\beta_k)$ can be treated as known, for it is estimated at a rate infinitely faster than $\sqrt{N}$. Sherman (1993) proves

$$\sqrt{N}(b_{kc} - \beta_{kc}) \cong N(0, V^{-1}\Delta V^{-1}); \qquad (9.12)$$

$$V \equiv E[\{x_{kc} - E(x_{kc} \mid x'\beta)\}\{x_{kc} - E(x_{kc} \mid x'\beta)\}'\psi(x'\beta) \cdot f_u(-x'\beta)],$$

$$\Delta \equiv E[\{x_{kc} - E(x_{kc} \mid x'\beta)\}\{x_{kc} - E(x_{kc} \mid x'\beta)\}'\{\psi(x'\beta)\}^2$$

$$\cdot F_u(-x'\beta)\{1 - F_u(-x'\beta)\}],$$

where $\psi(x'\beta)$ is the density of $x'\beta$, and $F_u(w) \equiv P(u \leq w)$ and $f_u = F_u'$.

Let $b_N \equiv (b_{kc}, 1)'$. In estimating the covariance matrix, the main difficulty is in getting $f_u(-x'\beta)$, for u is not estimable. But owing to $E(y \mid x_i'\beta) = 1 - F(-x_i'\beta)$, we can estimate $1 - F(-x_i'\beta)$ with

$$1 - F_N(-x_i'b_N) \equiv \sum_{j \neq i} M((x_j'b_N - x_i'b_N)/s)$$

$$\cdot y_j / \sum_{j \neq i} M((x_j'b_N - x_i'b_N)/s), \qquad (9.13)$$

where M is a kernel and s is a smoothing parameter. Then differentiating (9.13) wrt $x_i'b_N$, $f_u(-x'\beta)$ can be estimated by

$$f_N(-x_i'b_N) = -s^{-1}\sum_{j \neq i} M'((x_j'b_N - x_i'b_N)/s)y_j \bigg/ \sum_{j \neq i} M((x_j'b_N - x_i'b_N)/s)$$

$$+s^{-1}F_N(x_i'b_N) \cdot \sum_{j \neq i} M'((x_j'b_N - x_i'b_N)/s)/M((x_j'b_N - x_i'b_N)/s), \quad (9.14)$$

where $M' \equiv \partial M(w)/\partial w$. Thus V may be estimated by [estimating Δ is easier and so omitted; a numerical derivative can be used as in Sherman (1993)]

$$(1/N)\sum_i \{x_{ikc} - E_N(x_{kc} \mid x_i'b_N)\}\{x_{ikc}$$

$$- E(x_{kc} \mid x_i'b_N)\}'\psi_N(x_i'b_N)f_N(-x_i'b_N),$$

where

$$E_N(x_{kc} \mid x_i'b_N) \equiv \sum_{j \neq i} K_{ji}x_{jkc} \bigg/ \sum_{j \neq i} K_{ji}, \quad \psi_N(x_i'b_N) \equiv \{(N-1)h\}^{-1}\sum_{j \neq i} K_{ji},$$

and $K_{ji} \equiv K((x'_j b_N - x'_i b_N)/h)$; $K(\cdot)$ is an univariate kernel.

The advantage of RCE for the binary model is that RCE is the only $\sqrt{N}$-consistent method, which is smoothing-parameter-free at the present time, although its variance matrix has the density functions of u and $x'\beta$ which require a nonparametric estimation method as in CLAD.

9.10   Semiparametric Methods for Binary and Censored Panel Data

So far we have dealt only with cross-section data. In this section, we introduce two semiparametric estimators for panel data. The first estimator is Manski (1987)'s panel maximum score estimator (PMSC) extending the cross-section MSC. The second is Honoré (1992)'s trimmed LSE for censored panel data. See Chamberlain (1984) and Hsiao (1986) for panel data in general.

Consider a panel data model:

$$y_{it} = x'_{it}\beta + \alpha_i + u_{it}, \quad i = 1 \ldots N, \ t = 1 \ldots T, \tag{10.1}$$

where α_i is an unobserved time-invariant "unit-specific" effect. For instance, if y_{it} is the earnings of individual i at time t, then x_{it} is a vector of explanatory variables such as age (age_{it}), schooling (edu_{it}), and a dummy for city living (d_{it}), and α_i is the "ability" of the individual (or unit) i. Among x_{it}, edu_{it} is time-invariant (assuming only "post-education" adults), d_{it} is time-variant, and age_{it} is time-variant but in a deterministic way. There are many issues in panel data in relation to these characteristics. Also important is whether α_i is correlated to x_{it} ("fixed-effect" models) or not ("random effect").

For random-effect models, estimators for cross-section data can be applied, although there is an efficiency issue to take care of owing to the structure in the error term $\alpha_i + u_{it}$. The simplest approach for a random-effect model is estimating each "wave" (the data for each t) separately, and then combine the T sets of estimates using minimum distance estimation (MDE) under the restriction $\beta_1 = \beta_2 \ldots = \beta_T = \beta$, where β_t is the population parameter for the tth wave. Following the notations for MDE in Chapter 5, the restriction is [see (10.3) below for a simple example on $1_T \otimes I_k$]

$$\theta = (1_T \otimes I_k) \cdot \beta, \quad \text{where } \theta = (\beta'_1 \ldots \beta'_T)'. \tag{10.2}$$

For fixed-effect models, cross-section estimators are not directly applicable. We will assume fixed effect in this section, however, because fixed effect includes random effect as a special case, and ability to deal with fixed effect is an advantage allowed by panel data.

Suppose $T = 3$ and we can estimate β using two waves despite the fixed effect. Let β_{12} and β_{23} denote the population parameters for the first two and the last two waves, respectively. Then the estimate for β_{12} and β_{23} can be combined using the restriction $\beta_{12} = \beta_{23} = \beta$, which is

$$\begin{bmatrix} \beta_{12} \\ \beta_{23} \end{bmatrix} = \begin{bmatrix} I_k \\ I_k \end{bmatrix} \begin{bmatrix} \beta \\ \beta \end{bmatrix}. \tag{10.3}$$

So whether random or fixed effect, MDE is helpful in reducing the big data with T waves into smaller data with one or two waves.

Consider a fixed-effect binary response panel data (y_{it}, x'_{it}), where $y_{it} = 1[y^*_{it} > 0]$ and

$$y^*_{it} = x'_{it}\beta + \alpha_i + u_{it}, \quad i = 1\ldots N, \ t = 1, 2. \tag{10.4}$$

Define $y_i \equiv (y_{i1}, y_{i2})'$, $x_i \equiv (x'_{i1}, x'_{i2})'$, $u_i \equiv (u_{i1}, u_{i2})'$, and

$$\Delta y_i \equiv y_{i2} - y_{i1}, \quad \Delta x_i \equiv x_{i2} - x_{i1}, \quad \Delta u_i \equiv u_{i2} - u_{i1}. \tag{10.5}$$

Let $F_{u_j|x\alpha}$ denote the distribution of $u_j \mid (x', \alpha)$. The key assumptions for PMSC are the following:

(i) There is one component of Δx_i, say the kth, such that $\beta_k \neq 0$ and the conditional density of Δx_{ik} given almost every value of $(\Delta x_{i1} \ldots \Delta x_{i,k-1})$ is positive on R^1.

(ii) $F_{u_1|x\alpha} = F_{u_2 x\alpha}$, and the support of $u_1 \mid x\alpha$ is R^1 for almost every x and α.

An analog of the first assumption already appeared in MSC. In the second assumption, there is no restriction on the correlation between u_1 and u_2 given (x', α), nor on the dependence of u on (x', α); the unbounded support of $u_1 \mid x\alpha$ is to make sure $P(\Delta y \neq 0 \mid x, \alpha) > 0$ for almost every (x', α).

Since $P(\Delta y = -1 \mid x, \Delta y \neq 0) + P(\Delta y = 1 \mid \Delta y \neq 0) = 1$, $\mathrm{Med}(\Delta y \mid x, \Delta y \neq 0) = \pm 1$ depending on which is the greater between the two probabilities. Since

$$P(\Delta y = 1 \mid x, \Delta y \neq 0) = P(y_1 = 0, y_2 = 1 \mid x) / \{P(y_1 = 0, y_2 = 1 \mid x)$$

$$+ P(y_1 = 1, y_2 = 0 \mid x)\},$$

$$P(\Delta y = 1 \mid \Delta y \neq 0) \gtrless P(\Delta y = -1 \mid x, \Delta y \neq 0)$$

$$\Leftrightarrow P(y_1 = 0, y_2 = 1 \mid x) \gtrless P(y_1 = 1, y_2 = 0 \mid x).$$

Use $P(y_1 = 0, y_2 = 1 \mid x) = P(y_2 = 1 \mid x) - P(y_1 = 1, y_2 = 1 \mid x)$ and its like to get

$$P(\Delta y = 1 \mid x, \Delta y \neq 0) \gtrless P(\Delta y = -1 \mid x, \Delta y \neq 0)$$

$$\Leftrightarrow P(y_2 = 1 \mid x) \gtrless P(y_1 = 1 \mid x). \tag{10.6}$$

But $F_{u_1|x\alpha} = F_{u_2|x\alpha}$ implies

$$P(y_2 = 1 \mid x, \alpha) = P(y_2^* \geq 0 \mid x, \alpha) \overset{>}{\underset{<}{}} P(y_1^* \geq 0 \mid x, \alpha)$$

$$= P(y_1 = 1 \mid x, \alpha) \Leftrightarrow x_2'\beta \overset{>}{\underset{<}{}} x_1'\beta, \tag{10.7}$$

which also holds without α by integrating out α. Therefore, we get

$$\text{Med}(\Delta y_i \mid x_i, \Delta y_i \neq 0) = \text{sgn}(\Delta x_i'\beta). \tag{10.8}$$

Based on this, PMSC maximizes the following analog of (2.4):

$$(1/N) \sum_i \text{sgn}(\Delta x_i'b) \cdot \Delta y_i. \tag{10.9}$$

The observation that only Δy and Δx are informative conditional on $\Delta y \neq 0$ is predated by Chamberlain (1980), who proposed a parametric estimator for β under the assumptions that u is independent of x, and u_1 and u_i are iid logistic; a survey on the parametric literature for panel data with LDV can be found in Maddala (1987). Charlier et al. (1995a) smooth PMSC as Horowitz (1992) smooths MSC and then apply the smoothed PMSC to labor force participation data. Specifically, first they observe that an appropriate extension of PMSC for $T > 2$ is

$$(1/N) \sum_{i=1}^N \sum_{t=2}^T \sum_{s<t} \text{sgn}(\Delta x_{its}'b) \cdot (y_{it} - y_{is}) \cdot 1_{its}$$

or equivalently [compare to (2.4)]

$$(1/N) \sum_{i=1}^N \sum_{t=2}^T \sum_{s<t} 1[\Delta x_{its}'b > 0] \cdot (y_{it} - y_{is}) \cdot 1_{its}, \tag{10.10}$$

where $\Delta x_{its} \equiv x_{it} - x_{is}$, and $1_{its} = 1$ if (y_{it}, x_{it}') and (y_{is}, x_{is}') are observed and 0 otherwise. Then they replace the indicator function in (10.10) with a smooth function, the integral of a kernel, to obtain a practical asymptotic distribution.

Turning to a fixed-effect censored response model, consider (10.4) but with $y_{it} = \max(y_{it}^*, 0)$. For ease of exposition, we will discuss the main idea for $y_{it} > 0$ (truncated case). Assume that $u_1 \mid x\alpha$ and $u_2 \mid x\alpha$ are iid [no correlation between u_1 and u_2, although this may be relaxed somewhat as in Honoré (1992, p. 543)]. Since $y_j > 0 \Leftrightarrow u_j > -x_j'\beta - \alpha$, $j = 1, 2$, the different truncation points across j make the truncated error terms nonidentically distributed given (x', α). But if we can impose the conditions $u_1 > -x_2'\beta - \alpha$ and $u_2 > -x_1'\beta - \alpha$ additionally, we get

$$u_j > \max(-x_1'\beta, -x_2'\beta) - \alpha, \quad j = 1, 2, \tag{10.11}$$

restoring the iid condition due to the same truncation points.

Since

$$u_1 > -x_2'\beta - \alpha \Leftrightarrow y_1^* > -\Delta x'\beta, \quad u_2 > -x_1'\beta - \alpha \Leftrightarrow y_2^* > \Delta x'\beta,$$

$$u_1 > \max(-x_1'\beta, -x_2'\beta) - \alpha \Leftrightarrow y_1 > -\Delta x'\beta,$$

$$u_2 > \max(-x_1'\beta, -x_2'\beta) - \alpha \Leftrightarrow y_2 > \Delta x'\beta. \qquad (10.12)$$

On the set $A \equiv \{y_1 > -\Delta x'\beta, y_2 > \Delta x'\beta\}$, subtract $y_1 = x_1'\beta + \alpha + u_1$ from $y_2 = x_2'\beta + \alpha + u_2$ to get $\{\Delta y - \Delta x'\beta\}1[A] + (u_2 - u_1)1[A]$. The main idea is that $(u_2 - u_1)1[A]$ is symmetric around 0 given x, since $u_2 \mid (x, A)$ and $u_1 \mid (x, A)$ are iid. Thus we get $E\{(u_2 - u_1)1[A] \mid x\} = 0$ which implies

$$E\{(\Delta y - \Delta x'\beta) \cdot \Delta x'\beta \, 1[y_1 > -\Delta x'\beta, y_2 > \Delta x'\beta]\} = 0; \qquad (10.13)$$

the idea is reminiscent of STLS, yet distinct.

Honoré suggests two estimators (LAD and LSE versions) for the truncated case with the minimands having (10.13) as the first-order condition. Extending the above idea to the censored case, he also suggests two estimators (LAD and LSE version); the LSE version minimizes

$$(1/N) \sum_i [\{\max(y_{i2}, \Delta x_i'b) - \max(y_{i1}, -\Delta x_i'b) - \Delta x_i'b\}^2$$

$$- 2 \cdot 1[y_{i2} < \Delta x_i'b](y_{i2} - \Delta x_i'b)y_{i1} - 2 \cdot 1[y_{i1} < -\Delta x_i'b](y_{i1} + \Delta x_i'b)y_{i2}].$$
$$(10.14)$$

The first term becomes $(\Delta y - \Delta x)^2$ when $y_{i2} > \Delta x_i'b$ and $y_{i1} > -\Delta x_i'b$ as required by (10.13). The remaining two terms can be regarded as adjustment terms to make use of the observations with $y = 0$ and to make the minimand smooth when $y_{i2} < \Delta x_i'b$ or $y_{i1} < -\Delta x_i'b$. Note that the truncated model requires an additional assumption that the log of the density for $u_1 \mid xc$ be strictly concave as in QME. The estimator for (10.14) is $\sqrt{N}$-consistent and asymptotically normal with the variance

$$E^{-1}(1[-y_1 < \Delta x'\beta < y_2] \cdot \Delta x \Delta x') \cdot E[\{y_2^2 1[y_1 < -\Delta x'\beta]$$

$$+ y_1^2 1[y_2 < \Delta x'\beta] + (\Delta y - \Delta x'\beta)^2 1[-y_1 < \Delta x'\beta < y_2]\} \cdot \Delta x \Delta x'] \quad (10.15)$$

$$\cdot E^{-1}(1[-y_1 < \Delta x'\beta < y_2] \cdot \Delta x \Delta x').$$

An estimate for this is straightforward.

9.11 Specification Tests with Semiparametric Estimators

Semiparametric estimators are based on assumptions weaker than those used in parametric estimators. However, it is tempting to use parametric estimators, since they often are easy to compute and computer programs for semiparametric estimators are not readily available. It is possible to test parametric model specifications with semiparametric estimators or with their moment conditions. Also, the assumptions in one semiparametric estimator may be tested by another semiparametric estimator. In this section we examine the topic of specification tests with semiparametric methods; see Whang and Andrews (1993) for a more rigorous and diverse discussion. We will show a number of tests for censored (or truncated) models; testing ideas for other models can be easily inferred from our examples. Recall the three testing principles: Lagrangean multiplier (LM), likelihood ratio (LR) and Wald tests. Analogs of these are examined in this section: method of moments tests based on LM idea, an artificial regression test that is a LR test in a sense, and tests comparing two estimators as in Wald tests.

For the censored model $y = \max(x'\beta + u, 0)$, if $u \mid x$ is symmetric, then the WME moment condition (7.8) should hold for all w. With $r_i \equiv y_i - x_i' b_{\text{scls}}$, the following moment forms the basis of a symmetry test:

$$m(w, b_{\text{scls}}) \equiv (1/\sqrt{N}) \sum_i 1[x_i' b_{\text{scls}} > w]$$

$$\cdot \{1[|r_i| < w]r_i + (1[r_i > w] - 1[r_i < -w])w\} \cdot x_i. \tag{11.1}$$

Under the null hypothesis of symmetry, $m(w, b_{\text{scls}})$ follows an asymptotic normal distribution centered at zero. This yields a test statistic

$$m(w, b_{\text{scls}})' \cdot \Omega_{\text{scls}}^{-1} \cdot m(w, b_{\text{scls}}) \Rightarrow \chi_k^2, \tag{11.2}$$

where Ω_{scls} is the variance of $m(w, b_{\text{scls}})$ derived in the following.

Define two "trimmed error terms" λ and δ:

$$\lambda \equiv 1[x'\beta > w] \cdot \{1[|u| \leq w]u + (1[u > w] - 1[u < -w])w\},$$
$$\delta \equiv 1[x'\beta > 0] \cdot \{1[|u| \leq x'\beta]u + (1[u > x'\beta] - 1[u < -x'\beta]) \cdot x'\beta\}. \tag{11.3}$$

Then $m(w, b_{\text{scls}})$ follows $N(0, \Omega_{\text{scls}})$ (Lee, 1994) where

$$\Omega_{\text{scls}} \equiv E(xx'\lambda^2) + H_w S^{-1} E(xx'\delta^2) S^{-1} H_w$$

$$- H_w S^{-1} E(xx'\lambda\delta) - E(xx'\lambda\delta) S^{-1} H_w; \tag{11.4}$$

$$H_w \equiv E\{1[x'\beta > w, |u| < w] \cdot xx'\}; \quad S \equiv E\{1[|u| < x'\beta] \cdot xx'\}.$$

If we remove $1[x'b_{\text{scls}} > w]$ and replace b_{scls} with LSE in (11.1), and further remove $1[x'\beta > w]$ and $1[x'\beta > 0]$ in Ω_{scls}, then (11.2) can be used to test for symmetry in the fully observed linear model. We can combine several moment conditions corresponding to different w's in (7.8), but this does not guarantee improvement. Instead, using (11.1) for various w's separately may reveal the particular value at which the symmetry breaks down.

In the following, we present a test statistic based on two moment conditions corresponding to two values of w, w_1 and w_2. Cases of more moment conditions can be handled analogously.

Define $m(w_1, b_{\text{scls}})$ and $m(w_2, b_{\text{scls}})$ as in (11.1), and also define $m(w, b_{\text{scls}})$ as the $2k \times 1$ vector stacking $m(w_1, b_{\text{scls}})$ and $m(w_2, b_{\text{scls}})$. Further define λ_1 and λ_2 as

$$\lambda_1 \equiv 1[x'\beta > w_1] \cdot \{1[|u| \leq w_1]u + (1[u > w_1] - 1[u < -w_1])w_1\},$$

$$\lambda_2 \equiv 1[x'\beta > w_2] \cdot \{1[|u| \leq w_2]u + (1[u > w_2] - 1[u < -w_2])w_2\}; \quad (11.5)$$

let $\lambda \equiv (\lambda_1, \lambda_2)'$. Define η_1 and η_2 as

$$\begin{aligned} \eta &\equiv 1[x'\beta > w_1] \cdot \{1[|u| \leq w_1] \cdot u\}, \\ \eta_2 &\equiv 1[x'\beta > w_2] \cdot \{1[|u| \leq w_2] \cdot u\}; \end{aligned} \quad (11.6)$$

let $\eta \equiv (\eta_1, \eta_2)'$. Then the test statistic is

$$m(w, b_{\text{scls}})' \cdot \Lambda_{\text{scls}}^{-1} \cdot m(w, b_{\text{scls}}) \Rightarrow \chi^2_{2k}, \quad (11.7)$$

where

$$\begin{aligned} \Lambda_{\text{scls}} &= E(\lambda\lambda' \otimes xx') + H_{ww}S^{-1} \cdot E(xx'\delta^2) \cdot S^{-1}H'_{ww} \\ &\quad - H_{ww}S^{-1} \cdot E(\lambda'\delta \otimes xx') - E(\lambda\delta \otimes xx') \cdot S^{-1}H'_{ww}; \end{aligned}$$

$$H_{ww} \equiv E\{\eta \otimes xx'\}, \quad S = E\{1[|u| < x'\beta] \cdot xx'\}.$$

Powell (1986b, p. 155) proposes another symmetry test. Let $b_N(\alpha)$ denote the censored αth quantile estimator. The symmetry of $u \mid x$ around 0 implies, for a $\alpha > 1/2$,

$$\{b_N(\alpha) - b_N(1/2)\} - \{b_N(1/2) - b_N(1 - \alpha)\} =^p 0.$$

Using this test statistic is valid even when u is heteroskedastic. In fact, more specification tests can be devised comparing $b_N(\alpha_1) \ldots b_N(\alpha_J)$, where $0 < \alpha_1 < \cdots < \alpha_J < 1$. Analogously, comparing WME's (QME's) corresponding to different w's can yield more specification tests.

Suppose β and u_i are observed. Then

$$E_{u|x}(u^2 1[|u| < w] \cdot 1[x'\beta > w]) = 1[x'\beta > w] \cdot E_{u|x}(u^2 1[|u| < w]). \quad (11.8)$$

With heteroskedasticity, say $u = e \cdot g(x)$ where e is homoskedastic,

$$E_{u|x}(u^2 1[|u| < w] \cdot 1[x'\beta > w]) = 1[x'\beta > w] \cdot g(x)^2 \cdot E(e^2 1[|e| < w/g(x)]). \quad (11.9)$$

Unless $E(e^2 1[|e| < w/g(x)]) = g(x)^{-2}$, the right-hand side is $1[x'\beta > w]$ times a function of x. If u is homoskedastic, (11.9) is $1[x'\beta > w]$ times a constant, barring the unlikely event where $E_{u|x}(u^2) = \sigma^2$ but, for a $h(x)$,

$$E_{u|x}(u^2 1[|u| < w]) = (\sigma^2/2) + h(x)$$

and
$$E_{u|x}(u^2 1[|u| > 2]) = (\sigma^2/2) - h(x).$$

Set up the following artificial regression model:

$$u_i^2 \cdot 1[|u_i| < w, x_i'\beta > w] = 1[x_i'\beta > w] \cdot z_i'\alpha + v_i, \qquad (11.10)$$

where $z_i \equiv (1, z_2, \ldots, z_q)'$, and $z_2, \ldots, z_q$ are polynomial functions of the components of x. The null hypothesis is H_0: $\alpha_2 = \cdots = \alpha_q = 0$ ($\Leftarrow$ homoskedasticity). In view of (11.9), setting $z_2 \ldots z_q$ so that $z'\alpha \cong E_{u|x}(u^2 1[|u| < w])$ enhances the power of the test. Testing the significance of $\alpha_2 \ldots \alpha_q$ is equivalent to comparing the fitness of (11.10) with and without $z_2 \ldots z_q$, so the test is a LR test in this sense.

Since u_i and β are not observable in practice, they should be replaced by r_i and an estimate b_N, respectively. Rewrite (11.10) as

$$r_i^2 1[|r_i| < w, x_i'b_N > w] = 1[x_i'b_N > w] \cdot z_i'\alpha + v_i$$

$$+ (1[x_i'\beta > w] - 1[x_i'b_N > w]) \cdot z_i'\alpha \qquad (11.11)$$

$$+ (r_i^2 1[|r_i| < w, x_i'b_N > w] - u_i^2 1[|u_i| < w, x_i'\beta > w]).$$

Here α is estimated by a_N, the LSE of $r_i^2 1[|r_i| < w, x_i'b_N > w]$ on $1[x_i'b_N > w] \cdot z_i$. The error term in (11.11) has three components. The last one can be shown to be negligible under $E(u \cdot 1[|u| < w]) = 0$. Suppose $\beta_k \neq 0$ and x_k has a conditional density $\psi(x_k \mid x_c)$ with the continuous first derivative wrt x_k, where $x_c \equiv (1, x_2, \ldots, x_{k-1})'$. In the following, we show the asymptotic distribution of (11.11).

Assume $\beta_k > 0$ for a while. Denote zz' as $\zeta(x_c, x_k)$ to reflect the fact that the components of zz' are functions of x_c and x_k; note that ζ is a $q \times q$ matrix. Define $\beta_c \equiv (\beta_1, \ldots, \beta_{k-1})'$ and $L \equiv (L_c, L_k)$, where

$$L_c \equiv E[\zeta\{x_c, (w - x_c'\beta_c)/\beta_k\} \cdot \alpha \cdot x_c'\beta_k^{-1}$$

$$\cdot \psi\{(w - x_c'\beta_c)/\beta_k \mid x_c\}],$$

$$L_k \equiv [\zeta\{x_c, (w - x_c'\beta_c)/\beta_k\} \cdot \alpha \cdot (w - x_c'\beta_c)$$

$$\cdot \beta_k^{-2}\psi\{(w - x_c'\beta_c)/\beta_k | x_c\}]. \qquad (11.12)$$

Then it can be shown that (Lee, 1994)

$$\sqrt{N}(a_N - \alpha) =^d N(0, E^{-1}(1[x'\beta > w] \cdot zz') \cdot E(\xi\xi') \cdot E^{-1}(1[x'\beta > w]zz')),$$

$$\xi_i \equiv 1[x_i'\beta > w] \cdot z_i v_i - L \cdot S^{-1} x_i \delta_i; \qquad (11.13)$$

δ and S were defined in (11.3) and (11.4). If $\beta_k < 0$, the negative sign in ξ_i should be replaced by a positive sign. L_c and L_k can be estimated, respectively, by

$$(1/N) \sum_i \zeta_{Ni} a_N \cdot x_{ci}' b_{Nk}^{-1} \cdot 1[|x_{ki}$$

$$- (w - x'_{ci}b_{Nc})/b_{Nk}| < h]/(2h),$$

$$(1/N) \sum_i \zeta_{Ni}a_N(w - x'_{ci}b_{Nc})b_{Nk}^{-2} \cdot 1[|x_{ki}$$

$$- (w - x'_{ci}b_{Nc})/b_{Nk}| < h]/(2h), \qquad (11.14)$$

where

$$\zeta_{Ni} \equiv \zeta(w_{ci}, (w - x'_{ci}b_{Nc})/b_{Nk}),$$

and h is a smoothing parameter converging to 0 as $N \to \infty$.

Design a $(q - 1) \times q$ matrix A such that $H_0 : A \cdot \alpha = 0$ is equivalent to $H_0 : \alpha_2 = \cdots \alpha_q = 0$. Define Ω as the variance matrix in (11.13). Then, we get the following test statistic for homoskedasticity for the censored (and the truncated) model:

$$(Aa_N)' \cdot \{A\Omega_N^{-1}A'\}^{-1} \cdot (Aa_N) \Rightarrow \chi_{q-1}^2, \qquad (11.15)$$

where $\Omega_N =^p \Omega$. Ω_N can be obtained by replacing the true values and the error terms in Ω with the estimates and the residuals, respectively.

Recall the Hausman (1978) test idea comparing two estimators b_e and b_c, where b_e is efficient under H_0 but inconsistent under H_a, while b_c is inefficient under H_0 but consistent under both hypotheses. Consider setting $b_c = b_{scls}$ and $b_e = b_{mle}$. The test statistic is

$$(b_{mle} - b_{scls})' \cdot [V(b_{mle} - b_{scls})]^{-1} \cdot (b_{mle} - b_{scls}) \Rightarrow \chi_k^2. \qquad (11.16)$$

For this test, H_0 becomes the model assumption for MLE, and rejecting H_0 does not reveal specifically which aspect to H_0 is violated. Powell (1986b) suggests a similar test comparing MLE and a quantile estimator.

Quite often in practice, the middle inverted matrix in (11.16) fails to be p.d. If we have

$$\sqrt{N}(b_{scls} - \beta) =^p (1/\sqrt{N}) \sum_i \psi_i, \quad \sqrt{N}(b_{mle} - \beta) =^p (1/\sqrt{N}) \sum_i \mu_i,$$
$$(11.17)$$

$V(b_{mle} - b_{scls})$ can be estimated by

$$(1/N) \sum_i (\psi_i - \mu_i)(\psi_i - \mu_i)'/N. \qquad (11.18)$$

With the notations in (11.3) and (11.4), it can be shown that $\psi_i = S^{-1}x_i\delta_i$ and $\mu_i = E^{-1}(s_b^* s_b^{*\prime}) \cdot s_{bi}^*$, where s_b^* is the effective score for β (the residual of regressing the score function for β on the score function for σ_u). With (11.18), the Hausman test may be better called a Wald test, which is also based on the difference of two estimators without the assumption that one of them is efficient under H_0.

Let $E\{\mu(\beta)\} = 0$ denote the asymptotic first-order condition for SCLS in (6.5). Then b_{scls} satisfies

$$(1/\sqrt{N}) \sum_i \mu(b_{\text{scls}}) =^p 0. \tag{11.19}$$

Furthermore, with a Taylor's expansion,

$$0 =^p (1/\sqrt{N}) \sum_i \mu(b_{\text{scls}}) =^p (1/\sqrt{N}) \sum_i \mu(b_{\text{mle}})$$

$$+ A \cdot \sqrt{N}(b_{\text{scls}} - b_{\text{mle}}), \tag{11.20}$$

where $A =^p (1/N) \sum_i \partial E\{\mu(b_{\text{mle}})\}/\partial b$. Thus, $(b_{\text{mle}} - b_{\text{scls}})$ in (11.16) can be replaced by $A^{-1}(1/\sqrt{N}) \sum_i \mu(b_{\text{mle}})$, and the resulting test requires only b_{mle}. We can also design an equivalent test using the first-order condition of the MLE, which then requires only b_{scls}. Designing simpler tests equivalent to a Hausman test was suggested by Ruud (1984). In general, if a MLE and a semiparametric estimator are available for the same model, then we can construct a Wald-type test as in (11.16) with (11.18). Depending on which is easier to compute, we can choose an equivalent test computing only one estimator. Pagan and Vella (1989) provide applications and discussions on specification tests in general.

10
Semi-Nonparametrics

10.1 Introduction

In this chapter, we examine semiparametric estimators that require nonparametric techniques. In most cases, the estimators will be obtained in two stages, where the first stage is a nonparametric method. See Delgado and Robinson (1992) and Powell (1994) for surveys. In most cases, we will restrict our discussion to $\sqrt{N}$-consistent estimators. Although we defined the term "semiparametrics" in a narrow sense that obtaining the estimates does not require nonparametric techniques, in this chapter we will often use the term "semiparametrics" in its wide sense as the union of semiparametrics and semi-nonparametrics.

In Section 2, we introduce some asymptotic results useful for other sections. In Section 3, we show how to do weighted LSE efficiently without assuming the form of heteroskedasticity (Robinson, 1987). In Section 4, we review Robinson's (1988a) semilinear model as a compromise between linear and nonparametric models. In Section 5, we introduce three estimators for single-index models: Powell et al. (1989), Ichimura (1993), and Klein and Spady (1993). In Section 6, average derivative estimators in Härdle and Stoker (1989) and Stoker (1991) are reviewed, and multiple-index extensions of single-index estimators are discussed. In Section 7, we study nonparametric IVE for linear and nonlinear models of Newey (1990b) and Robinson (1991). In Section 8, we examine two estimators: Lee (1995b), for simultaneous equations with limited endogenous regressors, and Ahn and Manski (1993), for binary choice under uncertainty. In Section 9, a semi-nonparametric MLE with series expansion (Gallant and Tauchen, 1989) is studied. In Section 10, we review specification tests using nonparametric techniques. In Section 11, semiparametric efficiency is examined drawing upon Newey (1990a) and Bickel et al. (1993).

10.2   Some Useful Asymptotic Results

It will be helpful to collect some results for two-stage estimation (TSE) with a nonparametric first stage. The reader may choose to refer to this section when necessary, reading other sections first. Recalling our discussion on a two-stage estimator b_N with a finite dimensional nuisance parameter a_N ($=^p \alpha$), if b_N satisfies

$$(1/\sqrt{N}) \sum_i m(z_i, b_N, a_N) =^p 0, \tag{2.1}$$

then

$$\sqrt{N}(b_N - \beta) =^p \left\{-E^{-1}(m_b)\right\} \cdot (1/\sqrt{N}) \sum_i [m(z_i, \beta, \alpha) + E(m_a) \cdot \eta_i(\alpha)], \tag{2.2}$$

where

$$m_b \equiv \partial m(z, \beta, \alpha)/\partial b,$$
$$m_a \equiv \partial m(z, \beta, \alpha)/\partial a, \quad \sqrt{N}(a_N - \alpha) =^p (1/\sqrt{N}) \sum_i \eta_i(\alpha).$$

The results in the following hold when a high-order kernel taking negative as well as positive values is used in the first stage, and the bandwidth h is chosen to be smaller than the optimal bandwidth minimizing the asymptotic mean squared error; such a small h reduces asymptotic bias faster than the optimal bandwidth.

Suppose that the nuisance parameter is $\alpha_j \equiv E(w_j \mid q_j)$; assume that w is a scalar for ease of exposition. Let α_j be estimated by a "leave-one-out" kernel estimator (this is theoretically convenient)

$$a_j = \sum_{i \neq j} K((q_i - q_j)/h)w_j/K((q_i - q_j)/h), \tag{2.3}$$

where K is a kernel and h is a bandwidth. Denoting $m(z_i, \beta, \alpha)$ as m_i,

$$\sqrt{N}(b_N - \beta) =^p -E^{-1}(m_b) \cdot (1/\sqrt{N}) \sum_i [m_i + E(m_a \mid q_i)\{w_i - E(w \mid q_i)\}]. \tag{2.4}$$

If $\alpha_j = \nabla E(w \mid q_j) = \partial E(w \mid q_j)/\partial q_j$, then the $(1/\sqrt{N}) \sum_i$ term becomes

$$(1/\sqrt{N}) \sum_i [m_i - f(q)^{-1}[\nabla\{E(m_a \mid q_i)f(q_i)\}] \cdot \{w_i - E(w \mid q_i)\}]$$

$$= (1/\sqrt{N}) \sum_i [m_i - \{\nabla E(m_a \mid q_i) + E(m_a \mid q_i)\nabla f(q_i)/f(q_i)\}$$

$$\cdot \{w_i - E(w \mid q_i)\}], \tag{2.5}$$

where $f(q)$ is the density of q. With $\alpha_j = f(q_j)$, the $(1/\sqrt{N}) \sum_i$ term becomes

$$(1/\sqrt{N}) \sum_i [m_i + E(m_a \mid q_i)f(q_i) - E\{E(m_a \mid q)f(q)\}]. \tag{2.6}$$

If $\alpha_j = \nabla f(q_j)$, then the $(1/\sqrt{N}) \sum_i$ should be

$$(1/\sqrt{N}) \sum_i [m_i - \nabla\{E(m_a \mid q_i) \cdot f(q_i)\} + E[\nabla\{E(m_a \mid q_i) \cdot f(q_i)\}]], \tag{2.7}$$

where

$$\nabla\{E(m_a \mid q) \cdot f(q)\} = \nabla E(m_a(q) \mid q) \cdot f(q) + E(m_a(q)) \cdot \nabla f(q).$$

This is also applicable to $\alpha_j = \nabla f(q_j)/f(q_j)$ by treating $f(q_j)$ as known.

Call the terms following m_i in $(1/\sqrt{N}) \sum_i$ "correction terms," for they account for the first-stage estimation error; if α_j is known, no correction terms are necessary. In (2.4) to (2.7), if $m_a(z, \beta, \alpha) = 0$, then the correction term is zero. Since (2.4) to (2.7) do not depend on the kernel, it is likely that the choice of a nonparametric method does not matter. Results (2.4) to (2.7) were first derived by Newey (1994) using a series approximation for a_j. With the kernel estimator in (2.3), they can be proven using high-order kernels, U-statistic theories, and the uniform consistency theorem in the nonparametrics chapter; Robinson (1988a) seems to be the first to use high-order kernels to facilitate similar derivations. In (2.3) to (2.7), we assumed that w is a scalar. If there is more than one nuisance parameters (that is, if w is a vector), then the correction term is a sum of individual correction terms (Newey, 1994).

If there is no b_N in (2.1) but α is still infinite dimensional, then (2.4) to (2.7) gives $o_p(1)$ equivalent expressions to $(1/\sqrt{N}) \sum_i m(z_i, a_i)$. As an example, consider

$$(1/\sqrt{N}) \sum_i g(x_i) \cdot E_N(y \mid x_i), \tag{2.8}$$

where $\alpha_i \equiv E(y \mid x_i)$, $m(x_i, \alpha_i) = g(x_i)E(y \mid x_i)$, and $a_i = E_N(y \mid x_i)$. Using (2.4),

$$(1/\sqrt{N}) \sum_i g(x_i)E_N(y \mid x_i) =^p (1/\sqrt{N}) \sum_i g(x_i)E(y \mid x_i)$$

$$+ g(x_i)\{y_i - E(y \mid x_i)\} = (1/\sqrt{N}) \sum_i g(x_i) y_i, \qquad (2.9)$$

for $E(m_a \mid x) = E[\partial\{g(x)E(y \mid x)\}/\partial E(y \mid x) \mid x] = g(x)$. Also (2.9) implies

$$(1/\sqrt{N}) \sum_i g(x_i)\{E_N(y \mid x_i) - E(y \mid x)\} = (1/\sqrt{N}) \sum_i g(x_i)\{y - E(y \mid x)\}.$$
$$(2.10)$$

Sometimes we use nonparametric estimates obtained from subsamples. Imagine a discrete regressor π taking 0 or 1. Suppose we have two nuisance parameters $\alpha_{i\nu} = E(w \mid q_i, \pi_i = \nu)$, $\nu = 0, 1$. Then the $(1/\sqrt{N}) \sum_i$ term is (Lee, 1995c)

$$(1/\sqrt{N}) \sum_i \Bigg[m_i + \sum_{\nu=0,1} P(\pi = \nu \mid q_i)^{-1} E(m_{a_\nu} \mid q_i)\{y$$

$$- E(w \mid q_i, \pi_i = \nu)\} 1[\pi_i = \nu] \Bigg], \qquad (2.11)$$

which has two correction terms added.

10.3 Efficient Estimation with Unknown Form of Heteroskedasticity

We have seen that the variance matrix of an efficient estimator under the conditional moment condition $E(\psi(\beta) \mid x) = 0$ is $(\psi_b \equiv \partial\psi/\partial b)$

$$E_x^{-1}[E(\psi_b(\beta) \mid x) \cdot \{E(\psi(\beta)\psi(\beta)' \mid x)\}^{-1} \cdot E(\psi_b(\beta)' \mid x)]. \qquad (3.1)$$

Using this, with a given $k \times 1$ moment condition $E(u \mid x) = 0$ in the linear model $y = x'\beta + u$ [here $\psi(\beta) = y - x'\beta$], the efficiency bound is

$$E_x^{-1}\{x \cdot (V(u \mid x))^{-1} \cdot x'\} = E_x^{-1}[xx'\{V(u \mid x)\}^{-1}]. \qquad (3.2)$$

Under the known form of heteroskedasticity, the weighted LSE (WLS) attains (3.2). In this section, we present an estimator attaining this bound without specifying the form of heteroskedasticity.

Robinson (1987) shows how to implement an efficient estimation using a nonparametric estimate for $V(u \mid x)$ with LSE residuals; that is, estimate $V(u \mid x)$ nonparametrically with the LSE residuals and then apply WLS. Robinson suggests a nearest-neighbor method for estimating $V(u \mid x)$. But we use a kernel estimator $(r_j \equiv y_j - x_j' b_{\text{LSE}})$

$$V_N(u \mid x_i) = E_N(u^2 \mid x_i) = \sum_{j \neq i}^{N} K((x_j - x_i)/h) r_j^2 \Big/$$

$$\sum_{j\neq i}^{N} K((x_j - x_i)/h). \tag{3.3}$$

Another, perhaps better, alternative that Robinson (1987) suggests is

$$E_N(y^2 \mid x_i) - \{E_N(y \mid x_i)\}^2 = \sum_{j\neq i} K((x_j - x_i)/h)y_j^2 \Big/ \sum_{j\neq i} K((x_j - x_i)/h)$$

$$- \left\{ \sum_{j\neq i} K((x_j - x_i)/h)y_j \Big/ \sum_{j\neq i} K((x_j - x_i)/h) \right\}^2, \tag{3.4}$$

which does not depend on LSE and the linearly assumption $E(y \mid x) = x'\beta$; (3.4) is due to $V(u \mid x) = V(y \mid x) = E(y^2 \mid x) - \{E(y \mid x)\}^2$.

To see the effect of the nonparametric estimation on WLS, note that the moment condition for WLS b_N is $E\{(y - x'b)/V(u \mid x)\} = 0$ and b_N satisfies

$$(1/\sqrt{N}) \sum_i (y_i - x_i'b_N)/V_N(u \mid x_i) = 0. \tag{3.5}$$

Here, $m(z_i, b, a_i) = (y_i - x_i'b)/E_N(u^2 \mid x_i)$, $\alpha_i = E(u^2 \mid x_i)$, and

$$m_a(z_i, \beta, \alpha_i) = -u \cdot \{E(u^2 \mid x)\}^{-2} \Rightarrow E(m_a \mid x) = 0. \tag{3.6}$$

Thus, following (2.4), there is no effect of estimating $V(u \mid x)$ on WLS. That is, we can do as well as if we knew the functional form of $E(u^2 \mid x)$.

10.4 Semilinear Model

Robinson (1988a) considers the following model:

$$y = x'\beta + \theta(z) + u, \quad x \neq z, \quad E(u \mid x, z) = 0, \tag{4.1}$$

where θ is an unknown function of z, x is a $m \times 1$ vector and z is a $k \times 1$ vector. This is a mixture of a parametric component $x'\beta$ and a nonparametric one $\theta(z)$. Taking $E(\cdot \mid x)$ on (4.1), we get

$$E(y \mid z) = E(x \mid z)'\beta + \theta(z). \tag{4.2}$$

Subtracting this from (4.1),

$$y - E(y \mid z) = \{x - E(x \mid z)\}'\beta + u \Leftrightarrow y = E(y \mid z) + \{x - E(x \mid x)\}'\beta + u, \tag{4.3}$$

which does not have $\theta(z)$ any more. Here the deterministic part of y is decomposed into two: one is the effect of z on y $[E(y \mid z)]$ and the other is the effect on y of x net of z $[(x - E(x \mid z))'\beta]$.

In order to estimate β, first use kernel estimators

$$E_N(y \mid z_i) \equiv \sum_{j \neq i}^{N} K((z_j - z_i) \Big/ h) \cdot y_j \Big/ \sum_{j \neq i}^{N} K((z_j - z_i)/h), \qquad (4.4)$$

$$E_N(x \mid z_i) \equiv \sum_{j \neq i}^{N} K((z_j - z_i)/h_x) \cdot x_j \Big/ \sum_{j \neq i}^{N} K((z_j - z_i)/h_x), \qquad (4.5)$$

for $E(y \mid z)$ and $E(x \mid z)$, respectively. Substitute (4.4) and (4.5) into (4.3) to define a new dependent variable $y - E_N(y \mid z)$ and regressors $x - E_N(x \mid z)$. Apply LSE to the new model $y - E_N(y \mid z) \cong \{x - E_N(x \mid z)\}'\beta + u$ to get

$$b_N = \left[\sum_i \{x_i - E_N(x \mid z_i)\} \cdot \{x_i - E_N(x \mid z_i)\}' \right]^{-1}$$

$$\cdot \left[\sum_i \{x_i - E_N(x \mid z_i)\} \cdot \{y_i - E_N(y \mid z_i)\} \right]. \qquad (4.6)$$

The usual intercept term in LSE disappears due to the mean subtraction; so the intercept is not estimable in (4.6). Also if we allow variables to appear in both x and z, their coefficients in β are not estimable for the same reason. The LSE b_N has the following asymptotic distribution:

$$\sqrt{N}(b_N - \beta) =^d N(0, A^{-1}BA^{-1}), \qquad (4.7)$$

$$A =^p A_N \equiv (1/N) \sum_i \{x_i - E_N(x \mid z_i)\} \cdot \{x_i - E_N(x \mid z_i)\}',$$

$$B =^p B_N \equiv (1/N) \sum_i \{x_i - E_N(x \mid z_i)\} \cdot \{x_i - E_N(x \mid z_i)\}' \cdot v_i^2,$$

$$v_i \equiv \{y_i - E_N(y \mid z_i)\} - \{x_i - E_N(x \mid z_i)\}'b_N.$$

Despite the nonparametric first stage, the estimation errors do not affect the asymptotic distribution of b_N. This can be shown using (2.4); the steps are analogous to (3.5) and (3.6).

If the bandwidth is large enough so that $E_N(y \mid z) = \bar{y}$ (sample mean of y_i's) and $E_N(x \mid z) = \bar{x}$, then (4.3) is approximately equal to

$$y - \bar{y} = (x - \bar{x})'\beta + u \Leftrightarrow y = \bar{y} - \bar{x}'\beta + x'\beta + u, \qquad (4.8)$$

which is the linear model with $\theta(z)$ in (4.1) treated as the intercept term $\bar{y} - \bar{x}'\beta$. In this regard, the usual linear model $y = x'\beta + u$ (with $\bar{y} - \bar{x}'\beta$ absorbed into the intercept) is a special case of (4.1) when h is large, and it is a misspecified one if $\theta(z)$ indeed varies with z. If one wants to estimate $\theta(z)$ as well as β, then this can be done by a nonparametric regression of $y - x'b_N$ on z, for $E(y - x'\beta \mid z) = \theta(z)$ in (4.1).

The model (4.1) is applicable to a labor supply model with age $= z$, because the labor supply profile against age is likely to be nonlinear (increasing and then leveling off after a certain age). The usual practice of including age and age^2 to capture this pattern is not quite satisfactory, for it means declining labor supply after a peak age.

Chen (1988) and Speckman (1988) also consider semilinear models. Chamberlain (1992) shows that (4.6) attains the semiparametric efficiency bound under homoskedasticity (that is, the variance is the smallest possible under the given semiparametric assumptions). Staniswalis and Severini (1994) examine a semilinear model $E(y \mid x, z) = g\{x'\beta + \theta(z)\}$, where $g(\cdot)$ is a known function. For instance, this model is suitable for $y = 1[x'\beta + \theta(z) + u \geq 0]$ with $u =^d N(0,1)$, which implies $E(y \mid x, z) = \Phi\{x'\beta + \theta(z)\}$, where Φ is the $N(0,1)$ distribution function.

10.5   Estimators for Single-Index Models

Consider a single-index model

$$y = r(x) + u \quad \text{with} \quad E(y \mid x) = r(x) = G(x'\beta), \tag{5.1}$$

where $G(\cdot)$ is unknown. From this, we get

$$\nabla r(x) \ (= \partial r(x)/\partial x) = \{dG(x'\beta)/d(x'\beta)\} \cdot \beta.$$

For a weighting function $\omega(x)$,

$$E_x\{\omega(x)\nabla r(x)\} = E_x\{\omega(x)dG(x'\beta)/d(x'\beta)\} \cdot \beta \equiv \gamma_\omega \beta, \tag{5.2}$$

which is proportional to β provided $\gamma_\omega \neq 0$; β is identified only up to the unknown scale factor γ_ω. In this section, we introduce three estimators for single-index models. The first two [Powell et al. (1989) and Ichimura (1993)] are good for single-index models in general, and the third [Klein and Spady (1993)] is good only for binary response models.

If we choose $\omega(x) = f(x)$, the density of x, we have the density weighted *average derivative* δ:

$$\delta \equiv E\{f(x)\nabla r(x)\} = E\{f(x)dG(x'\beta)/d(x'\beta)\} \cdot \beta \equiv \gamma_f \beta. \tag{5.3}$$

Suppose x has a continuous density function $f(x)$ which is zero at the boundary "∂X" of its support on R^k, [this excludes unity (1) and functionally dependent regressors such as $x_3 = x_2^2$]. With $r(x)$ bounded, integration by parts yields

$$\delta = \gamma_f \beta = E\{f(x)\nabla r(x)\} = \int \nabla r(x) \cdot f(x)^2 dx$$

$$= r(x)f(x)^2|_{\partial X} - 2\int [r(x)\nabla f(x) \cdot f(x)]dx = -2 \cdot E[\{r(x) + u\} \cdot \nabla f(x)] \tag{5.4}$$

$$= -2 \cdot E\{y\nabla f(x)\}. \tag{5.5}$$

We can estimate δ by plugging a kernel nonparametric estimate into $\nabla f(x)$ and using a sample moment for $E\{y\nabla f(x)\}$ will be shown subsequently. The reason why we use $f(x)$ as the weight in (5.3) is to prevent $f_N(x)$ from appearing in the denominator of the nonparametric estimate, because having $1/f_N(x)$ is troublesome when $f_N(x) \cong 0$, although this problem can be overcome by "trimming" [using observations with $f_N(x_i) > \varepsilon$ for some $\varepsilon > 0$]. In the usual linear model with $G(x'\beta) = x'\beta$, δ becomes $Ef(x) \cdot \beta$ for $dG(x'\beta)/d(x'\beta) = 1$ in (5.3). Hence, if we want to make (5.3) more comparable to LSE, we can estimate $\delta^* \equiv \delta/Ef(x)$.

Let $f_N(x_i)$ denote a (leave-one-out) kernel estimator for $f(x_i)$:

$$f_N(x_i) \equiv \{(N-1)h^k\}^{-1} \sum_{j \neq i} K((x_j - x_i)/h). \tag{5.6}$$

Then, a natural estimator for $\nabla f(x_i)$ is

$$\nabla f_N(x_i) = \{(N-1)h^{k+1}\}^{-1} \sum_{j \neq i} \nabla K((x_i - x_j)/h), \tag{5.7}$$

since $-\nabla K(z) = \nabla K(-z)$ from the symmetry of K. Following (5.5), define a density-weighted average derivative estimator (ADE) for δ:

$$\delta_N \equiv (-2/N) \sum_i y_i \nabla f_N(x_i)$$

$$= \{2/(N(N-1))\} \sum_i \sum_{j \neq i} (1/h^{k+1})\nabla K((x_j - x_i)/h)y_i \tag{5.8}$$

$$= \{2/(N(N-1))\} \sum_{i<j} (1/h^{k+1})\nabla K((x_j - x_i)/h)(y_i - y_j),$$

again using $-\nabla K(z) = \nabla K(-z)$.

Recall the U-statistic theory in the preceding chapter; for a U-statistic U_N and its parameter θ such that

$$U_N \equiv \{2/(N(N-1))\} \sum_{i<j} g(z_i, z_j) \quad \theta \equiv E\{g(z_i, z_j)\}, \tag{5.9}$$

we get

$$\sqrt{N}(U_N - \theta) =^p (2/\sqrt{N}) \sum_i [E\{g(z_i, z_j \mid z_i)\} - \theta]$$

$$=^d N(0, 4 \cdot E[E\{g(z_i, z_j \mid z_i)\} - \theta]^2). \tag{5.10}$$

Apply this to (5.8) to obtain

$$\sqrt{N}(\delta_N - \delta) =^d N(0, 4\Omega); \tag{5.11}$$

Ω can be estimated by its sample analog Ω_N:

$$\Omega_N = (1/N) \sum_i [E_N\{g(z_i, z_j) \mid z_i\}][E_N\{g(z_i, z_j) \mid z_i\}]' - \delta_N \delta_N',$$

where

$$E_N\{g(z_i, z_j) \mid z_i\} \equiv (N-1)^{-1} \sum_{j \neq i} (1/h)^{k+1} \cdot \nabla K((x_j - x_i)/h) \cdot (y_i - y_j).$$

For a $\sqrt{N}$-consistent estimator b_N for β, if $\sqrt{N}(b_N - \beta) =^p (1/\sqrt{N}) \sum_i \psi_i$, then ψ_i is called the *influence function* and b_N is said to be *asymptotically linear*; the influence function shows the contribution of each observation to the total estimation error $\sqrt{N}(b_N - \beta)$. For the U-statistic U_N in (5.9), the influence function is $2 \cdot [E\{g(z_i, z_j \mid z_i)\} - \theta]$. But this does not show well the sources of the estimation error. To see this better, observe

$$\sqrt{N}(\delta_N - \delta) = (-2/\sqrt{N}) \sum_i [y_i \nabla f_N(x_i) - E\{y \nabla f(x)\}]. \tag{5.12}$$

Invoking (2.7) with $E(m_a \mid x) = E(y \mid x) = r(x)$ and $\nabla E(m_a \mid x) = \nabla r(x)$, the first term on the right-hand side (rhs) is

$$(1/\sqrt{N}) \sum_i [y_i \nabla f(x_i) - \nabla r(x_i) \cdot f(x_i) - r(x_i) \nabla f(x_i)$$

$$+ E\{\nabla r(x) \cdot f(x) + r(x) \nabla f(x)\}]$$

$$= (1/\sqrt{N}) \sum_i [u_i \nabla f(x_i) - \nabla r(x_i) \cdot f(x_i)$$

$$+ E\{\nabla r(x) \cdot f(x) + r(x) \nabla f(x)\}]. \tag{5.13}$$

Substituting this into (5.12), the rhs of (5.12) becomes

$$(-2/\sqrt{N}) \sum_i [u_i \nabla f(x_i) - [\nabla r(x_i) \cdot f(x_i) - E\{\nabla r(x) \cdot f(x)\}]], \tag{5.14}$$

which breaks down the influence function into two sources of error. Since the covariance between the two terms is zero due to $E(u \mid x) = 0$, the variance is a sum of two individual variances.

Powell et al. (1989) further suggest an estimator for $\delta^* = \delta/Ef(x)$:

$$\delta_N^* \equiv \left[\sum_i \nabla f_N(x_i) \cdot x_i' \right]^{-1} \cdot \left[\sum_i \nabla f_N(x_i) \cdot y_i \right]. \tag{5.15}$$

The reason why this is a legitimate estimator for δ^* can be seen in the next section. Equation (5.15) has the form of an instrumental variable estimator with $\nabla f_N(x)$ as an instrument in the regression of y on x. In the next section, we will examine more of this kind of estimator. In the Monte Carlo study of Powell et al. (1989), δ_N^* performed better than δ_N. Härdle and Tsybakov (1993) discuss how to choose a smoothing parameter for the density weighted ADE.

One shortcoming of ADE is requiring x to have a continuous density and its components to be functionally independent of one another. Relaxing these, Ichimura (1993) proposes a semiparametric LSE (SLS) for single-index models under independence of u from x by minimizing

$$Q_N(b) \equiv (1/N) \sum_i \{y_i - E_N(y \mid x_i'b)\}^2 \qquad (5.16)$$

wrt b, where $E_N(y \mid x_i'b)$ is a nonparametric estimator for $E(y \mid x_i'b)$:

$$E_N(y \mid x_i'b) = \sum_{j \neq i} K((x_j'b - x_i'b)/h)y_j \Big/ \sum_{j \neq i} K((x_j'b - x_i'b)/h). \qquad (5.17)$$

The SLS b_{SLS} is not a two-stage estimator for $E_N(y \mid x_i'b)$ depends on b, but the fact that $E(y \mid x'\beta)$ is estimated makes SLS a two-stage estimator in a sense; so it may be better to call SLS a "1.5-stage estimator."

There should be at least one continuous regressor with nonzero coefficient for SLS, and β is identified only up to a scale if Assumption 4.2(4) of Ichimura holds, which is rather engaging to be discussed here. If we do a grid-search in implementing SLS, first fix b and evaluate $Q_N(b)$ at

$$b', E_N(y \mid x_1'b_N), \ldots, E_N(y \mid x_N'b); \qquad (5.18)$$

repeating this for all possible b's, we get b_{SLS} minimizing (5.16). Alternatively, one may use a Gauss–Newton-type algorithm with $\partial Q_N(b)/\partial b$.

b_{SLS} is $\sqrt{N}$-consistent and asymptotically normal with the variance

$$E^{-1}\{\nabla E(y \mid x'\beta) \cdot \nabla E(y \mid x'\beta)'\} \cdot E\{V(u \mid x) \cdot \nabla E(y \mid x'\beta)$$

$$\cdot \nabla E(y \mid x'\beta)'\} \cdot E^{-1}\{\nabla E(y \mid x'\beta) \cdot \nabla E(y \mid x'\beta)'\}, \qquad (5.19)$$

where

$$\nabla E(y \mid x'\beta) \cong (\partial E_N(y \mid x'b)/\partial b\big|_{b=\beta})$$

$$= \{x - E(x \mid x'\beta)\} \cdot dG(x'\beta)/d(x'\beta). \qquad (5.20)$$

The variance has the same form as that of a nonlinear LSE with $G(\cdot)$ known except that the regressor part of (5.19) is $x - E(x \mid x'\beta)$ not x. Technically, this is due to the b in the kernel for $E_N(y \mid x'b)$: when $E_N(y \mid x'b)$ is differentiated wrt b, $x_j - x$ comes out and x_j yields $E(x \mid x'\beta)$. Other than this, the fact that $G(\cdot)$ is estimated by a nonparametric method does not show up in (5.19); this is because $G(\cdot)$ has been "concentrated out" [see Newey (1994, (3.11) and (3.12))].

Consider a binary response model

$$y = 1[y^* \geq 0] = 1[x'\beta + u \geq 0] \tag{5.21}$$

with $f(u \mid x) = f(u \mid x'\beta)$ and $E(u \mid x) = 0$, which implies $E(u \mid x'\beta) = 0$ and $V(u \mid x) = V(u \mid x'\beta)$. Denoting the distribution function of $u \mid x'\beta$ as $F(u)$ and its density as $f(u)$, a MLE would maximize

$$(1/N) \sum_{i=1}^{N} y_i \ln\{1 - F(-x_i'b)\} + (1 - y_i) \ln\{F(-x_i'b)\} \tag{5.22}$$

wrt b. Since $V(u \mid x'\beta)$ can be also a function of $x'\beta$, b in the likelihood function may appear in the variance function as well as in the mean function. Klein and Spady (1993) propose a quasi-MLE (QMLE) maximizing

$$Q_N(b) \equiv (1/N) \sum_{i=1}^{N} y_i \ln\{1 - F_N(-x_i'b)\} + (1 - y_i) \ln\{F_N(-x_i'b)\} \tag{5.23}$$

wrt b, where $F_N(-x_i'b)$ is an estimate for $F(-x_i'b)$, say

$$F_N(-x_i'b) = \sum_{j \neq i} K((x_j'b - x_i'b)/h)(1 - y_j) \Big/ \sum_{j \neq i} K((x_j'b - x_i'b)/h), \tag{5.24}$$

which is motivated by $F(-x_i'\beta) = E(1 - y \mid x_i'\beta)$. QMLE is also a "1.5-stage estimator" as SLS. Klein and Spady (1993) also allow nonlinear indices.

Assume that x_k is continuous with $\beta_k \neq 0$ and that $u \mid x'\beta$ has the unbounded support. Then as in RCE and MSC of the previous chapter, only $\beta_1/|\beta_k| \cdots \beta_{k+1}/|\beta_k|$, $\mathrm{sgn}(\beta_k)$ are identified. For asymptotic distribution, treat $\mathrm{sgn}(\beta_k)$ as known, since it is estimated at a rate infinitely faster than $\sqrt{N}$ as in RCE. When the scale of y is observed, ADE and SLS estimate all k parameters but there is an unknown scale factor attached to all parameters. When the scale of y is not observed as in binary models, we can estimate only $k - 1$ parameters and $\mathrm{sgn}(\beta_k)$. So "up-to-scale-identifiability" can have different meanings.

The population version of (5.22) can be written as

$$E_x[\{1 - F(-x'\beta)\} \cdot \ln\{1 - F(-x'b)\} + F(-x'\beta) \cdot \ln\{F(-x'b)\}] \tag{5.25}$$

$$\equiv E_x[t \cdot \ln(s) + (1 - t) \cdot \ln(1 - s)], \tag{5.26}$$

where $t \equiv 1 - F(-x'\beta)$ and $s \equiv 1 - F(-x'b)$; (5.26) is maximized at $t = s \Leftrightarrow \beta = b$, which is the basis for identifying β in the Klein–Spady estimator b_{QMLE}, as SLS is based on the fundamental fact that the squared loss function is minimized at $E(y \mid x)$.

The $(k-1) \times 1$ b_{QMLE} is $\sqrt{N}$-consistent and asymptotically normal with $V[\sqrt{N}(b_{\text{QMLE}} - \beta_c)]$ being

$$E^{-1}[\{x_c - E(x_c \mid x'\beta)\}\{x_c - E(x_c \mid x'\beta)\}' f^2(-x'\beta)/$$

$$\{F(x'\beta)(1 - F(x'\beta))\}]. \tag{5.27}$$

This is reminiscent of the variance of probit except that the regressor part is $x_c - E(x_c \mid x'\beta)$ not x. As in SLS, estimating F by F_N does not affect the asymptotic distribution of b_{QMLE} except in the regressor part. Cosslett (1987) and Chamberlain (1986) prove that (5.27) is the semiparametric efficiency bound under the independence of u from x, which is stronger than the "index sufficiency" condition used by QMLE.

10.6   Average Derivative Estimators (ADE) and Multiple Indices

In the first part of the previous section, we examined the density-weighted average derivative estimator (ADE) for single-index models. Consider the "unweighted" ADE $E\{\nabla r(x)\}$ instead of (5.2). Since

$$E\{\nabla r(x)\} = E\{dG(x'\beta)/d(x'\beta)\} \cdot \beta, \tag{6.1}$$

$E\{\nabla r(x)\} = \beta$ if $G(x'\beta) = x'\beta$. That is, the (unweighted) average derivative becomes β if the model is indeed linear. In this sense, the average derivative is more natural than weighted ones.

Observe that

$$E\{\nabla r(x)\} = E\{\lambda(x) \cdot y\}, \text{ where } \lambda(x) \equiv -\nabla f(x)/f(x) = -\nabla \ln f(x), \tag{6.2}$$

which follows from [recall (5.4) and (5.5)]

$$E\{\nabla r(x)\} = \int \nabla r(x) \cdot f(x) dx = r(x)f(x)\big|_{\partial X} - \int \nabla f(x) \cdot r(x) dx$$

$$= -\int [\{\nabla f(x)/f(x)\} \cdot r(x)] \cdot f(x) dx = E\{\lambda(x)r(x)\} = E\{\lambda(x)y\}.$$

Further observe that (Stoker, 1991), using (6.2) and $E\{\lambda(x)\} = 0$,

$$\mu \equiv E\{\nabla r(x)\} = \text{COV}\{\lambda(x), y\} \tag{6.3}$$

$$= \{\text{COV}(\lambda(x), x)\}^{-1} \cdot \text{COV}(\lambda(x), y), \tag{6.4}$$

where (6.4) follows from

$$I_k = E\{\partial x/\partial x\} = x \cdot f(x)\big|_{\partial X} - \int \nabla f(x) \cdot x dx$$

$$= \int \{\lambda(x)x\} f(x) dx = \text{COV}(\lambda(x), x). \tag{6.5}$$

Let $\lambda_N(x_i) \equiv -\nabla f_N(x_i)/f_N(x_i)$, which are defined in (5.7) and (5.6). Also let $\nabla r_N(x_i)$ be given by

$$\nabla r_N(x_i) = h^{-1} \sum_{j \neq i} \nabla K((x_i - x_j)/h) y_j / f_N(x_i) - r_N(x_i) \nabla f_N(x_i)/f_N(x_i),$$

which is the gradient of the leave-one-out kernel estimator $r_N(x_i)$. From (6.1), (6.3), and (6.4), respectively, we get the following three ADE's for μ:

$$\text{Direct ADE: } (1/N) \sum_i \nabla r_N(x_i), \tag{6.6}$$

$$\text{Indirect ADE: } (1/N) \sum_i \lambda_N(x_i)(y_i - \bar{y}), \tag{6.7}$$

$$\text{Indirect IVE ADE: } \left[(1/N) \sum_i \lambda_N(x_i)(x_i - \bar{x})' \right]^{-1}$$

$$\cdot (1/N) \sum_i \lambda_N(x_i)(y_i - \bar{y}), \tag{6.8}$$

where $\bar{x}$ and $\bar{y}$ may be omitted $E\lambda(x) = 0$. (6.8) is similar to δ_N^* in the previous section; δ_N^* has $\nabla f_N(x_i)$ in place of $\lambda_N(x_i)$ due to the weighting by $f(x_i)$. Under certain conditions, (6.6) to (6.8) are asymptotically equivalent and $\sqrt{N}$-consistent (Stoker, 1991), although different regularity conditions are needed for different estimators. In small samples, however, it seems that (6.8) behaves better than (6.6) and (6.7).

Denoting the indirect ADE as m_N, Härdle and Stoker (1989) prove that $\sqrt{N}(m_N - \mu)$ is asymptotically normal with the variance

$$V[u\lambda(x) + \{\nabla r(x) - E\nabla r(x)\}]; \tag{6.9}$$

compare this to (5.14) which differs only by $f(x)$ [(6.9) can be proven easily using (2.7)]. The variance is estimable with

$$(1/N) \sum_i r_{Ni} r_{Ni}' - \left\{ (1/N) \sum_i r_{Ni} \right\} \cdot \left\{ (1/N) \sum_i r_{Ni} \right\}', \tag{6.10}$$

where

$$r_{Ni} \equiv \{(N-1)h^k\}^{-1} \sum_{j \neq i} [\{\nabla K((x_i - x_j)/h)/h$$

$$- K((x_i - x_j)/h)\lambda_N(x_j)\} y_j / f_N(x_j) + \lambda_N(x_i) y_i].$$

This has been applied to a demand analysis by Härdle et al. (1991). Härdle and Stoker (1989) also estimate $G(\cdot)$ by nonparametrically regressing y_i on $x_i'b_N$.

Newey and Stoker (1993, (3.6)) show that the semiparametric efficiency bound for a weighted average derivative $E\{\omega(x)\nabla r(x)\}$ is the expected value of the outer-product of

$$\{-\nabla\omega(x) - \omega(x)\nabla f(x)/f(x)\} \cdot u + \{\omega(x)\nabla r(x) - E(\omega(x)\nabla r(x))\}. \quad (6.11)$$

With $\omega(x) = 1$, we get the influence function for m_N in (6.9), which means m_N is an efficient estimator for $E\nabla r(x)$.

Single-index models provide a sensible compromise between the parametric and purely nonparametric models. The single-index model limits the nonlinearity into the indexing function $G(\cdot)$, preserving the linearity in $x'\beta$. Consequently, it is more "manageable" than the purely nonparametric model. One major drawback however is the up-to-scale identification that more or less limits its application to cases where the scale of the response variable is not observed.

The "direct" ADE $(1/N)\sum_i \nabla r_N(x_i)$, which is also treated in Rilstone (1991), is interesting even if the model is not a single-index model. In a nonparametric model $y = r(x) + u$, $\nabla r(x_0)$ measures the marginal effect of x on y at $x = x_0$; $\nabla r(x) = \beta$ for all x if $r(x) = x'\beta$. But since we have N different marginal effects $\nabla r(x_i)$, $i = 1 \ldots N$, we may want to use $(1/N)\sum_i \nabla r_N(x_i)$ as a measure representing the N marginal effects; this is analogous to using a location measure to represent N observations.

Using (2.5) with $m_i(x_i, a_i) = a_i = \nabla r_N(x_i)$ and $u_i = y_i - r(x_i)$,

$$(1/\sqrt{N})\sum_i \nabla r_N(x_i) =^p (1/\sqrt{N})\sum_i [\nabla r(x_i) - \{\nabla f(x_i)/f(x_i)\} \cdot u_i].$$

Thus,

$$\sqrt{N}\left\{(1/N)\sum_i \nabla r_N(x_i) - E\nabla r(x)\right\}$$

$$=^p (1/\sqrt{N})\sum_i [\{\nabla r(x_i) - E\nabla r(x)\} - \{\nabla f(x_i)/f(x_i)\}u_i], \quad (6.12)$$

where the influence function is the same as that in (6.9). Note that one estimate for $V[\nabla r(x)]$ is the "sample" variance

$$(1/N)\sum_i \{\nabla r_N(x_i)\}^2 - \{(1/N)\nabla r_N(x_i)\}^2, \quad (6.13)$$

which, however, accounts only for the first term in (6.12). Obviously, the first-stage error of estimating $\nabla r(x)$ with $\nabla r_N(x)$ appears in the second term of (6.12). (6.13) is zero if either $r(x) = 0$ or $r(x) = x'\beta$. Thus (6.13) may be used as a measure for the nonlinearity in $r(x)$ as x varies.

Single-index models can be extended to double- or multiple-index models. As an example of triple-index model (an example of double index model is given in the preceding chapter), consider a switching regression model with uknown sample separation:

$$
\begin{aligned}
y &= x_1'\beta_1 + u_1 \quad \text{if } z'\gamma + v > 0, \\
&= x_2'\beta_2 + u_2 \quad \text{if } z'\gamma + v \le 0;
\end{aligned}
\tag{6.14}
$$

y is a mixture of two populations depending on $z'\gamma + v$. Denoting the joint density of (u_1, u_2, v) as $f(u_1, u_2, v)$, which is independent of x,

$$
E(y \mid x, z) = E(y \mid x, v \ge -z'\gamma) \cdot P(v \ge -z'\gamma \mid z) + E(y \mid x, v < -z'\gamma)
$$

$$
\cdot P(v < -z'\gamma \mid z)
$$

$$
= \int\int\int_{-z'\gamma} (x_1'\beta_1 + u_1) f(u_1, u_2, v) dv du_1 du_2
$$

$$
+ \int\int\int^{-z'\gamma} (x_2'\beta_2 + u_2) \cdot f(u_1, u_2, v) dv du_1 du_2
\tag{6.15}
$$

$$
= x_1'\beta_1 \cdot P(v > -z'\gamma \mid z) + x_2'\beta_2 \cdot P(v \le -z'\gamma \mid z)
$$

$$
+ \int\int\int_{-z'\gamma} u_1 f(u_1, u_2, v) dv du_1 du_2
$$

$$
+ \int\int\int^{-z'\gamma} u_2 f(u_1, u_2, v) dv du_1 du_2,
\tag{6.16}
$$

where ∞ and $-\infty$ are omitted in $\int$. This is a triple index model:

$$
E(y \mid x, z) = G(x_1'\beta_1, x_2'\beta_2, z'\gamma).
\tag{6.17}
$$

If $f(u_1, u_2, v) = f(u_1)f(u_2)f(v)$ and $Eu_1 = Eu_2 = 0$, then the last two terms in (6.16) disappear. Another example of multiple-index model can be seen in multinomial choice models where the regression functions in the choice utility equations become the indices. See Ichimura and Lee (1991) for an extension of Ichimura (1993) to multiple-index models. Also see L.F. Lee (1995) for a multiple-index extension of Klein and Spady (1993) for multinomial choice models.

A multiple-index model can also arise as a simplification of nonparametric models. One such method is "projection pursuit regression (PPR)" in Friedman and Stuetzle (1981) with

$$
E(y \mid x) = \sum_{j=1}^{p} G_j(x'\beta_j),
\tag{6.18}
$$

where p is a integer to be chosen by the researcher, β_j is a $k \times 1$ parameter vector, and $G_j(\cdot)$ is an unknown univariate function, $j = 1 \ldots p$. To implement this, first find b_1 and $G_{N1}(\cdot)$ minimizing $(1/N) \sum_i \{y_i - G_{N1}(x_i'b_1)\}^2$.

Then find b_2 and $G_{N2}(\cdot)$ minimizing $(1/N)\sum_i\{y_i-G_{N1}(x_i'b_1)-G_{N2}(x_i'b_2)\}^2$. This continues until the reduction in the sum of squared errors becomes small, which then determines p. The main idea of PPR is replacing a high-dimensional nonparametric problem with many one-dimensional nonparametric subproblems. In practice, one can try various values for β_j and estimate $G_j(\cdot)$ with a kernel method for a given $x'\beta_j$ at the jth step. Alternatively, we may approximate $G_j(s)$ with a polynomial function, say $\alpha_{j1}s+\alpha_{j2}s^2+\alpha_{j3}s^3$, and estimate α_{j1}, α_{j2}, α_{j3}, and β_j jointly. Either way, estimating β_j and $G_j(\cdot)$ is a very difficult computational problem, almost countless local minima can be found. It is necessary for p to go to infinity as $N\to\infty$ for (6.18) to approximate an arbitrary $E(y\mid x)$.

A special case of PPR (at least from our econometric viewpoint) is a artificial neural network model of "a single hidden layer with p neurons" [see Kuan and White (1994), for instance] in which

$$E(y\mid x)=\alpha_0+\sum_{j=1}^{p}\alpha_j\cdot G(x'\beta_j), \tag{6.19}$$

where $G(\cdot)$ is a known nonlinear function, for instance, $G(s)=(1+e^{-s})^{-1}$. Here $\alpha_0,\ \alpha_j,\ \beta_j,\ j=1\ldots p$, are the parameters to be estimated; computationally, this seems as hard as PPR due to too many local minima. As in PPR, p should go to ∞ as $N\to\infty$; in fact, β_j's is not identified in this case (Bierens, 1994, p. 94), which explains the computational problem.

10.7   Nonparametric Instrumental Variable Estimation

Consider the following model:

$$y_1=\alpha y_2+x_1'\beta_1+u_1,\quad E(u_1\mid x)=0, \tag{7.1}$$

which is one of two simultaneous equations, where y_2 is the endogenous regressor and x is the exogenous variables of the system that includes at least one variable not in x_1. Suppose $E(y_2\mid x)$ is known. Then we can estimate α and β_1 applying IVE with $(E(y_2\mid x),x_1')'$ as the instrument. Defining

$$z\equiv(E(y_2\mid x),x_1')',\quad w\equiv(y_2,x_1')',\quad \gamma\equiv(\alpha,\beta_1')',\quad g_N\equiv(a_N,b_{1N}')',$$

the variance matrix of $\sqrt{N}(g_N-\gamma)$ is

$$E^{-1}(wz')\cdot E(u_1^2zz')\cdot E^{-1}(zw'). \tag{7.2}$$

Since

$$E(wz')=E_x[E(w\mid x)\cdot z']=E_x(zz')=E(zz'),$$

(7.2) becomes

$$E^{-1}(zz') \cdot E(u_1^2 zz') \cdot E^{-1}(zz'). \qquad (7.3)$$

Under homoskedasticity, this becomes $E(u_1^2) \cdot E^{-1}(zz')$.

In general, we do not know $E(y_2 \mid x)$, which means that the above IVE is not feasible. If we know that the other part of the simultaneous system (that is, the y_2 equation) is linear, then $E(y_2 \mid x)$ is a linear function of x. But even if $E(y_2 \mid x)$ is not linear, or even if the y_2 equation is unspecified, still we can have a feasible version by replacing $E(y_2 \mid x)$ with a nonparametric estimate $E_N(y_2 \mid x)$. Let $z_{Ni} \equiv (E_N(y_2 \mid x_i), x_i')'$. Then the feasible IVE, still denoted as g_N, satisfies

$$(1/\sqrt{N}) \sum_i (y_{1i} - w_i g_N) z_{Ni} = 0. \qquad (7.4)$$

Using (2.4), $E(m_a \mid x) = E(u \mid x) = 0$: there is no effect of using an estimated instrument (in fact, this holds very generally). Thus, the feasible IVE has the same asymptotic distribution as the infeasible IVE. The feasible IVE is equivalent to the 2SLSE for (7.1); to see this, rewrite (7.1) as

$$y_1 = \alpha E(y_2 \mid x) + x_1' \beta_1 + u_1 + \alpha\{y_2 - E(y_2 \mid x)\}$$
$$= \alpha E_N(y_2 \mid x) + x_1' \beta_1 + u_1$$
$$+ \alpha\{y_2 - E(y_2 \mid x)\} - \alpha\{E_N(y_2 \mid x) - E(y_2 \mid x)\},$$

where the error term has three components. When LSE is applied in the second stage of 2SLSE, the last two error terms cancel each other due to (2.10), and only u_1 is left as the error term in the second-stage LSE. Thus, (7.2) is also the asymptotic variance of the 2SLSE.

Deriving the efficient covariance matrix (3.1) under the moment condition $E\{(y - \alpha_1 y_2 - x_1' \beta_1) \mid x\} = E(u_1 \mid x) = 0$, we get

$$E_x^{-1}\{E(w \mid x) \cdot E(w' \mid x) \cdot (V(u_1 \mid x))^{-1}\} = E_x^{-1}\{zz' \cdot (V(u_1 \mid x))^{-1}\}. \qquad (7.5)$$

Under homoskedasticity, this becomes

$$E(u_1^2) \cdot E^{-1}(zz').$$

Thus the IVE (2SLSE), feasible or infeasible, is efficient under homoskedasticity. In the rest of this section, we discuss IVE for nonlinear models.

Suppose we have a nonlinear $s \times 1$ moment condition

$$E\{\rho(z_i, \gamma) \mid x_i\} = E\{u_i \mid x_i\} = 0, \qquad (7.6)$$

where z_i includes some endogenous regressors and some exogenous regressors x_i. Assume homoskedasticity:

$$E\{\rho(z_i, \gamma) \cdot \rho(z_i, \gamma)' \mid x_i\} \equiv \Omega \quad \text{(a constant matrix).} \qquad (7.7)$$

For instance, let $\gamma = (\alpha, \beta')'$ be a $k \times 1$ vector, $z_i \equiv (d_i, x_i', y_i)'$ and

$$\rho(z_i, \gamma) = y_i - \alpha \cdot d_i - r(x_i, \beta) = u_i, \tag{7.8}$$

where d_i is an endogenous dummy variable and $r(\cdot)$ is a known nonlinear function of β.

Recall the efficient covariance matrix (3.1):

$$E_x^{-1}[E(\rho_g(\gamma) \mid x) \cdot \{E(\rho(\gamma)\rho(\gamma)' \mid x\}\}^{-1} \cdot E(\rho_g(\gamma)' \mid x)], \tag{7.9}$$

where $\rho_g \equiv \partial\rho/\partial g$ is a $k \times s$ matrix. Under homoskedasticity, this becomes

$$E_x^{-1}\{E(\rho_g(\gamma) \mid x) \cdot \Omega^{-1} \cdot E(\rho_g(\gamma)' \mid x)\}. \tag{7.10}$$

If we have a consistent estimate for γ, then Ω can be estimated with the residuals. The problem is in getting $E(\rho_g(\gamma) \mid x) = E(\rho_g(z_i, \gamma) \mid x)$. In the example (7.8), it is ($r_b \equiv \partial r/\partial b$)

$$E\{(d_i, r_b(x_i, \beta)')' \mid x_i)\} = (E(d_i \mid x_i), r_b(x_i, \beta)')'. \tag{7.11}$$

Unless we know the conditional distribution of $d \mid x$, there is no way of getting $E(d_i \mid x_i)$ parametrically. However, the result for nonparametric IVE with linear models suggests that we may replace $E(d_i \mid x_i)$ with a kernel nonparametric estimator

$$E_N(d_i \mid x_i) \equiv \sum_{j \neq i}^{N} K((x_j - x_i)/h)d_j \Big/ \sum_{j \neq i}^{N} K((x_j - x_i)/h) \tag{7.12}$$

and attain the efficiency bound (7.10) under homoskedasticity. This is indeed the case as proved by Newey (1990b).

Specifically, the following is the procedure for efficient IVE of nonlinear (simultaneous) equations under homoskedasticity:

(i) Use inefficient instruments to estimate γ by g_0 consistently; for (7.8), we may use k linearly independent functions of x, assuming that d is correlated with x.

(ii) Estimate Ω with $\Omega_N \equiv (1/N) \sum_i \rho(z_i, g_0) \cdot \rho(z_i, g_0)'$. \hfill (7.13)

(iii) Obtain $E(\rho_g(z_i, \gamma) \mid x_i)$ using a nonparametric estimate such as (7.12). Denote the estimates as $D_i \equiv D_i(z_i, g_0)$ ($k \times s$ matrix).

(iv) Take one step from g_0 to get the efficient estimate g_N:

$$g_N = g_0 - \left\{\sum_i D_i \Omega_N^{-1} D_i'\right\}^{-1} \cdot \sum_i D_i \Omega_N^{-1} \rho(z_i, g_0). \tag{7.14}$$

A practical problem associated with any semi-nonparametric estimation is choosing a bandwidth, which also appears in the above nonparametric IVE of Newey (1990b). Robinson (1991) proposes another way to estimate $E(\rho_g \mid x)$ without the bandwidth problem. Robinson (1991) assumes independence between the error term and the exogenous variables in the system. Then using the residuals obtained from an initial $\sqrt{N}$-consistent estimate, $E(\rho_g \mid x)$ is estimated analogously to the method of simulated moments as presented in the following.

Let $z \equiv (x', y)'$. A key assumption in the Robinson approach is that $\rho(z, \gamma) = u$ is solvable for y to yield

$$y = R(x, \gamma, u). \tag{7.15}$$

Substitute this into $\rho_g(y, x, u, \gamma)$ to get $\rho_g\{R(x, \gamma, u), x, u, \gamma\}$. Recall that what we want to estimate is

$$E[\rho_g\{R(x, \gamma, u), x, u, \gamma\} \mid x_i]. \tag{7.16}$$

Let $\hat{u}_i$, $i = 1, \ldots, N$, denote the residuals using an initial $\sqrt{N}$-consistent estimate g_0. Then we can estimate (7.16) with

$$\{1/(N-1)\} \sum_{j \neq i} \rho_g\{R(x_i, g_0, \hat{u}_j), x_i, \hat{u}_j, g_0\}; \tag{7.17}$$

it is possible not to use all $N - 1$ observations. Replacing D_i in (7.13) with (7.17), we can attain the same efficiency as in Newey (1990b).

10.8 Nonparametric Estimation for Limited Endogenous Regressors

In this section we examine two estimators. The first is a two-stage estimator for simultaneous equations with limited endogenous regressors by Lee (1995b). The second is a binary choice problem under uncertainty by Ahn and Manski (1993). Both have a common aspect of replacing the conditional mean of a limited variable with a nonparametric estimate.

Consider a censored structural form (SF) equation

$$y_1 = \max(\alpha \cdot y_2 + x_1'\beta + \varepsilon, 0), \tag{8.1}$$

where $y_2 = \tau_2(y_2^*)$ is a transformation of the latent continuous variable y_2^* with $E(y_2^*\varepsilon) \neq 0$, and x_1 is a $k_1 \times 1$ vector of exogenous variables. As an example, y_1 may be female labor supply, while y_2 is a dummy variable for a labor union membership (then τ_2 is an indicator function). As another example, y_1 may be expenditure on durable goods, while y_2 $(= y_2^*)$ is household income (then τ_2 is an identity function). More generally than (8.1), we may consider

$$y_1 = \tau_1(y_1^*) = \tau_1(\alpha \cdot \tau_2(y_2^*) + x_1'\beta + \varepsilon) = \tau_1(\alpha \cdot y_2 + x_1'\beta + \varepsilon), \qquad (8.2)$$

where τ_1 is a function whose form is known; $\tau_1(\cdot) = \max(\cdot, 0)$ in (8.1). If both τ_1 and τ_2 are indicator functions, then we get simultaneous equations with binary-dependent variables. The regressor y_2 in (8.1) may have its own SF with the regressors y_1 and x_2. Define a $k \times 1$ exogenous regressor vector x as the variables in x_1 and x_2.

Rewrite (8.1) as

$$y_1 = \max(\alpha \cdot E(y_2 \mid x) + x_1'\beta + \varepsilon + \alpha\{y_2 - E(y_2 \mid x)\}, 0). \qquad (8.3)$$

Define $v \equiv y_2 - E(y_2 \mid x)$. Then α and β can be estimated in two stages. The first step is estimating $E(y_2 \mid x)$ with a kernel nonparametric estimate $E_N(y_2 \mid x)$. Further rewriting (8.3) as

$$y_1 = \max(\alpha \cdot E_N(y_2 \mid x) + x_1'\eta + \varepsilon + \alpha v$$

$$+ \alpha\{E(y_2 \mid x) - E_N(y_2 \mid x)\}, 0), \qquad (8.4)$$

the second step is estimating (8.4) with a semiparametric method that requires weak assumptions on $\varepsilon + \alpha v$. This strategy is applicable not only to (8.1) but to (8.2) in general. The first nonparametric step is uniform regardless of the form of τ_2, for we need only $E_N(y_2 \mid x)$. The second step, however, is τ_1-specific. If $\tau_1(\cdot) = \max(\cdot, 0)$ as in (8.1), the second step needs a censored model estimator, and if τ_1 is an indicator function, the second step needs a binary model estimator. For (8.1), we will use Powell's (1986) SCLS in the second stage, although almost all $\sqrt{N}$-consistent semiparametric censored model estimator can be used. Lee (1995b) also shows a simpler version under the assumption $E(y_2 \mid x) = x'\eta$, as well as extensions to the general model (8.2) and to cases with more than one endogenous regressors.

Define the following to simplify notations:

$$\gamma \equiv (\alpha, \beta')', \qquad u \equiv \varepsilon + \alpha v,$$
$$z \equiv (E(y_2 \mid x), x_1')', \qquad \tilde{z} \equiv (E_N(y_2 \mid x), x_1')',$$

where x and z are, respectively, $k \times 1$ and $(k_1 + 1) \times 1$ vectors. We will denote a generic element for γ as $g \equiv (a, b')'$. Now (8.3) can be written as

$$y_1 = \max(z'\gamma + u, 0). \qquad (8.5)$$

Minimize the following SCLS minimand wrt g:

$$Q_N(g, h_N) \equiv (1/N) \sum_i [\{y_{1i} - \max(y_{1i}/2, \tilde{z}_i'g)\}^2$$

$$+ 1[y_{1i} > 2\tilde{z}_i'g] \cdot [(0.5y_{1i})^2 - \{\max(\tilde{z}_i'g, 0)\}^2], \tag{8.6}$$

to get the nonparametric two stage SCLS g_N. Then,

$$\sqrt{N}(g_N - \gamma) \cong N(0, E^{-1}\{1[|u| < z'\gamma] \cdot zz'\} \cdot D \cdot E^{-1}\{1[|u| < z'\gamma] \cdot zz'\}), \tag{8.7}$$

$$D \equiv E[1[z'\gamma > 0] \cdot \min\{u^2, (z'\gamma)^2\} \cdot zz']$$

$$+ \alpha^2 E[1[z'\gamma > 0] \cdot v^2 \cdot \{F_{u|x}(z'\gamma) - F_{u|x}(-z'\gamma)\}^2 \cdot zz']$$

$$- 2\alpha \cdot E[\{1[|u| < z'\gamma] \cdot u + (1[u > z'\gamma] - 1[u < -z'\gamma]) \cdot z'\gamma\}$$

$$\cdot 1[z'\gamma > 0] \cdot v \cdot \{F_{u|x}(z'\gamma) - F_{u|x}(-z'\gamma)\} \cdot zz'],$$

where $f_{u|x}$ is the density of $u \mid x$ and $F_{u|x}$ is the distribution function. As for estimating D, the second and third terms can be estimated by

$$a_N^2(1/N) \sum_i 1[\tilde{z}_i'g_N > 0] \cdot \tilde{v}_i^2 \cdot \{F_N(\tilde{z}_i'g_N)\}^2 \cdot \tilde{z}_i\tilde{z}_i'$$

$$- 2a_N(1/N) \sum_i \{1[|\tilde{u}_i| < \tilde{z}_i'g_N] \cdot \tilde{u}_i + (1[\tilde{u}_i > \tilde{z}_i'g_N]$$

$$- 1[\tilde{u}_i < -\tilde{z}_i'g_N]) \cdot \tilde{z}_i'g_N\}$$

$$\cdot 1[\tilde{z}_i'g_N > 0] \cdot \tilde{v}_i \cdot F_N(\tilde{z}_i'g_N) \cdot \tilde{z}_i\tilde{z}_i', \tag{8.8}$$

where

$$\tilde{v}_i \equiv y_{2i} - m_N(x_i), \quad \tilde{u}_i \equiv y_{1i} - \tilde{z}_i'g_N,$$

$$F_N(\tilde{z}_i'g_N) \equiv \{1/((N-1)s^k)\} \sum_{j \neq i} 1[|\tilde{u}_j| < \tilde{z}_i'g_N]$$

$$\cdot K((x_j - x_i)/s)/f_N(x_i);$$

is a smoothing parameter. Under the independence between x and u, F_N becomes

$$F_N(\tilde{z}_i'g_N) \equiv \{1/(N-1)\} \sum_{j \neq i} 1[|\tilde{u}_j| < \tilde{z}_i'g_N]. \tag{8.9}$$

If there is no censoring, then the indicator functions and $F_{u|x}$ in (8.7) drop out to make the nonparametric SCLS equal to the nonparametric 2SLSE in the preceding section with the asymptotic variance

$$E^{-1}(zz') \cdot E(\varepsilon^2 zz') \cdot E^{-1}(zz'). \tag{8.10}$$

Consider a binary choice problem where the utility s from the choices 0 and 1 are

$$s_0 = z_0'\alpha + w_0'\beta + u_0, \quad s_1 = z_1'\alpha + w_1'\beta + u_1,$$
$$w_0 = E(w \mid x, 0) + v_0, \quad w_1 = E(w \mid x, 1) + v_1, \tag{8.11}$$

where $E(v_j \mid x, j) = 0$, x may overlap with z_0 and z_1, $E(w \mid x, j)$ is the conditional mean of w given x and j is chosen for $j = 0, 1$, z_0 is a $k_z \times 1$ vector, and w is a $k_w \times 1$ vector. Here, w is realized only after the agent makes his or her choice; thus, although w is a single vector, w as relevant to his or her decision becomes two different vectors w_0 and w_1. Substitute w_0 and w_1 equations into the first line of (8.11) to get

$$
\begin{aligned}
s_0 &= z_0'\alpha + E(w \mid x, 0)'\beta + v_0'\beta_0 + u_0, \\
s_1 &= z_1'\alpha + E(w \mid x, 1)'\beta + v_1'\beta_1 + u_1.
\end{aligned}
\tag{8.12}
$$

The agent chooses 1 if $s_1 > s_0$, which is

$$
u_1 - u_0 + (v_1 - v_0)'\beta > -(z_1 - z_0)'\alpha - \{E(w \mid x, 1) - E(w \mid x, 0)\}'\beta. \tag{8.13}
$$

Denoting the choice as y and assuming that the lhs of (8.13) follows $N(0, \sigma^2)$, where σ is an unknown constant, we have

$$
P(y = 1 \mid z, x) = \Phi[z'\alpha/\sigma + \{E(w \mid x, 1) - E(w \mid x, 0)\}'\beta/\sigma], \tag{8.14}
$$

where $z \equiv z_1 - z_0$ and $\Phi(\cdot)$ is the $N(0, 1)$ distribution function.

If we specify $E(w \mid x, 0)$ as a function of x, then a parametric estimation is possible for (8.14). But Ahn and Manski (1993) suggest the following two-stage method. First, replace $E(w \mid x, j)$, $j = 0, 1$, with

$$
E_N(w \mid x, j) = \sum_i K((x_i - x)/h)1[y_i = j] \cdot w_i \bigg/ \sum_i K((x_i - x)/h)1[y_i = j].
\tag{8.15}
$$

Second, apply probit to (8.14) with $E(\cdot \mid \cdot)$ replaced by $E_N(\cdot \mid \cdot)$. Note that "rational expectation" is assumed: the agent's subjective expectation is the same as the objective expectation governing the realization of w.

Define

$$
\omega \equiv E(w \mid x, 1) - E(w \mid x, 0), \quad \gamma \equiv (\alpha'/\sigma, \beta'/\sigma)',
$$

where γ is a $(k_z + k_w) \times 1$ vector. Then the TSE g_N has the following asymptotic distribution:

$$
\sqrt{N}(g_N - \gamma) =^d N(0, I_f^{-1}(I_f + \Omega)I_f^{-1}); \tag{8.16}
$$

$$
\Omega \equiv E[\mu(x) \cdot \{V(w \mid x, 0)/P(y = 0 \mid x) + V(w \mid x, 1)/P(y = 1 \mid x)\} \cdot \mu(x)'],
$$

$$
\mu(x) \qquad = E[\partial^2\{y \cdot \ln \Phi + (1 - y) \cdot \ln \Phi\}/\partial g \partial \omega \mid x],
$$
$$
(k_z + k_w) \times k_w
$$

where I_f is the probit information matrix when ω is known, and Ω has two factors: one is the variance $V(w \mid x, j)$ of the first-stage estimate, and the other is the link $\mu(x)$ through which the first-stage error affects the second stage. Ahn and Manski (1993) assume that, given x and $y = j$, v_j is independent of z to exclude the covariance term between the first-stage and second-stage errors. But this is unnecessary, since we can easily allow the covariance term using (2.11) with m being the score function of probit.

10.9   Semi-Nonparametric MLE with Series Expansion

In this section, we examine "semi-nonparametric MLE" with series expansion in Gallant and Tauchen (1989); see also Gallant and Nychka (1987). To be coherent with our focus on iid data, we present the (time series) work of Gallant and Tauchen in the iid framework.

Suppose that y_i is a $m \times 1$ response variable vector and x_i is a $k \times 1$ regressor vector independent of u_i, and

$$
\begin{array}{ccc}
y_i & = & \beta \cdot x_i & + & u_i. \\
m \times 1 & & m \times k \; k \times 1 & & m \times 1
\end{array}
\tag{9.1}
$$

If $u_i =^d N(0, \Omega) = N(0, CC')$, where C is the lower-triangular Cholesky decomposition, then $v_i \equiv C^{-1}u_i =^d N(0, I_m)$. Denoting the $N(0, I_m)$ density as $\phi(\cdot)$, the joint density for u_i is

$$
\phi(C^{-1}u_i)/\text{DET}(C),
\tag{9.2}
$$

where $\text{DET}(C)$ denotes the determinant of C. Under MLE, we can estimate β and C by maximizing the following wrt b and C:

$$
(1/N) \sum_i \ln[\phi\{C^{-1}(y_i - bx_i)\}/\text{DET}(C)].
\tag{9.3}
$$

But the assumptions for this MLE may be too restrictive. In the following, we show how to relax the assumptions in three aspects: nonnormality, dependence of x and v, and heteroskedasticity.

Recall our discussion on series expansion in Chapter 8 where we introduced a variation of the Hermite polynomial. Consider approximating the density function of $v_i \equiv C^{-1}u_i$ with

$$
h(v) = \left\{ \sum_{|\alpha|=0}^{A} a_\alpha v^\alpha \right\}^2 \cdot \phi(v)/M,
\tag{9.4}
$$

where

$$
\alpha \equiv (\alpha_1, \ldots, \alpha_m)', \quad \alpha_j\text{'s are nonnegative integers,}
$$

$$
|\alpha| \equiv \sum_{j=1}^{m} \alpha_j, \quad v^\alpha \equiv \prod_{j=1}^{m} v_j^{\alpha_j},
$$

$$
M = \int \left\{ \sum_{|\alpha|=0}^{A} a_\alpha v^\alpha \right\}^2 \cdot \phi(v)dv.
$$

To see what $a_\alpha v^\alpha$ looks like, let $A = 2$. Then

$|\alpha| = 0 \Leftrightarrow$ all α_j's are zero.

$|\alpha| = 1 \Leftrightarrow$ only one of α_j's is 1 and the others are 0.

$|\alpha| = 2 \Leftrightarrow$ either only one of α_j's is 2 and the others are 0,
or two of α_j's are 1.

Thus, (9.4) becomes

$$\left\{ a_0 + \sum_{j=1}^{m} a_j v_j + \sum_{j=1}^{m}\sum_{q=1}^{m} a_{jq} v_j v_q \right\}^2 \cdot \phi(v) \Big/ M. \tag{9.5}$$

The density function $h(v)$ includes $\phi(v)$ as a special case when only a_0 is nonzero and the other α_j's are all zero in (9.5).

To allow dependence between x and v, let a_α, $|\alpha| = 1, \ldots, A$ ($|\alpha| = 0$ is excluded), be functions of x,

$$a_\alpha(x) \equiv \sum_{|\gamma|=0}^{G} a_{\alpha\gamma} \cdot x^\gamma,$$

$\gamma \equiv (\gamma_1, \ldots, \gamma_k)'$, γ_j's are nonnegative integers,

$$|\gamma| \equiv \sum_{j=1}^{k} \gamma_j, \quad x^\gamma \equiv \prod_{j=1}^{k} x_j^{\gamma_j}. \tag{9.6}$$

For instance, with $G = 2$, a_{12} in (9.5) is

$$a_{12,0} + \sum_{j=1}^{k} a_{12,j} \cdot x_j + \sum_{j=1}^{k}\sum_{q=1}^{k} a_{12,jq} \cdot x_j x_q. \tag{9.7}$$

If all the coefficients other than the intercepts in $a_\alpha(x)$'s are zero, then this indicates independence between x and v.

If we want to allow heteroskedasticity in addition to (9.6), this can be accommodated by letting C be a function of x as in

$$\underset{m \times m}{C(x_i)} \equiv \underset{m \times m}{C_0(x_i, \eta)}, \tag{9.8}$$

where $C(x_i, \eta)$ is a matrix of functions whose functional form is known and parametrized by η such that $C(x_i, 0)$ is a constant matrix. Then heteroskedasticity will be captured by η in $C(x_1)$.

Combining (9.4), (9.6), and (9.8), the conditional likelihood is

$$h\{C(x_i)^{-1}(y_i - bx_i) \mid x_i\} \equiv \left[\sum_{|\alpha|=0}^{A} a_\alpha(x_i) \right.$$

$$\cdot \{C(x_i)^{-1}(y_i - bx_i)\}^\alpha \Bigg]^2 \cdot \phi\{C(x_i)^{-1}(y_i - bx_i)\} \Bigg/ \tag{9.9}$$

$$\left[\text{DET}\{C(x_i)\} \cdot \int \left\{ \sum_{|\alpha|=0}^{A} a_\alpha(x_i) \cdot v^\alpha \right\}^2 \phi(v) dv \right].$$

Thus, we maximize

$$(1/N) \sum_i h\{C(x_i)^{-1}(y_i - bx_i) \mid x_i\} \tag{9.10}$$

wrt b, η [in $C(x_i)$], a_0, and $a_{\alpha\gamma}$. (9.9) can be simplified in a number of ways. For instance, if we believe that x and u are independent, then only (9.4) is necessary. If we believe that u depends on x only through $V(u \mid x)$, then (9.4) and (9.8) are enough. Maximizing (9.10) may be done using numerical derivatives in practice, since deriving the analytic derivatives can be rather cumbersome.

Recall that series expansion idea is also a nonparametric method, and the smoothing parameter is the number of the terms in the expansion; in the semi-nonparametric MLE, A and G are the smoothing parameters. If we have a function penalizing high A and G, this can be used for the choice of A and G. Owing to the smoothing aspect, the estimators maximizing (9.10) are not $\sqrt{N}$-consistent in principle, although this is often ignored in practice.

10.10 Nonparametric Specification Tests

It is possible to employ nonparametric techniques to test model specifications. In this section, we examine those tests.

Consider $y = E(y \mid x) + u$, where x is a $k \times 1$ vector and

$$H_0 \colon P\{E(y \mid x) = r(x, \beta)\} = 1 \text{ vs. } H_a \colon P\{E(y \mid x) = r(x, \beta)\} < 1, \tag{10.1}$$

where $r(x, \beta)$ is a known (non)linear function of β. One easy way to test H_0 is examining the correlation between $y - r(x, \beta)$ and $E(y \mid x)$, for $y - r(x, \beta) = u + E(y \mid x) - r(x'\beta)$ includes $E(y \mid x)$ if H_0 is wrong. Replacing u_i and $E(y \mid x_i)$, respectively, by $v_i \equiv y_i - r(x_i, b_N)$ and $E_N(y \mid x_i)$, where b_N is the nonlinear LSE, a test statistic is $(1/\sqrt{N}) \sum_i v_i E_N(y \mid x_i)$, which is equal to

$$(1/\sqrt{N}) \sum_i u_i E_N(y \mid x_i) + (1/\sqrt{N}) \sum_i E_N(y \mid x_i)$$

$$\cdot \{E(y \mid x_i) - r(x_i, b_N)\}. \tag{10.2}$$

Applying the mean value theorem to $r(x_i, b_N)$, the second term is

$$(1/\sqrt{N}) \sum_i E_N(y \mid x_i)\{E(y \mid x_i) - r(x_i, \beta) - \nabla r(x_i, b_N^*)'(b_N - \beta)\}$$

$$=^p \sqrt{N} \cdot E[E(y \mid x)\{E(y \mid x) - r(x, \beta)\}]$$
$$- (1/N) \sum_i E(y \mid x_i)\nabla r(x_i, \beta)' \cdot \sqrt{N}(b_N - \beta), \qquad (10.3)$$

where $b_N \in (b_N, \beta)$ and $\nabla r \equiv \partial r/\partial b$. Define λ_i such that $\sqrt{N}(b_N - \beta) =^p$ $(1/\sqrt{N}) \sum_i \lambda_i$. Substitute λ_i into (10.3) and use this to rewrite (10.2) as

$$(1/\sqrt{N}) \sum_i [u_i E(y \mid x_i) - E\{E(y \mid x)\nabla r(x, \beta)'\} \cdot \lambda_i] \qquad (10.4)$$

$$+ \sqrt{N} \cdot E[E(y \mid x) \cdot \{E(y \mid x) - r(x, \beta)\}]; \qquad (10.5)$$

(2.4) is envoked for the first term of (10.4) with $E(m_a \mid x) = E(u \mid x) = 0$.

Define ζ_i such that (10.4) $= (1/\sqrt{N}) \sum_i \zeta_i \Rightarrow N(0, \sigma^2)$ where $\sigma^2 =^p$ $(1/N) \sum_i \zeta_{iN}\zeta_{iN}'$; ζ_{iN} is ζ_i with its unknowns replaced by estimates. (10.5) is zero under H_0. Under H_a however, there is no guarantee that (10.5) is nonzero despite $E(y \mid x) - r(x, \beta) \neq 0$ for some x. In the following, we introduce tests avoiding this pitfall; a test statistic analogous to $(1/\sqrt{N}) \sum_i v_i E_N(y \mid x_i)$ was proposed by Wooldridge [1992, (3.4)] although its asymptotic distribution was derived under homoskedasticity.

The conditional moment condition $E(u \mid x) = 0$ implies $E\{u \cdot g(x)\} = 0$ for any square-integrable function of x. Method-of-moments tests with a finite number of functions $g_1(x) \ldots g_m(x)$ have been discussed already: the test statistic is $g_N' C_N^{-1} g_N =^d \chi_m^2$, where $C_N =^p V(g_N)$ and (v_i's are residuals)

$$g_N \equiv \left\{ (1/\sqrt{N}) \sum_i v_i g_1(x_i) \ldots (1/\sqrt{N}) \sum_i v_i g_m(x_i) \right\}'. \qquad (10.6)$$

De Jong and Bierens (1994) show that if the number m of functions goes to infinity and if the sequence of functions spans the space of square-integrable functions, then a method-of-moments test using the infinite moment conditions can detect any kind of violation of H_0, that is, the test is "consistent." The test statistic they propose is

$$\sqrt{2m}\{g_N' C_N^{-1} g_N - m) =^d N(0, 1), \qquad (10.7)$$

which is a CLT applied to m-many χ_1^2 random variables, not a χ^2 test with an infinite degree of freedom. In practice, one typically uses low-order polynomial functions of x for g_N; unless m is very large, it is likely that χ_m^2 still provides a better approximation than (10.7).

The main step leading to (10.7) is rewriting $g'_N C_N^{-1} g_N$ as a degenerate U-statistic and then applying a CLT in Hall (1984). Since degenerate U-statistics appear frequently in this literature, as a digression, we present a CLT in De Jong (1987). For $W_N \equiv \sum_{i<j} w_{ijN}(z_i, z_j)$, where $w_{ijN}(z_i, z_j) = w_{jiN}(z_j, z_i)$, $E\{w_{ijN}(z_i, z_j) \mid z_i\} = 0$, z_i's are independent, w_{ijN} is square-integrable, $\sigma_N^2 \equiv V(W_N)$, and $\sigma_{ij}^2 \equiv V(w_{ijN})$, $W_N/\sigma_N =^d N(0,1)$ if (Theorem 2.1 of De Jong)

$$\sigma_N^{-2} \max_{1 \leq i \leq N} \left\{ \sum_{j=1}^{N} \sigma_{ij}^2 \right\} \to 0, \quad \text{and}$$

$$\sigma_N^{-4} E(W_N^4) \to 3 \quad \text{as } N \to \infty. \tag{10.8}$$

De Jong (1987) shows another CLT (Theorem 5.3) useful for degenerate U-statistics without the fourth moment condition in (10.8); the theorem is more general than the CLT in Hall (1984). Let μ_{iN}, $i = 1 \ldots N$, be the eigenvalues of a symmetric matrix $[a_{ijN}]$, and let $W_N \equiv \sum_{i<j} a_{ijN} \cdot w_N(z_i, z_j)$, where z_i's are iid, $w_N(z_i, z_j) = w_N(z_j, z_i)$,

$$E\{w_N(z_i, z_j) \mid z_i\} = 0, \quad E\{w_N(z_i, z_j)\}^2 = 1, \quad \sigma_N^2 \equiv V(W_N).$$

Then $W_N/\sigma_N =^d N(0,1)$, if for a sequence of real numbers $\{A_N\}$, as $N \to \infty$,

$$A_N^2 \cdot \sigma_N^{-2} \max_{1 \leq i \leq N} \left\{ \sum_{j=1}^{N} a_{ij}^2 \right\} \to 0,$$

$$E\{w_N^2(z_i, z_j) \cdot 1[|w_N(z_i, z_j)| > A_N]\} \to 0 \tag{10.9}$$

$$\sigma_N^{-2} \max_{1 \leq i \leq N} \mu_{iN}^2 \to 0$$

or, instead of the last condition of (10.9)

$$E\{w_N(z_1, z_2) w_N(z_1, z_3) w_N(z_4, z_2) w_N(z_4, z_3)\} \to 0.$$

The above De Jong and Bierens (1994) test for a conditional moment condition is predated by Bierens' (1990) "consistent conditional moment test" (CCM). CCM is based upon the key fact that, if x is bounded and $E(\varepsilon \mid x) \neq 0$ for some x, then $t \in R^k$ such that $E(\varepsilon \cdot e^{t'x}) \neq 0$ is "dense" in R^k; that is, for almost all t in R^k, $E(\varepsilon \cdot e^{t'x}) \neq 0$. Thus, if H_0 is wrong, then $E(\varepsilon \mid x) \neq 0$ for some x where $\varepsilon \equiv y - r(x, \beta)$. We can use

$$\sqrt{N} \cdot M_N(t) \equiv \sqrt{N} \cdot (1/N) \sum_i \{y_i - r(x_i, b_N)\} \cdot e^{t'x_i} =^d N(0, s(t)^2) \tag{10.10}$$

as a test statistic where $s(t)^2$ and its estimate are in Bierens (1990, (13) to (16)). If x_i is not bounded, then replace x_i in $e^{t'x_i}$ with a one-to-one function $\psi(x)$ that is bounded [for instance, $\tan^{-1}(x)$].

Let $W_N(t) = N \cdot M_N(t)^2/s_N(t)^2$, where $s_N(t) =^p s(t)$. To implement CCM, t should be chosen. It is tempting to try different t's and choose one, say $\hat{t}$, that maximizes $W_N(t)$, but then $W_N(\hat{t})$ will no longer follow χ_1^2. Instead, Bierens (1990) suggests the following. First, fix a set T in which t lies, and choose an infinite sequence of t's, $t_1, t_2, \ldots$, such that the sequence is dense in T. Second, choose the optimal $\hat{t}$ maximizing $W_N(t)$ from the sequence. Third, choose $\gamma > 0$, $\rho \in (0, 1)$ and $t_0 \in T$, and let

$$\tilde{t} = t_0 \text{ if } W_N(\hat{t}) \leq W_N(t_0) + \gamma N^\rho, \quad \tilde{t} = \hat{t} \text{ otherwise.} \tag{10.11}$$

Then under H_0, $W_N(\tilde{t}) =^d \chi_1^2$, although $W_N(\tilde{t})$ no longer follows a non-central χ_1^2 under H_a. (10.11) makes choosing $\hat{t}$ over t_0 difficult with the "penalty function" γN^ρ. In practice, one has to choose a dense sequence $t_1, t_2, \ldots, \gamma, \rho$ and t_0, which is sometimes cumbersome. A quick review of CCM can be found in the appendix of Bierens and Pott–Buter (1990), who apply CCM to an Engel curve estimation.

Gozalo (1993) proposes a test analogous to CCM. Instead of searching for t as in CCM, he suggests a comparison of the parametric regression function with a nonparametric one at a number of selected points of x to test (10.1). Let $E(y \mid x) = r(x)$, $r_N(x) = E_N(y \mid x)$, and

$$\sigma_N(x)^2 \equiv V\big[(Nh^k)^{0.5}\{r_N(x) - r(x)\}\big]. \tag{10.12}$$

Consider a number of points in the range of x, say $(x^{(1)} \ldots x^{(d)})$. Define $T_N \equiv (T_N^{(1)} \ldots T_N^{(d)})'$ such that

$$T_N^{(j)} \equiv (Nh^k)^{0.5}\{r_N(x^{(j)}) - r(x^{(j)}, b_N)\}/\sigma_N(x^{(j)}). \tag{10.13}$$

Since $r(x^{(j)}, b_N) =^p r(x^{(j)}, \beta)$ under H_0 at the rate $\sqrt{N}$, while $r_N(x^{(j)}) =^p r(x^{(j)})$ at the rate $(Nh^k)^{0.5}$, under H_0,

$$T_N^{(j)} =^p (Nh^k)^{0.5}\{r_N(x^{(j)}) - r(x^{(j)}, \beta)\}/\sigma_N(x^{(j)}) =^d N(0,1); \tag{10.14}$$

that is, asymptotically, estimating β is as good as knowing β. Using the asymptotic independence among $r_N(x^{(j)})$, $j = 1 \ldots d$, the test statistic is

$$\sum_{j=1}^{d}\{T_N^{(j)}\}^2 =^d \chi_d^2. \tag{10.15}$$

This test is also applicable for omitted variable tests. Suppose that z may be omitted in a null model $y = E(y \mid x) + u$. Then $E_N(y \mid x, z) - E_N(y \mid x)$ can be used to test the possible omission. As in (10.13) to (10.14), the key point is that the convergence rate of $E_N(y \mid x) =^p E(y \mid x)$ should be faster than that of $E_N(y \mid x, z) =^p E(y \mid x, z)$, which then excludes discrete z. In both testing (10.1) and omitted variables, the difficulty is in selecting the evaluation points. Since the degree of freedom is m, selecting points where H_0 holds more or less decreases the power of the test. To overcome this, Gozalo suggests a "power-search" as in (10.11).

In a parametric context with the error term distribution specified, Staniswalis and Severini (1991) propose two tests analogous to (10.15). Rather than trying to select evaluation points, they propose plotting the test statistic vs. evaluation points to detect the set of x on which H_0 does not hold. They also consider letting $m \to \infty$ as $N \to \infty$: using (10.8), a properly centered and normalized test statistic converges to $N(0, 1)$.

Instead of selecting evaluation points for (10.15), Härdle and Mammen (1993) propose the use of a weighted integrated quadratic difference between the parametric and nonparametric regression functions:

$$Nh^{k/2} \cdot \int [r_N(x) - \tau\{r(x, b_N)\}]^2 \cdot w(x)dx, \qquad (10.16)$$

where $w(x)$ is a weighting function and

$$\tau\{g(x)\} \equiv \sum_i K((x_i - x)/h)g(x_i) \Big/ \sum_i K((x_i - x)/h). \qquad (10.17)$$

The purpose of $\tau(\cdot)$ is to center $r_N(x)$ properly under H_0; this also appears in Staniswalis and Severini (1991). Using (10.8), they show (10.16) minus $h^{-k/2} \cdot K^{(2)}(0) \cdot \int \sigma^2(x)w(x) \cdot f(x)^{-1}dx$ converges to (under H_0)

$$N\left(0, 2 \cdot K^{(4)}(0) \cdot \int \sigma^4(x) \cdot w(x)^2 \cdot f(x)^{-2}dx\right), \qquad (10.18)$$

where $K^{(j)}$ denotes the j-time convolution product of K and $\sigma^2(x) = V(u \mid x)$ is assumed to be continuous. Härdle and Mammen (1993) however do not recommend using (10.18) which depends on a stochastic expansion with $O_p(N^{-1/10})$ error terms; rather they recommend bootstrap methods.

Delgado and Stengos (1994) consider ($x \cap z \neq \emptyset$)

$$H_0: E(y \mid x, z) = r(x, \beta) \quad \text{vs.} \quad H_a: E(y \mid x, z) = E(y \mid z), \qquad (10.19)$$

where the two hypotheses are non-nested. Following Davidson and MacKinnon (1981), they set up an artificial regression model

$$y = (1 - \delta) \cdot r(x, \beta) + \delta \cdot E(y \mid z) + u \qquad (10.20)$$

and propose to test (10.19) extending the "J-test" of Davidson and MacKinnon: replace $E(y \mid z)$ with $E_N(y \mid z)$, and estimate δ and β jointly (J is from the word "jointly") treating x and $E_N(y \mid z)$ as regressors. H_0 is then equivalent to $\delta = 0$. They also suggest another test "P-test" for (10.20).

For x fixed ("fixed design," not random regressors), Müller (1992) proposes to test (10.1) by comparing derivatives as well as the regression functions, because the difference can be sharper in derivatives.

Consider the following model with heteroskedasticity:

$$y_i = x_i'\beta + u_i, \quad u_i^2 = z_i'\alpha + \varepsilon_i, \tag{10.21}$$

where x is a $k \times 1$ vector, z is a $k_z \times 1$ vector that may overlap with x, $E(u_i^2 \mid z_i) = z_i'\alpha$, and $\varepsilon_i \equiv u_i^2 - z_i'\beta$. Here we assume $z_i'\alpha > 0$ for all i, although it may be better to use a nonnegative function of z and α. With the first component of z_i being 1, we can test homoskedasticity with

$$H_0: \alpha_2 = \cdots = \alpha_{k_z} = 0, \quad H_a: \text{not } H_0, \tag{10.22}$$

by doing LSE of $(y_i - x_i' b_{\text{lse}})^2$ on z_i. But this test can be misleading when $E(y \mid x) \neq x'\beta$. B.J. Lee (1992) suggests a nonparametric heteroskedastic-ity test regressing v_i^2 where $v_i = y_i - E_N(y \mid x_i)$ on z_i without specifying the regression function. B.J. Lee (1992) in fact suggests running LSE of $v_i^2 f_N(x_i)$ on $z_i f_N(x_i)$ to remove the denominator $f_N(x_i)$ in $E_N(y \mid x_i)$ and simplify the asymptotic distribution theory.

10.11   Semiparametric Efficiency

Consider a model $P \equiv \{P_\theta : \theta \in \Theta\}$ where P_θ is a probability distribution indexed by θ and $\Theta \subset R^m$. The mapping $\theta \to P_\theta$ is a "parametrization." For instance, $P_\theta = N(\mu, \sigma^2)$, where $\theta = (\mu, \sigma) \in R \times (0, \infty) \subset R^2$. Let z_i, $i = 1 \ldots N$, be iid observations drawn from P_θ. Suppose that the model is parametrized by $\theta = (\beta', \eta')'$, where β is a $k \times 1$ parameter vector of main interest and η is a nuisance parameter.

If η is finite dimensional, then the efficiency bound for β is well known under some regularity conditions. In this section, we discuss the efficiency bound for a finite-dimensional β when η is nonparametric (infinite dimen-sional); for instance, $\eta(x) = V(y \mid x)$ where $z = (y, x')'$. This section draws mainly upon Bickel, Klaassen, Ritov and Wellner (1993, BKRW from now on). Cases with both β and η being nonparametric are not discussed. Of-ten certain statements need to hold only at the true value of θ and not at all possible θ. Then, we will use θ_0 (and β_0) to denote the true value. We discuss finite-dimensional nuisance parameters first.

Suppose η is of dimension $m-k$. With $\theta = (\beta', \eta')'$, $\beta = (I_k, 0_{k \times (m-k)}) \cdot \theta$, but we will use $\beta(\theta)$ or $\beta(P_\theta)$, for β can depend on θ in a more complex way in general. Note that $P_\theta \neq P_\lambda$ does not necessarily imply $\beta(\theta) \neq \beta(\lambda)$: e.g., imagine $\lambda = (\beta', \xi')'$. Let $p(z, \theta)$ denote the density of z. Define

$$I^{-1}(\beta_0 \mid \theta_0) \equiv E_{\theta_0}^{-1}(\tilde{S}_{\beta_0} \tilde{S}_{\beta_0}')$$

as the "(parametric) information bound" for β where

$$\tilde{S}_{\beta_0} \equiv S_{\beta_0} - \{E_{\theta_0}(S_{\beta_0} S_{\eta_0}') \cdot E_{\theta_0}^{-1}(S_{\eta_0} S_{\eta_0}') \cdot S_{\eta_0}\},$$

$$S_{\beta_0} \equiv \partial \ln\{p(z,\theta)\}/\partial\beta\big|_{\beta=\beta_0}, \quad S_{\eta_0} \equiv \partial \ln\{p(z,\theta)\}/\partial\eta\big|_{\eta=\eta_0}; \quad (11.1)$$

$$S_\beta \equiv 0 \equiv S_\eta \quad \text{if} \quad p(z,\theta) = 0.$$

P_θ is a "regular parametric model," if Θ is open, $p(z,\theta)$ is continuously differentiable in θ for almost all z, and $I(\theta) \equiv E_\theta(S_\theta S_\theta') < \infty$ is nonsingular and continuous in θ where $S_\theta \equiv \partial \ln\{p(z,\theta)\}/\partial\theta$ $[S_\theta \equiv 0$ if $p(z,\theta) = 0]$; see BKRW (p. 12) for a precise definition for which the preceding conditions are sufficient.

An estimator sequence $\{b_N\}$ for β is (locally) "regular" at θ if, with $\delta_N \equiv \sqrt{N}\{b_N - \beta(\theta_N)\}$ and $\sqrt{N}(\theta_N - \theta)$ being bounded,

$$E_{\theta_N} g(\delta_N) \to E_\theta g(\delta) \quad \text{for some rv } \delta, \quad (11.2)$$

for any continuous and bounded function $g(\cdot)$, and if the distribution of δ does not depend on $\{\theta_N\}$. If $\{b_N\}$ is regular at all $\theta \in \Theta$, then $\{b_N\}$ is regular.

Showing that $I^{-1}(\beta \mid \theta)$ is the (parametric) efficiency bound is done by a "convolution theorem": the limit distribution of any regular estimator for β can be represented as a sum of two random variables, say ν_θ and Δ_θ, where $\nu_\theta =^d N(0, I^{-1}(\beta \mid \theta))$ and ν_θ and Δ_θ are independent. From (11.1), if $E(S_\beta S_\eta') = 0$, then the efficiency bound for β with η unknown agrees with the efficiency bound with η known. In this case β is said to be "adaptively estimable."

An estimator b_N is asymptotically linear with an influence function $\zeta_i \equiv \zeta(z_i, \theta)$ if

$$\sqrt{N}\{b_N - \beta(\theta)\} =^p (1/\sqrt{N}) \sum_i \zeta_i,$$

$$E_\theta\{\zeta_i \zeta_i'\} \text{ is continuous in } \theta; \quad (11.3)$$

e.g., the influence function of the MLE for β is $I^{-1}(\beta \mid \theta) \cdot \tilde{S}_\beta$, called the "efficient influence function." For such a b_N (BKRW, p. 38–39):

$$b_N \text{ is regular}$$

$$\Leftrightarrow \zeta(z,\theta) - I^{-1}(\beta \mid \theta) \cdot \tilde{S}_\beta \text{ is orthogonal to } [S_\beta, S_\eta], \quad (11.4)$$

where $[S_\beta, S_\eta]$ is the closed linear span of S_β and S_η. Thus, any influence function of a regular asymptotically linear estimator must be a sum of $\psi \equiv I^{-1}(\beta \mid \theta) \cdot \tilde{S}_\beta$ and a term orthogonal to it. Once we have a $\zeta(z,\theta)$, we can obtain ψ as $\zeta(z,\theta) - \Pi\{\zeta(z,\theta) \mid [S_\beta, S_\eta]\}$, where Π is the (linear) projection.

It is easy to get an asymptotically linear estimator. More difficult is verifying its regularity. According to Newey (1990a, Theorem 2.2), an asymptotically linear estimate is regular iff

$$\partial\beta(\theta_0)/\partial\theta = E_{\theta_0}\{\zeta(\theta_0) \cdot S_{\theta_0}'\}. \quad (11.5)$$

For instance, consider the MLE with $\zeta = \psi = E^{-1}(\tilde{S}_\beta \tilde{S}'_\beta) \cdot \tilde{S}_\beta$. Since $\beta = (I_k, 0_{k \times (m-k)}) \cdot \theta$, the lhs is $\partial \beta(\theta)/\partial \theta = (I_k, 0_{k \times (m-k)})$, which is also the rhs due to $E(\tilde{S}_\beta S'_\beta) = E(\tilde{S}_\beta \tilde{S}'_\beta)$ and $E(\tilde{S}_\beta S'_\eta) = 0$. In fact, we can add any term, say, μ, to $\tilde{S}_\beta$ in $E^{-1}(\cdot) \cdot \tilde{S}_\beta$ without disturbing (11.5), so long as μ is orthogonal to $[S_\beta, S_\eta]$. This observation shows that (11.4) is equivalent to (11.5) for the parameterization $\beta = (I_k, O_{k \times (m-k)}) \cdot \theta$.

Turning to nonparametric nuisance parameters, denote the nuisance parameter by α now. Any subset Q of $P = \{P_\theta, \theta \in \Theta\}$ is called "a regular parametric submodel" of P, if Q is parametrized by a finite-dimensional parameter η and the parametrization is regular. For instance, consider the following semiparametric model:

$$ y = x'\beta + u \quad \text{and} \quad E(u \mid x) = 0, \tag{11.6} $$

where the distribution of x and $u \mid x$ is α and $\theta = (\beta', \alpha)'$. Under independence of u from x, a regular parametric submodel for this is

$$ \{P_\theta(y, x) : \theta = (\beta', \sigma)', \ y = x'\beta + u, \ u =^d N(0, \sigma^2), \ x =^d N(0, 1)\}. \tag{11.7} $$

Replacing $N(0, \sigma^2)$ with a logistic distribution, we get another parametric submodel. If we replace $u =^d N(0, \sigma^2)$ with u symmetric around 0, then this is not a parametric submodel, for the symmetric distribution is not parametrized.

Let β be a scalar. Define the "semiparametric information bound" as

$$ \sup_Q I^{-1}(\beta \mid Q), \tag{11.8} $$

since, not knowing α, we should not be able to do better than $I^{-1}(\beta \mid Q)$ for any parametric Q. Ideally, we calculate the parametric information bound for each Q, and then take the supremum to obtain (11.8). Any Q^* such that $I^{-1}(\beta \mid Q^*) = \sup_Q I^{-1}(\beta \mid Q)$ is called a "least favorable submodel." If we know a least favorable submodel, then we can get (11.8) and possibly construct an estimator attaining the bound.

With the nuisance parameter nonparametric, we need to generalize the score function for the finite-dimensional nuisance parameter. The corresponding concept is "tangent," "tangent set," and "tangent space," which is a subset of the space $L_2^0(P_0)$ of mean-zero square integrable functions wrt $P_0 \equiv P_{\theta_0}$. Let η be a scalar and $|\eta| < 1$. A tangent $h(z)$ for a one-dimensional regular parametric submodel P_η is

$$ \partial \ln\{p_\eta(z)\}\big|_{\eta=0}. $$

The tangent set $\dot{P}^0$ is the union of the tangents. The tangent space $\dot{P}$ is the closed linear span of $\dot{P}^0$ (i.e., $\dot{P}$ includes linear combinations of tangents and their limit points).

For example, suppose $\beta = p_0(0)$, the density of z at $z = 0$; α is the rest of $p(z)$. A one-dimensional parametric submodel P_η with the density $p_\eta(z)$ is (BKRW, p. 48)

$$p_\eta(z) = p_0(z)\{1 + \eta h(z)\}, \quad \text{where}$$

$$\sup_z |h(z)| < 1, \quad \int h(z)p_0(z)dz = 0, \tag{11.9}$$

which goes through $p_0(z)$ when $\eta = 0$. The restrictions on h ensure that $p_\eta(z)$ is a proper density and they make P_η regular; the boundedness assumption on h implies $h \in L_2^0(P_0)$. Since $\partial \ln\{p_\eta(z)\}/\partial\eta|_{\eta=0} = h(z)$, $h(z)$ is a tangent. The tangent space is the closed linear span of $h(z)$'s subject to the restrictions in (11.9). Another one-dimensional parametric submodel passing through $p_0(z)$ is

$$p_\eta(z) = p_0(z)\Psi\{\eta h(z)\} \bigg/ \int \Psi\{\eta h(s)\}dP_0(s) \text{ where } h \in L_2^0(P_0), \tag{11.10}$$

$\Psi: R \to (0, \infty)$ is bounded and continuously differentiable with bounded derivative Ψ', $\Psi(0) = \Psi'(0) = 1$, and bounded Ψ'/Ψ. For this,

$$\beta(P_\eta) = p_\eta(0) = p_0(0)\Psi\{\eta h(0)\} \bigg/ \int \Psi\{\eta h(s)\}dP_0(s). \tag{11.11}$$

In this example $h \in L_2^0(P_0)$ is enough to make P_η regular, and $\dot{P} = L_2^0(P_0)$.

If there are two parameters in a parametric submodel, then, typically, the tangent space is the sum of the two tangent spaces. For instance, let $P = \{P_{(\alpha,\beta)}: \alpha \in A, \beta \in B\}$ with $P_0 = P_{(\alpha_0,\beta_0)}$. Also let $\dot{P}_1$ be the tangent space with $\beta = \beta_0$ fixed, and $\dot{P}_2$ be the tangent space with $\alpha = \alpha_0$ fixed. Then $\dot{P} = \dot{P}_1 + \dot{P}_2$.

The scalar parameter $\beta(P_\eta)$ is "pathwise differentiable" on P at P_0 if there exists a linear (and bounded) mapping $\psi(h): \dot{P} \to R$ such that

$$\beta(P_\eta) = \beta(P_0) + \eta \cdot \psi(h) + o(|\eta|), \tag{11.12}$$

where P_η is any regular parametric submodel and h is the tangent for P_η; i.e., $\psi(h)$ is the derivative of $\beta(P_\eta)$ wrt η at $\eta = 0$ (ψ depends on P_0). Intuitively, $\psi(h)$ can be viewed as a "product" of two derivatives $\psi(z)$ and $h(z)$ (chain rule): $h(z)$ is the score function reflecting the effect of η on P_η, and $\psi(z)$ reflects the effect of a change in P_η on β. One example is (11.5) with $\zeta = \psi$ and $\psi(h) = E(\psi \cdot S'_{\theta_0})$. BKRW (p. 61, Theorem 1) show that the bound (11.8) is achieved by $\bar{E}_{\theta_0}\psi^2$ if $\psi(z)$ belongs to the tangent set.

If β is a $k \times 1$ vector, we get $\psi = (\psi_1 \ldots \psi_k)'$, where each component satisfies (11.12). BKRW (p. 62) show that $E(\psi\psi') \geq I^{-1}(\beta \mid Q)$, and if the closed linear space $[\psi_1 \ldots \psi_k]$ is a subset of the tangent set, then there exists a one-dimensional regular parametric submodel Q such that

$$c'I^{-1}(\beta \mid Q)c = c'E(\psi\psi')c$$

for each $k \times 1$ vector c. Thus, $E(\psi \psi')$ is appropriately defined as the multidimensional semiparametric efficiency bound when $[\psi_1 \ldots \psi_k]$ is a subset of the closure of the tangent set. In view of this, define ψ as the "efficient influence function." The justification of this definition is shown by a convolution theorem (BKRW, p. 63) analogous to the parametric convolution theorem under the assumption of a regular estimator and $[\psi_1 \ldots \psi_k]$ being a subset of the closure of the tangent set.

(11.12) can be rewritten as $d\beta(P_n)/d\eta\big|_{\eta=0} = \psi(h)+o(1) = E\{\psi(z)h(z)\}+o(1)$, which is analogous to (11.5) with ζ replaced by ψ. Also analogously to (11.4), for any asymptotically linear estimator b_N with the influence function ζ (BKRW, p. 65),

$$b_N \text{ is regular } \Leftrightarrow \zeta - \psi \text{ is orthogonal to } \dot{P}. \tag{11.13}$$

This shows a way to obtain an efficient influence function (so the bound): first get ζ, and then $\psi = \zeta - \Pi(\zeta \mid \dot{P})$, where Π is the (Hilbert space) projection on $\dot{P}$.

As an example, recall $\beta(P_\eta) = p_\eta(0)$ with (11.9). Since

$$\beta(P_\eta) = p_o(0) \cdot \{1 + \eta h(0)\} = \beta(P_0) + p_0(0)h(0) \cdot \eta, \tag{11.14}$$

$\psi(h) = p_o(0)h(0)$ for the submodel (11.9). Even if we use a different submodel (11.10), Taylor-expanding the rhs of (11.11) around $\eta = 0$,

$$p_0(z) + p_0(z)h(z)\eta \Rightarrow p_0(0)\{1 + \eta h(0)\} \text{ at } z = 0, \tag{11.15}$$

which is the same as (11.14) except that there is no restriction on h other than $h \in L_2^0(P_0)$. This suggests $\psi(h) = p_0(0)h(0)$ with $h \in L_2^0(P_0)$. Since $h(0)$ can be made to be arbitrarily big, the semiparametric information bound is 0. An easy example with a nonzero bound is $\beta(P_\eta) = \int z dP_\eta(z)$.

The preceding pathwise differentiability approach is good for finding the efficiency bound when β can easily be expressed as a function of P_η. More frequently, β is defined only implicitly through a restriction as in extremum estimators. In such cases, another approach based upon the efficient score is more convenient. With A denoting a subset of a function space, let the model be

$$P = \{P_{(\beta,\alpha)} : \beta \in B \in R^k, \alpha \in A\},$$

and define P_1 and P_2 as

$$P_1 = \{P_{(\beta,\alpha_0)} : \beta \in B\}, \quad P_2 = \{P_{(\beta_0,\alpha)} : \alpha \in A\}; \tag{11.16}$$

denote the tangent space for P_2 as $\dot{P}_2$. Let $S_{\beta_0} \equiv \partial \ln P_1/\partial \beta\big|_{\beta=\beta_0}$. Then assuming a regular parametrization for P_1, the efficiency bound is

$$E^{-1}(\tilde{S}_{\beta_0}\tilde{S}'_{\beta_0}) \text{ where } \tilde{S}_{\beta_0} \equiv S_{\beta_0} - \Pi(S_{\beta_0} \mid \dot{P}_2). \tag{11.17}$$

In practice, determining the tangent set $\dot{P}_2$ exactly can be difficult. Instead, we may get a candidate T for $\dot{P}_2$. Then either $T \subseteq \dot{P}_2$ or $T \supseteq \dot{P}_2$ (since the largest of $\dot{P}_2$ is the set of the mean-zero square integrable functions, all cases other than $T \subseteq \dot{P}_2$ can be turned into $T \supseteq \dot{P}_2$ by expanding T). First, suppose $T \subseteq \dot{P}_2$. Then the efficient score is too large and the resulting efficiency bound is too small (too optimistic). But if we can show a regular estimator achieving this bound, then the bound is legitimate. Also the bound is "sharp" in the sense that there indeed exists a regular estimator achieving the bound. Second, suppose $T \supseteq \dot{P}_2$. Then the efficient score is too small and the resulting efficiency bound is too large (too pessimistic). But if we can show a regular least favorable submodel, then the bound is legitimate; we may be able to design an estimator on the submodel to make the bound sharp.

In the rest, we show two simple examples of semiparametric efficiency bounds. In the first example, the orthocomplement of the tangent set is easily determined, while the tangent set is easier to see in the second. More examples can be seen in Newey (1990a) and BKRW.

With a $s \times 1$ moment condition $Em(z, \beta) = 0$, the efficiency bound is

$$\{E(S_\beta m') \cdot E^{-1}(mm') \cdot E(mS'_\beta)\}^{-1}. \tag{11.18}$$

If a given semiparametric restriction can be expressed as a vector moment condition, then (11.18) will be the semiparametric efficiency bound. Here the tangent set is $L_2^0(P_0)$ minus $[m]$, so that (11.18) is the projection of S_β on the orthocomplement $[m]$ of the tangent set; the tangent space has the "co-dimension" s [see Luenberger (1969, p. 65)]. If $y = x'\beta + u$ and $E(u \mid x) = 0$, then the efficiency bound is

$$E^{-1}\{xx' \cdot \sigma(x)^{-2}\} = E^{-1}\{(xu\sigma(x)^{-2})(xu\sigma(x)^{-2})'\}, \tag{11.19}$$

where $\sigma^2(x) = V(u \mid x)$. We will show this by verifying that the efficient score for β is $xu\sigma(x)^{-2}$. Since $E(u \mid x) = 0$ with $E(u^2) < \infty$ is equivalent to $E\{u \cdot g(x)\} = 0$ for any square-integrable $g(x)$, the orthocomplement of the tangent set must be $E\{u \cdot g(x)\}$ for any square-integrable $g(x)$. The score vector for β in the linear model is $-xf'(u \mid x)/f(u \mid x)$, where $f' = \partial f(u \mid x)/\partial u$. Hence, showing

$$E[\{-xf'(u \mid x)/f(u \mid x) - ux\sigma(x)^{-2}\} \cdot ug(x)]$$

$$= E_x[xg(x) \cdot E_{u|x}\{-uf'(u \mid x)/f(u \mid x) - u^2\sigma(x)^{-2}\}] = 0 \tag{11.20}$$

is enough for any square-integrable $g(x)$. The first term in $E_{u|x}$ is

$$-\int u\{f'(u \mid x)/f(u \mid x)\}f(u \mid x)du = -\int uf'(u \mid x)du$$

$$= -uf(u \mid x)|_{-\infty}^{\infty} + \int f(u \mid x)du = 1, \tag{11.21}$$

assuming $f(\pm\infty \mid x)$ is zero for all x. The second term in $E_{u|x}$ is 1. Therefore, (11.20) holds.

In the linear model with u independent of x, denote a parametric submodel by $f(u; s)$, where $f(u; \sigma)$ is the true density. The log-likelihood is $\ln\{f(y - x'b, s)\}$. The score vectors are

$$S_\beta = -x \cdot f_b(u; \sigma)/f(u; \sigma), \quad S_\sigma = f_s(u; \sigma)/f(u; \sigma), \tag{11.22}$$

where the subscript denotes the partial derivatives. Since there is no restriction on $f(u; \sigma)$, the tangent space should be the square integrable functions of u in view of (11.10). Then the projection of S_β is equal to the conditional mean:

$$E(S_\beta \mid u) = E(-x \cdot f_b(u; \sigma)/f(u; \sigma) \mid u) = -E(x) \cdot f_b(u; \sigma)/f(u; \sigma). \tag{11.23}$$

Thus the efficient score is $-\{x - E(x)\} \cdot f_b(u; \sigma)/f(u; \sigma)$. The semiparametric efficiency bound from this agrees with the Cramer–Rao bound in the MLE for the slope coefficients. This suggests that the slope coefficients may be estimated adaptively under the independence between x and u, which is indeed the case.

Appendix: Gauss Programs for Selected Topics

1. Introduction

In this appendix, we present GAUSS programs for selected topics which are relatively easy to program. Although the programs are simple and short, they can be used in real applications with some fine tuning. Some programs use simulated data which are not provided, and the others use a small data set for female labor supply which are given in the last section of this chapter.

The female labor supply data with $N = 200$ were drawn from the 1987 Michigan Panel Study of Income Dynamics (PSID). The data set was named "lab200.dat". The labor supply is censored at 0 with the 22% censoring % (that is, 44 individuals record zero labor supply). The variables are

    col 1: wife labor supply in hours per year
    col 2: the other household income
    col 3: wife age (age)
    col 4: wife schooling in years (edu)
    col 5: dummy, 1 if husband is a professional or manager (hus)
    col 6: number of pre-school children, age 0 to 5 ($pkid$)
    col 7: number of primary school children, age 6 to 12 ($skid$)
    col 8: dummy, 1 if house is on mortgate ($mort$).

The dependent variable (y) is the column 1 divided by 12, which is the labor supply in hours per month. The other household income divided by 1000 (inc) will be treated as an exogenous regressor except when we discuss simultaneous equations where y and inc are the endogenous variables. As for the exogenous variables, we use 1, age, age2 (=age^2), edu, pkid, and skid (and inc); hus and mort will be used as instuments when inc is treated as endogenous. The variable age2 is included to pick up the possible up and down pattern of labor supply as age increases. When the labor data set is used for an estimator, our discussion on the numerical estimation result will be rather brief, because our goal here is not in discovering any new facts from the data set.

The summary statistics of the data are (LQ is the lower quartile, and UQ is the upper quartile)

	MIN	LQ	MED	UQ	MAX
y	0	29.42	121.00	160.00	395.17
inc	0	18.00	27.82	41.80	268.00
age	18	28	35	44	64
edu	7	12	12	14	17
pkid	0	0	0	1	4
skid	0	0	0	1	4
hus	0	0	0	1	1
mort	0	0	1	1	1

The following is the list of the programs (the chapter number at the end of the line shows the relevant chapter for the program):

Section 2: LSE and GMM for a linear model (Ch. 2)
Section 3: Method-of-moments test for omitted variables (Ch. 3)
Section 4: Censored model MLE (Ch. 4)
Section 5: Two-stage estimation for a selection model (Ch. 5)
Section 6: GMM for linear simultaneous equations (Ch. 6)
Section 7: Kernel nonparametric regression (Ch. 8)
Section 8: SCLS for censored model and symmetry test (Ch. 9)
Section 9: MDE for a simultaneous censored model with SCLS (Ch. 5, 9)
Section 10: Semilinear model (Ch. 10)

Section 9 has not been discussed previously, although the relevant theory was covered in Chapters 5 and 9; Lee (1995a) shows a related GAUSS program.

The reader is assumed to have some basic knowledge of GAUSS. Each program is more or less self-contained. But the earlier ones, particularly Section 2, have more comments (in GAUSS, comments are given inside /* ... */). The (data) input and (result) output processes are not treated in detail; the input process is more complicated. The output of a matrix AAA to a file named "result.out" can be done with (the last period is not a part of the command)

output file = result.out reset;   aaa;   .

The word "reset" means that, if the file result.out already exists, it will be overwritten. If the reader desires to attach the output at the end of result.out, then simply replace "reset" with "on." Since the version of GAUSS used here is somewhat old, one may be able to shorten the programs considerably now when using a newer version.

One way to verify if a program is correct is to do a simulation (or Monte Carlo) study. Suppose we write a program on LSE for the linear model. To do a simulation study, suppose we set up a simple model

$$y_i = 2.5 \cdot x_i + u_i, \quad x_i =^d N(0,1), \quad \text{and} \quad u_i =^d N(0,1), \quad N = 200,$$

where x_i and u_i are independent. Generate 500 data sets where each set has 200 observations. We will get 500 estimates for the slope along with the 500 standard deviation estimates for the slope estimates; denote them as $b_1, \ldots, b_{500}$ and $s_1, \ldots, s_{500}$, respectively. If the program is correct, then it must be the case that

(i) $2.5 \cong \bar{b} \equiv (1/500) \sum_{j=1}^{500} b_j$,

(ii) $(1/500) \sum_{j=1}^{500} s_j \cong \left\{ (1/500) \sum_{j=1}^{500} (b_j - \bar{b})^2 \right\}^{1/2}$.

If (i) holds but not (ii), then the estimation part of the program is right but the variance part of the program is wrong. One caveat is that, in semi- and nonparametric methods where both estimates and variance matrix depend on smoothing parameters, (i) and (ii) may not hold as well as in parametric models.

2. LSE and GMM for a Linear Model

In this section, we ignore the censoring problem and apply LSE and GMM to the labor data. The LSE result with inc treated as exogenous is [the numbers in $(\cdot)$ are the t-values]:

1	age	age^2	edu	inc	pkid	skid
49.03	3.76	−0.07	4.33	−0.24	−37.44	−12.37
(0.83)	(1.38)	(−2.28)	(1.75)	(−1.90)	(−4.54)	(−1.59)

where $R^2 = 0.23$.

Now treating inc as endogenous, we use hus and mort as the instruments. The GMM result is

1	age	age^2	edu	inc	pkid	skid
111.94	1.48	−0.05	2.04	0.27	−38.32	−12.90
(1.33)	(0.38)	(−1.17)	(0.70)	(0.45)	(−4.37)	(−1.64)

where $R^2 = 0.21$ is measured by the squared correlation between y and the predicted value of y from the GMM estimates; but such R^2 should not be taken seriously when some regressors are correlated with the error term. Testing $H_0: \beta_{edu} = 0$ and $\beta_{inc} = 0$ with the GMM variance matrix, the p-value for the Wald test statistic following χ_2^2 is 0.42, which fails to reject H_0. To see how good the instruments are, we calculated the correlation coefficients between inc and the instruments, which are CORR(inc,hus) = 0.34 and CORR(inc,mort) = 0.22.

Examining the LSE and GMM, the age profile is quadratic; increasing at a decreasing rate. One year increase in schooling increases labor supply by 4.3 hours per month in LSE and 2 hours in GMM. One pre-school child decreases the labor supply by about 38 hours per month which is around 1/4 of the monthly labor supply. One primary school child decreases the labor supply by about 12 hours per month, which is about 1/3 of b_{pkid}. Overall, comparing LSE and GMM estimates, the signs are alike but the magnitudes of the estimates are rather different, except those for pkid and skid. Moreover, edu and inc both appeared statistically rather significant in LSE and insignificant in GMM. The following is the GAUSS program where all variance matrices are in heteroskedasticity-consistent form.

```
          /* LSE-BIV-GMM example program for a linear model */

new;                    /* beginning of the program */
format /m1/rd 7,2;      /* output format command */
n=200;                  /* n = # observations */
one=ones(n,1);          /* one is n many 1's for intercept */

/* read in the data and make data columns */

load dat[n,8]=lab200.dat;    /* The data file "lab200.dat" is read
                                into the n*8 matrix "dat" */

y=dat[.,1]/12; inc=dat[.,2]/1000; age=dat[.,3];
edu=dat[.,4]; hus=dat[.,5]; pkid=dat[.,6];
skid=dat[.,7]; mort=dat[.,8]; age2=age∧2;

/* make x matrix of dimension n*k, where k = # columns of x */

x=one~age~age2~edu~inc~pskid~skid; k=cols(x);
```

```
/* LSE */     /* py=predicted y, ulse=residual, r2lse=r−square */
ixx=invpd(x'*x);   lse=ixx*(x'*y);   pylse=x*lse;
ulse=y−pylse; u2lse=ulse∧2; r2lse=rsquare(y,pylse);
covlse=ixx*(x'*(x* ~u2lse))*ixx;   tvlse=tv(lse,covlse);
/* It is possible to print the variable names using the command
   "printfm", although we do not do it in the following. */

"LSE and t-values"; lse'|tvlse';
?; "LSE r-square "   r2lse;

proc rsquare(z1,z2):     /* z1 and z2 are the inputs for the proc */
   local cov, corr;      /* cov, corr defined only within the proc */
   cov=meanc((z1−meanc(z1)).*(z2−meanc(z2)));
   corr=cov/(stdc(z1)*stdc(z2));
   retp(corr∧2);          /* output from the procedure */
endp;                     /* end of the procedure */


proc tv(b,cov);          /* procedure for t-values */
   retp(b./sqrt(diag(cov)));
endp;


/* BIV */        /* z is a n*s instrument matrix */
z=one~age~age2~edu~pkid~skid~hus~mort;     izz=invpd(z'*z);
biv=indpd(x'*z*izz*z'*x)*z'*z*izz*z'*y;   u2ive=(y-x*biv)∧2;

/* GMM */
izdz=invdp(z'*(z*~u2ive));   covgmm=invpd(x'*z*izdz*z'*x);
gmm=covgmm*x'*z*izdz*z'*y;   tvgmm=tv(gmm,covgmm);
pygmm=x*gmm;   r2gmm=rsquare(y,pygmm);

?; "GMM and t-values"; gmm'|tvgmm';
?; "GMM r-square   "   r2gmm;

/* Check the correlations to see how good the instruments are */

?; "corr01   "   sqrt(rsquare(inc,hus));
   "corr02   "   sqrt(rsquare(inc,mort));

/* Wald test for edu=0 and inc=0 which are the 4th and 5th in x.
   g is the dof; H0: capr*beta=0. GMM, not LSE, is used */

iden=eye(k); capr=(iden[.,4]~iden[.,5])'; g=rows(capr);
d=capr*gmm; wald=d'*invpd(capr*covgmm*capr')*d;
?; "pvalue   "   cdfchic(wald,g);

end;              /* end of the program */
```

3. Method of Moments Test for Omitted Variables

In this section we show a method-of-moments test for omitted variables. In the previous section, using GMM, we failed to reject H_0: $\beta_{edu}=0$ and $\beta_{inc}=0$. So, in this section, we use only age, age2, pkid, and skid as the regressors, and test if pkid^2 and skid^2 are omitted. First, running LSE, we get (compare this to LSE in Section 2)

1	age	age2	pkid	skid	
115.28	3.11	-0.07	-38.71	-13.43	
(2.27)	(1.15)	(-2.14)	(-4.88)	(-1.72)	$R^2=0.21$.

Second, testing

$$H_0: \text{pkid}^2 \text{ and skid}^2 \text{ are not omitted,}$$

the p-value of the test statistic is 0.61, so we fail to reject H_0. The following is the GAUSS program where the LSE part overlaps with the program in Section 2.

```
/* Method of Moment omitted variable test for a linear model */

new;
format /m1/rd 7,2;        /* output format command */
n=200;                    /* n = # observatins */
one=ones(n,1);            /* one is n many 1's for intercept */

load dat[n,8]=lab200.dat;
y=dat[.,1]/12; inc=dat[.,2]/1000; age=dat[.,3];
edu=dat[.,4]; hus=dat[.,5]; pkid=dat[.,6];
skid=dat[.,7]; mort=dat[.,8]; age2=age^2;

x=one~age~age2~pkid~skid;    k=cols(x);

ixx=invpd(x'*x); lse=ixx*(x'*y); pylse=x*lse;
ulse=y-pylse; u2lse=ulse^2; r2lse=rsquare(y,pylse);
covlse=ixx*(x'*(x*~u2lse))*ixx; tvlse=tv(lse,covlse);

"LSE and t-values"; lse'|tvlse';
?; "LSE r-square   "   r2lse;

proc rsquare(z1,z2);
   local cov,corr;
   cov=meanc((z1-meanc(z1)).*(z2-meanc(z2)));
corr=cov/(stdc(z1)*stdc(z2));
retp(corr^2); endp;

proc tv(b,cov);
   retp(b./sqrt(diag(cov)));
```

endp;

/* testing if pkid∧2 and skid∧2 are omitted */

z=(pkid∧2)~(skid∧2);   g=cols(z);
testvec=sumc(z*~ulse)/sqrt(n);

/* calculate the compoents of the var(test stat), capc */

capa=(z'*x)*ixx;   var1=z'*(z*~u2lse)/n;
cov=(z'*(x*~u2lse)/n)*capa';
var2=capa*(x'*(x*~u2lse)/n)*capa';

capc=var1−cov−cov'+var2;

chi2=testvec'*invpd(capc)*testvec;   /* chi-square test stat */
?; "p-value   "   cdfchic(chi2,g);
end;

4. Censored MLE

In the preceding sections we ignored the censoring problem in the data. In this section, censored MLE is applied to the labor supply data, treating inc as exogenous. The estimation result is

1	age	age2	edu	inc	pkid	skid
−6.91	5.78	−0.11	6.64	−0.29	−46.24	−16.93
(−0.08)	(1.38)	(−2.23)	(2.52)	(−1.16)	(−5.29)	(−2.39)

where $\sigma_N = 77.59$ with the t-value being 16.47, and $\sum_i \inf = -941.14$. Overall, the magnitude of the MLE and the t-values are higher than those in LSE and GMM. For a later use, we also obtained b_N/σ:

$$-0.09 \quad 0.07 \quad -0.00 \quad 0.09 \quad -0.00 \quad -0.60 \quad -0.22.$$

The following is the GAUSS program.

/* Censored MLE program */

```
new; format /m1 /rd 7,2;
n=200; cr=0.0001;                        /* cr=stopping criterion */
step=0.5;                        /* step size in iteration, a # in (0,1] */
one=ones(n,1);

load dat[n,8]=lab200.dat;

y=dat[.,1]/12; inc=dat[.,2]/1000; age=dat[.,3];
edu=dat[.,4]; hus=dat[.,5]; pkid=dat[.,6];
```

skid=dat[.,7]; mort=dat[.,8]; age2=age∧2;

/* ind is the n*1 indicator column for uncensored y */

ind = y.>0;

/* x is n*k. kx is # of columns in x. k is # of estimates including sigma
which is a "nuisance parameter" */

x=one∼age∼age2∼edu∼ind∼pkid∼skid;
kx=cols(x); k=kx+1; /* kx is # of b's, and k is kx +1 for sigma */

/* procedure for log-likelihood and gradients */

```
proc like(b,s); local xb,ss,xbs,d,c,r,l1,l2;
  xb=x*b; xbs=xb/s; ss=s∧2; d=pdfn(xbs); c=cdfn(xbs); r=y−xb;
  l1=ind.*(−0.5*ln(2*pi) − 0.5*ln(ss) − (r∧2)/(2*ss) );
  l2=(1−ind).*ln(cdfn(−xbs));
retp(sumc(l1+l2)); endp;
```

```
proc first(b,s); local xb,ss,xbs,d,c,r,f1,f2;
  xb=x*b; ss=s∧2; xbs=xb/s; d=pdfn(xbs); c=cdfn(xbs); r=y−xb;
  f1=−x*∼( (d./(1−c)).*(1−ind)/s − (r.*ind/ss) );
  f2=−(ind/s) + ( xb.*((d./(1−c)).*(1−ind))/ss )
       + (r∧2).*ind/(s∧3);
retp(f1∼f2); endp;
```

```
lse=(x'y)/(x'x);          /* this is a GAUSS special command for LSE */
res=y−x*lse;
a0=lse|sqrt(meanc(res²));          /* a0 is an initial estimate for MLE */
```

routine:

```
b0=a0[1:kx,1]; s0=abs(a0[k,1]);
grad=first(b0,s0); hessi=−grad'*grad;
a1=a0+invpd(−hessi)*sumc(grad)*step;          /* key iteration routine */
```

/* stopping criterion; we could also use (sumc(grad))'*(sumc(grad))
of the absolute difference of the likelihoods for a0 and a1. */

if meanc(abs(a1−a0))<cr; goto final; endif;

```
b1=a1[1:kx,1]; s1=abs(a1[k,1]);
?; "log-likelihood   "  like(b1,s1); "estimates   "  a1';
a0=b1|s1; goto routine;
```

final:

```
b=a1[1:kx,1]; s=a1[k,1]; mle=b|s; covmle=invpd(−hessi);
tvmle=mle./sqrt(diag(covmle));
```

```
?;?; "log-likelihood   "  like(b,s);
"Final Estimates and t-values"; mle'|tvmle';
```

?; "b/sigma   " b'/s;   /* standardized estimate to be compared
                               with probit later */

/* wald test for age=0 and edu=0 which are the 2nd and 4th in x.
  g is the dof for the test; H:capr*alpha=0. */

iden=eye(k); capr=(iden[.,2]~iden[.,4])'; g=rows(capr);
d=capr*mle; wald=d'*invpd(capr*covmle*capr')*d;
?; "pvalue   " cdfchic(wald,g);
end;

5. Two-Stage Estimation for a Selection Model

In applying censored MLE to the labor data, we assumed implicitly that the decision to work and how many hours to work are determined by the same equation. If the binary decision of whether to work or not is based on a different equation from how much to work, then Heckman's two-stage estimator (TSE) for selection models is appropriate.

First, create a dummy variable "ind," which is 1 if a woman works and 0 otherwise. Then applying probit to the data with the regressors, age, age2, edu, inc, pkid, and skid, we get the estimates and t-values

1	age	age2	edu	inc	pkid	skid
-0.14	0.02	-0.00	0.19	-0.00	-0.73	-0.38
(-0.07)	(0.20)	(-0.95)	(3.12)	(-0.40)	(-4.54)	(-2.43).

The signs of the estimates are the same as those of b_N/s in the previous section, but the magnitudes of the estimates are somewhat different.

Second, applying LSE with the selection correction term to the data with $y > 0$, we get [γ is the coefficient of the correction term, and the italic numbers in $(\cdot)$ are the t-values ignoring the first stage error]

1	age	age2	edu	inc	pkid	skid	γ
122.29	2.27	-0.04	-0.04	-0.25	-25.79	-3.82	16.03
(1.89)	(0.80)	(-1.07)	(-0.01)	(-1.71)	(-2.32)	(-0.39)	(0.51)
(1.93)	*(0.82)*	*(−1.09)*	*(−0.01)*	*(−1.73)*	*(−2.33)*	*(−0.40)*	*(0.54)*

The selection correction term is not significant. Comparing this result to the LSE in Section 2 applied to all data with $y=0$ as well as $y>0$, the signs are similar but the magnitude and the statistical significance are lower in the TSE. The same can be said in comparing censored MLE with the TSE. By comparing the two t-value lines with and without considering the first-stage error, there is very little difference for our data set.

The following is the GAUSS program; recall that the first part of the program includes probit for binary models which could have been provided in an earlier section.

/* Heckman's Two Stage Estimation for a selection model */

```
new; format/m1/rd 7,2;
nn=200; one=ones(nn,1);
cr=0.000001; step=0.5;
eps=0.00001;                              /* eps is to prevent division by 0 */

load dat[nn,8]=lab200.dat;

y=dat[.,1]/12; inc=dat[.,2]/1000; age=dat[.,3];
edu=dat[.,4]; hus=dat[.,5]; pkid=dat[.,6];
skid=dat[.,7]; mort=dat[.,8]; age2=age∧2;

ind=y.>0; x=one~age~age2~edu~inc~pkid~skid; k=cols(x);

proc lse(x,y);
retp(x'y/(x'x)); endp;

proc tv(a,cov);
retp( a./sqrt(diag(cov)) ); endp;

proc like(a); local prob;
   prob=cdfn(x*a);
retp(ind'*ln(prob) + (1−ind)'*ln(1−prob)); endp;

proc first(a); local xa,den, prob,e,denom;
   xa=x*a; den=pdfn(xa); prob=cdfn(xa); denom=prob.*(1−prob);
   denom=denom+eps*(denom.<eps);
   e=(ind−prob).*den./denom;
retp(x*~e); endp;

a0=lse(x,ind);                           /* initial estimate for probit */

STEP1:                                   /* probit step with all data */

grad=first(a0); hessi=−grad'*grad;
a1=a0+invpd(−hessi)*sumc(grad)*step;
if meanc(abs(a1−a0))<cr; alp=a1; goto step2; endif;

?; "log-likelihood    "   like(a1);
"estimates"; a1'; a0=a1; goto step 1;

STEP2:

grad=first(alp);                         /* alp is probit MLE */
info=grad'*grad; covmle=invpd(info);
?;?; "first stage probit log-likelihood    "   like(alp);
"probit estimates and t-values"; alp'|(tv(alp,covmle))';
```

```
y1=selif(y,ind); x1=selif(x,ind);          /* select data with ind>0 */
xalp=x1*alp; n=rows(y1);                    /* n is # of selected data */
pd=pdfn(-xalp); cd-cdfn(xalp);

lam=pd./cd;                                 /* lam is n*1 correctin column */
xlam=x1~lam;                                /* regressor in 2nd stage LSE */
b=lse(xlam,y1);                             /* second stage LSE */
py1=xlam*b;                                 /* predicted y */

/* - - - - second stage variance estimation using the whole data - - - - */

xalp=x*alp; pd=pdfn(-xalp); cd=cdfn(xalp);
lam=pd./cd; lam2=lam^2; xlam=x~lam;

py=xlam*b; e=y-py;
gam=b[k+1, 1];                              /* gam is the coefficient for lam */

qbb=xlam'*(xlam*~ind);                      /* second order matrix for cov */
iqbb=invpd(qbb);

capa=-gam*( xlam'*( x*~(ind.*(-xalp.*lam-lam2)) ) );
qb=xlam*~(ind.*e) + grad*covmle*capa';
qbqb=qb'*qb;                                /* first order matrix for cov */
covtse=iqbb*qbqb*iqbb';                     /* second stage cov matrix */

qblse=xlam*~(ind.*e);
covlse=iqbb*(qblse'*qblse)*iqbb';
tvtse=tv(b,covtse);                         /* covlse ignores the 1st stage error,
                                               under-estimating covtse */
?; "second stage LSE, correct t-value and incorrect t-value";
b'|tvtse'; ?; (tv(b,covlse))';
?; "r-square    "  rsquare(y1,py1);

proc square(z1,z2);
   local cov,corr;
   cov=meanc((z1-meanc(z1)).*(z2-meanc(z2)));
   corr=cov/(stdc(z1)*stdc(z2));
retp(corr^2); endp;

end;
```

6. GMM for Linear Simultaneous Equations

The following is a program for GMM for linear simultaneous equations. As can be seen in the program, the data are simulated with the true model and parameters given in the program. We use two simultaneous equations with three exogeneous variables. The first equation excludes one regressor and the second equation excludes two regressors. So far we used matrices to get estimates in GAUSS programs. In this section, since we have another dimension (the number of equation), we will use vector, not matrix, to prevent confusion. That is, in getting $\sum_i x_i x_i'$, we will be using a "do loop" along with $k \times 1$ vector x_i's, not $X'X$ where X is a $N \times k$ matrix. More appropriately, since we will need $\sum_i (I_2 \otimes x_i)(I_2 \otimes x_i)'$, we will be using a do loop along with the $2k \times 2$ matrix $I_2 \otimes x_i$.

```
/* GMM for lnear simultaneous equations with simulated data */

new; format /m1 /rd 5,2;
n=200; one=ones(n,1);

/* SF model: y1 = alp1*y2 + bet11 + bet12*x12 + u1,
             y2 = alp2*y1 + bet21 + bet23*x23 + bet24*x24 + u2;

   x23 and x24 are excluded from y1, x12 is excluded from y2 */

bet11=0.5; bet12=0.5;                        /* setting SF parameters */
bet21=1; bet23=1; bet24=1;
alp1=0.5; alp2=0.5; theta=1−alp1*alp2;       /* theta = 0.25 */

/* x and u generation */

x12=2*rndu(n,1); x23=rndn(n,1); x24=rndn(n,1).>0;
x1=one~x12; x2=one~x23~x24;
x=one~x12~x23~x24;                   /* x is the system exo. var's */
k1=cols(x1); k2=cols(x2); k=cols(x);

u1=rndu(n,1)−0.5;
u2=sqrt(0.2)*rndn(n,1)+sqrt(0.8)*u1;       /* u2 is correlated with u1 */

/* y1 and y2 generation with RF */

y1=(bet21+bet23*x23+bet24*x24)*alp1/theta
   + (bet11+bet12*x12)/theta + (u1+alp1*u2)/theta;
y2=(bet11+bet12*x12)*alp2/theta
   + (bet21+bet23*x23+bet24*x24)/theta + (u2+alp2*u1)/theta;

proc tv(a,cov);
retp(a./sqrt(diag(cov))); endp;

/* GMM */    /* caphj is SF regressors */

caph1=y2~one~x12; caph2=y1~one~x23~x24;
id=eye(2); ck1=cols(caph1); ck2=cols(caph2);
ck=ck1+ck2;

sumwz=zeros(ck,2*k);                          /* initialization for the do loop */
sumzz=zeros(2*k,2*k); sumzy=zeros(2*k,1); i=1;

do until i>n;
```

```
  xi=x[i,.]'; zi=id.*.xi;
  wi=( ( caph1[i,.]~zeros(1,ck2) )|( zeros(1,ck1)~caph2[i,.]) )';
  yi=y1[i,1]|y2[i,1];
  sumwz=sumwz+wi*zi'; sumzz=sumzz+zi*zi'; sumzy=sumzy+zi*yi;
i=i+1; endo;

invz=invpd(sumzz);
biv=invpd(sumwz*invz*sumwz')*sumwz*invz*sumzy;   /* biv obtained */

sumwz=zeros(ck,2*k); sumzdz=zeros(2*k,2*k);
sumzy=zeros(2*k,1); i=1;

do until i>n;
  xi=x[i,.]'; zi=id.*.xi;
  wi=( ( caph1[i,.]~zeros(1,ck2) )|( zeros(1,ck1)~caph2[i,.]) )';
  yi=y1[i,1]|y2[i,1]; vi=yi-wi'*biv;
  sumwz=sumwz+wi*zi'; sumzdz=sumzdz+zi*vi*vi'zi'; sumzy=sumzy+zi*yi;
i=i+1; endo;

invzdz=invpd(sumzdz); covgmm=invpd(sumwz*invzdz*sumwz');
gmm=covgmm*sumwz*invzdz*sumzy; tvgmm=tv(gmm,covgmm);

sf1=gmm[1:ck1,1]; tvsf1=tvgmm[1:ck1,1];
sf2=gmm[ck1+1: ck,1]; tvsf2=tvgmm[ck1+1:ck,1];

"GMM estimates and t-values for linear simultaneous system";
?; "first equation"; sf1'|tvsf1';
?; "second equation"; sf2'|tvsf2';
end;
```

7. Kernel Nonparametric Regression

A simple kernel regression program using the $N(0,1)$ kernel is given in this section. As can be seen in the program, the data set is simulated with the true regression function being the normal density function going up and down. In the procedure for the kernel regression function, we get two outputs: one is the density function for x and the other is the regression function. The former is for pointwise confidence intervals; homoskedasticity is assumed to simplify the program. Note that the program does not take advantage of the symmetry of $K(a - b) = K(b - a)$, which the reader may use to save time. For an automatic choice of the smoothing parameter h with the least squares cross validation, see Section 10.

/* Kernel Nonparametric Regression Program */

```
new; format /m1/rd 6,2;
n=200;
```

```
h=0.35;                                    /* h is the smoothing parameter */

/* data (x,y) generation */

x=rndn(n,1); rho=pdfn(x/0.5)/0.5;
y=rho+rndn(n,1); z=x~y~rho;
z=sortc(z,1);                              /* rearrange data in x-increasing order */
x=z[.,1]; y=z[.,2]; rho=z[.,3];    /* rho is not a part of the data */

proc (2) = npr(y,x); local reg,i,f;
   i=1; reg=zeros(n,1); f=zeros(n,1);
   do until i>n;
        f[i,1]=sumc( pdfn((x−x[i,1])/h) )/(n*h);
        reg[i,1]=sumc( pdfn((x−x[i,1])/h).*y)/(n*h) /f[i,1];
   i=i+1; endo;
retp(f,reg); endp;

{f,reg}=npr(y,x);                          /* reg is the kernel estimate */
s=sqrt(meanc((y−reg)∧2));

v=pdfn(rndn(500,1)); mk=meanc(v);
adj=( 1.96*s*(mk/(n*h))∧0.5 )./sqrt(f);
lcb=reg−adj; ucb=reg+adj;                  /* confidence interval */

z=x~lcb~reg~rho~ucb; med=int(n/2);        /* true value rho included
*/
z[med−20;med−1,.]~z[med+1:med+20,.];    /* part of z printed */

/* the following is to plot a part of the estimate with |x|<1.5 */

ind=(abs(x).<1.5); xreg=x~reg;
sxreg=selif(xreg,ind); selx=sxreg[.,1]; selreg=sxreg[.,2];
library pgraph; xy(selx,selreg);

end;
```

8. SCLS for Censored Model and Symmetry Test

In this section we apply symmetrically censored LSE (SCLS) to the labor data and test the symmetry. For the readers not interested in the test, we present two programs: one is for SCLS only, and the other is for the symmetry test, which, in fact, includes SCLS routine.

Applying SCLS to the labor data, we obtained the following result:

1	age	age2	edu	inc	pkid	skid
−10.80	7.52	−0.13	4.62	−0.20	−52.19	−20.48
(−0.14)	(1.65)	(−2.03)	(1.87)	(−1.65)	(−7.16)	(−2.58)

This result is not much different from that of MLE.

In the symmetry test, we give two versions: one uses two moment conditions and the other uses only one moment condition. The former is put between the comments sign so that the reader can activate it if interested in it. Also included is the Hausman specification test comparing the MLE and SCLS. The symmetry specification test for censored models can be applied to noncensored models with a simple adjustment.

Using one moment condition with various "tuning constant" w between 20 and 110, we obtained

w	20	30	40	50	60	70	80	90	100	110
p-value	0.10	0.13	0.33	0.28	0.29	0.18	0.06	0.00	0.00	0.00

which suggests that the symmetry may not hold beyond ± 80 in the range of the error term. Recall that $\sigma_N \cong 78$ in MLE; thus, the symmetry appears to hold up one standard deviation distance from 0. But, since SCLS is inconsistent under asymmetry, the preceding diagnosis should be taken with caution. In choosing w, the estimate for σ in MLE was helpful; the preceding range for w (20 to 110) was built around $\sigma_N \cong 78$. In fact, we could not calculate the test statistic beyond $w=110$ which is close to 1.4 times σ_N.

Using two w's with a least one w close to 100, the p-values of the tests were almost 0. The Hausman test gave a p-value, which is 1; the reason is clear: above SCLS result is not much different from the MLE result.

*/ SCLS program for censored models */

```
new; format /m1 /rd 7,2;
cr=0.0001;
n=200; one=ones(n,1);

load dat[n,8]=lab200.dat;

y=dat[.,1]/12; inc=dat[.,2]/1000; age=dat[.,3];
edu=dat[.,4]; hus=dat[.,5]; pkid=dat[.,6];
skid=dat[.,7]; mort=dat[.,8]; age2=age^2;

ind=y.>0; x=one~age~age2~edu~inc~pkid~skid;
kx=cols(x); k=kx+1;
lse=(x'y)/(x'x); b0=lse;                        /* initial value for scls */

/* scls iteration routine */

proc obj(b); local xb,regy,reg0,indyx,err;              /* scls minimand */
   xb=x*b; regy=maxc((0.5*y')|xb'); indyx=y.>(2*xb);
   reg0=maxc(zeros(1,n)|xb'); err=((0.5*y)^2)-(reg0^2);
retp(meanc((y-regy)^2)+meanc(err.*indyx)); endp;

sclsjob:
```

```
xb=x*b0; indx=xb.>0;
b1=invpd(x'*(x*~indx))*(x'*( indx.*minc(y'|(2*xb')) ));
"scls and minimand"; b1'~obj(b1);

if meanc(abs(b1-b0))<cr; scls=b1; goto final; endif;
b0=b1; goto sclsjob;

final;

/* scls variance matrix calculation */

xb=x*scls; indx=xb.>0; u=y-xb; indu=abs(u).<xb;
h=x'*(x*~(indx.*indu)); invh=invpd(h);
lam=( (u.*indu) + ((u.>xb)-(u.< -xb)).*xb ).*indx;
covscls=invh*(x'*(x*~(lam∧2)))*invh;

tvscls=scls./sqrt(diag(covscls));
?; "final scls and t-values"; scls'|tvscls';
?; "scls-minimand value    "  obj(scls);
end;
```

```
                /* Symmetry specification test for censored model */
new; format /m1 /rd 7,2;
cr=0.00001; step=0.5;
n=200; one=ones(n,1);
w1=50; w2=100;                          /* two w's for symmetry test */

load dat[n,8]=lab200.dat;

y=dat[.,1]/12; inc=dat[.,2]/1000; age=dat[.,3];
edu=dat[.,4]; hus=dat[.,5]; pkid=dat[.,6];
skid=dat[.,7]; mort=dat[.,8]; age2=age∧2;
ind=y.>0; x=one~age~age2~edu~inc~pkid~skid;
kx=cols(x); k=kx+1;

lse=(x'y)/(x'x); b0=lse;                /* initial values for scls and mle */
s0=sqrt(meanc((y-x*lse)∧2));

sclsjob:

sb=x*b0; xind=xb.>0;
b1=invpd(x'*(x*~xind))*(x'*( xind.*minc(y'|(2*xb')) ));
"scls-step"; b1';

if meanc(abs(b1-b0))<cr; scls=b1;
    a0=lse|s0; ?; goto mlejob; endif;
b0=b1; goto sclsjob;

proc first(b,s); local xb,ss,xbs,d,c,r,f1,f2;
    xb=x*b; ss=s∧2; xbs=xb/s; d=pdfn(xbs); c=cdfn(xbs); r=y-xb;
```

```
   f1=−x*~( (d./(1−c)).*(1−ind)/s − (r.*ind/ss) );
   f2=−(ind/s) + ( xb.*((d./(1−c)).*(1−ind))/ss )
        + (r∧2).*ind/(s∧3);
retp(f1~f2); endp;

mlejob:

score=first(a0[1:kx,1],a0[k,1]);
a1=a0+invpd(score'score)*sumc(score)*step;
"mle-step"; a1';

if meanc(abs(a0−a1))<cr; mle=a1; ?; goto done; endif;
a0=a1; goto mlejob;

done:

/*
/* symmetry test with two w's */

xb1=x*scls; indx1=xb1.>w1; u1=y−xb1; indu1=abs(u1).<w1;
lam1=( (u1.*indu1) + ((u1.>w1)−(u1.<−w1))*w1 ).*indx1;
h1=x'*(x*~(indx1.*indu1));

xb2=x*scls; indx2=xb2.>w2; u2=y−xb2; indu2=abs(u2).<w2;
lam2=( (u2.*indu2) + ((u2.>w2)−(u2.<−w2))*w2 ).*indx2;
h2=x'*(x*~(indx2.*indu2));

hw=h1|h2;

xbs=x*scls; indxs=xbs.>0; us=y−xbs; indus=abs(us).<xbs;
hs=x'*(x*~(indxs.*indus)); invhs=invpd(hs);
lams=( (us.*indus) + ((us.>xbs)−(us.<−xbs)).*xbs ).*indxs;
covscls=invhs*(x'*(x*~(lams∧2)))*invhs;

rep=((x*~lam1)~(x*~lam2))−((x*~ lams)*invhs)*hw';
cov=rep'*rep/n;

tv=sumc((x*~lam1)~(x*~lam2))/sqrt(n);
wmets=tv'*invpd(cov)*tv;
pvwme=cdfchic(wmets,2*kx);
/*

/* symmetry test with one w, w1 */

xb1=x*scls; indx1=xb1.>w1; u1=y−xb1; indu1=abs(u1).<w1;
lam1=( (u1.*indu1) + ((u1.>w1)−(u1.<−w1))*w1 ).*indx1;
h1=x'*(x*~(indx1.*indu1)); hw=h1;

xbs=x*scls; indxs=xbs.>0; us=y−xbs; indus=abs(us).<xbs;
hs=x'*(x*~(indxs.*indus)); invhs=invpd(hs);
lams=( (us.*indus) + ((us.>xbs)−(us.<−xbs)).*xbs ).*indxs;
covscls=invhs*(x'*(x*~(lams∧2)))*invhs;
```

```
rep=(x*~lam1)−(x*~lams)*invhs*hw';
cov=rep'*rep/n;

tv=sumc((x*~lam1))/sqrt(n);
wmets=tv'*invpd(cov)*tv;
pvwme=cdfchic(wmets,kx);

/* specification test comparing MLE and SCLS */

bmle=mle[1:kx,1];
score=first(bmle,mle[k,1]); sh=score[.,1:kx]; ss=score[.,k];
esh=sh−ss*invpd(ss'ss)*(ss'sh); iinf=invpd(esh'esh);
repmle=esh*iinf;

rep=−(x*~lams)*invhs−repmle;
covhaus=rep'*rep;
haus=(scls−bmle)'*invpd(covhaus)*(scls−bmle);
pvhaus=cdfchic(haus,kx);

?; "scls and mle (except the scale factor)"; scls'|bmls'; ?;
"pv−wme    "  pvwme; "pv-haus    "  pvhaus;
end;
```

9. MDE for a Simultaneous Censored Model

In this section we consider two simultaneous equations with the endogenous variables y and inc in the labor data. We will estimate the y equation only excluding hus and mort from the y-equation; there is not enough information to specify the inc equation. Applying the minimum distance estimation (MDE) along with SCLS and LSE applied to the reduced form (RF) y and inc equations, respectively, we obtained the following for the y-equation:

1	age	age2	edu	inc	pkid	skid
43.21	5.55	−0.10	2.78	−0.00	−50.00	−18.02
(0.35)	(0.86)	(−1.20)	(0.69)	(−0.01)	(−6.04)	(−1.79).

This result showing insiginficant inc is somewhat different from the earlier SCLS where inc was treated as exogenous with $b_{inc} = -0.20$ with the t-value -1.65.

The MDE program with SCLS is simpler than a MDE using censored MLE for the y-equation. The reason is that the censored MLE estimates σ as well as β, while SCLS estimates only β. Owing to the nuisance parameter σ, the MDE with MLE requires one more step than the MDE with SCLS. Since the program for the former can be inferred from that for the latter, we did not present the MDE program with censored MLE.

/* MDE Program with SCLS applied to censored equation */

```
new; format /m1 /rd 7,2;
cr=0.00001; n=200;
one=ones(n,1);

load dat[n,8]=lab200.dat;

y=dat[.,1]/12; inc=dat[.,2]/1000; age=dat[.,3];
edu=dat[.,4]; hus=dat[.,5]; pkid=dat[.,6];
skid=dat[.,7]; mort=dat[.,8]; age2=age∧2;

x=one~age~age2~edu~pkid~skid~hs~mort; k=cols(x);

proc obj(b);
   local xb,maxyxb, max0xb, indyxb;
   xb=x*b; maxyxb=maxc((0.5*y')|xb');
   max0xb=maxc(zeros(1,n)|xb'); indyxb=y.>(2*xb);
retp( meanc((y−maxyxb)∧2) + meanc(indyxb.*((0.5*y)∧2 − max0xb∧2))
);
endp;

proc scls(b,y,x);
   local nb,txb,indx,small,invx;
   indx=(x*b).>0; invx=inv(x'*(x*~indx));
   txb=2*x*b; small=minc(y'|(txb'));
   nb=invx*(x'*(indx.*small));
rept(nb); endp;

b=(x'y)/(x'x);

sclsjob:

nb=scls(b,y,x);
if meanc(abs(b−nb))<cr; goto final; endif;
b=nb, b'~obj(b); goto sclsjob;

final:

b=nb; xb=x*b; u=y−xb; ind=(xb.>0);
indu=(abs(u).<xb); small=minc((u∧2)'|(xb∧2)');
c=x'*(x*~indu); invc=invpd(c); d=x'*(x*~(ind.*small));
cov=invc*d*invc; tratio=b./sqrt(diag(cov));

?; "Estimates b, tratio"; b'|tratio'; "minimand    "  obj(b);

/* MDE step */

h1=b; h2=(x'inc)/(x'x); v2=inc−x*h2;
id=eye(k); j1=id[.,1:k−2];
ch1=h2~j1; ig1=(ch1'h1)/(ch1'ch1); a1=ig1[1,1];

xb=x*h1; indx=xb.>0; u=y−xb; indu=abs(u).<xb;
```

```
small=minc((u∧2)'|(xb∧2)');
m=x'*(x*~(indx.*indu)); invm=invpd(m);
g=x'*(x*~(indx.*small)); invxx=invpd(x'*x); xxv2=x'*(x*~(v2∧2));

xe=x'*( x*~( ( (u.*indu) + ((u.>xb)-(u.<−xb)).*xb ).*v2.*indx ) );
c11=invm*g*invm; c12=(a1∧2)*invxx*xxv2*invxx;
c13=a1*invm*xe*invxx; c1=c11+c12−c13−c13'; ic1=invpd(c1);

cov1=invpd(ch1'*ic1*ch1);
g1=cov1*(ch1'ic1*h1);
tratio=g1./sqrt(diag(cov1));

?; "MDE estimates for SF"; g1'|tratio';
end;
```

10. Semilinear Model

In this section, we present a program for semilinear model $y = x'\beta+\theta(z)+u$, where $\theta(z)$ is an unknown function of z. A simulated data set is used where $z = (x1, x2)'$ and $\theta(z) = (x1 + x2 + x1 \cdot x2)/2$. In the program we choose the smoothing parameter with the least squares cross-validation, which was not used in the kernel regression program given earlier. Also to reduce the bias in the estimates, a third-order kernel derived from the standard normal kernel is used.

```
                /* Semilinear model with a third order kernel */

new; format /m1 /rd 7,2;
n=100; adj=0.0001;                        /* adj is to prevent division by 0 */

hlow=0.1; hhigh=5; hinc=0.3;                           /* search area for h */
hrep=int((hhigh−hlow)/hinc)+1;

u=rndn(n,1); x0=rndn(n,1).>0;                          /* data generation */
x1=rndn(n,1); x2=2*rndu(n,1); x2=x2/stdc(x2);
y=1 + x0 + (x1+x2+x1.*x2)/2 +u;

proc nrp(y,x1,x2,h);
   local reg,i,x1i,x2i,yi,k1ij,k2ij,kn1,kn2,denom;
   i=1; reg=zeros(n,1);

do until i>n;

   if i==1; x1i=x1[2:n,1]; x2i=x2[2:n,1]; yi=y[2:n,1];
      elseif i==n; x1i=x1[1:n−1,1]; x2i=x2[1:n−1,1]; yi=y[1:n−1,1];
      else; x1i=x1[1:i−1,1]|x1[i+1:n,1]; x2i=x2[1:i−1,1]|x2[i+1:n,1];
            yi=y[1:i−1,1]|y[i+1:n,1];
   endif;
```

```
    kn1=pdfn((x1[i,1]−x1i)/h); kn2=pdfn((x2[i,1]−x2i)/h);
    k1ij=(3/2)*kn1−0.5*kn1.*((x1[i,1]−x1i)/h)∧2;
    k2ij=(3/2)*kn2−0.5*kn2.*((x2[i,1]−x2i)/h)∧2;
    denom=sumc(k1ij.*k2ij);
    reg[i,1] = sumc((k1ij.*k2ij).*yi)/(denom+adj*(abs(denom)<adj));
i=i+1; endo;

retp(reg); endp;

i=1; comp=zeros(hrep,2); h=hlow;
do until i>hrep;
    comp[i,1]=meanc((y−nrp(y,x1,x2,h))∧2); comp[i,2]=h;
    i=i+1; h=h+hinc;
endo;

hy=comp[minindc(com[.,1]),2]; my=nrp(y,x1,x2,hy);

i=1; comp=zeros(hrep,2); h=hlow;
do until i>hrep;
    comp[i,1]=meanc((x0−npr(x0,x1,x2,h))∧2); comp[i,2]=h;
    i=i+1; h=h+hinc;
endo;

hx0=comp[minindc(comp[.,1]),2]; mx0=npr(x0,x1,x2,hx0);

mdx0=x0−mx0; mdy=y−my;

invmdx0=invpd(mdx0'mdx0);
lse=invmdx0*(mdx0'mdy); res=mdy−mdx0*lse; res2=res∧2;

cov=invmdx0*(mdx0'*(mdx0*~res2))*invmdx0;
tvlse=lse./sqrt(diag(cov));
"estimate and t-value   "  lse'~tvlse;'
?; "optimal hy and hx0   "   hy~hx0;
end;
```

11. Data

y*12	inc*1000	age	edu	hus	pkid	skid	mort
1896	32600	33	12	0	0	0	1
0	23578	37	13	0	0	1	1
882	31000	37	12	1	0	1	1
1596	19387	36	9	0	0	2	1
1834	12347	35	11	0	0	0	1
727	41800	46	16	1	0	1	1
1936	16000	26	14	0	1	0	0
0	15720	63	8	0	0	0	0
0	24460	62	12	0	0	0	0
1214	21500	22	15	0	0	0	0
0	43179	33	12	0	0	1	1
0	32000	28	14	1	2	0	0
0	80000	26	13	1	2	0	1
0	23965	62	12	0	0	0	0
0	27404	54	10	0	0	0	1
0	23700	31	12	0	1	1	1
1764	37000	33	13	0	0	3	0
2008	24000	23	8	0	1	1	0
1880	19000	35	11	0	0	1	1
1715	11000	32	10	0	0	2	0
0	9925	26	10	0	1	1	0
1312	43140	36	12	0	0	2	1
1720	33600	25	16	1	0	0	0
1790	10000	21	13	1	0	0	0
1920	21750	29	12	0	1	0	0
2046	22000	38	12	0	0	2	1
2390	7000	28	13	0	1	1	0
1920	38040	32	12	1	0	0	0
0	14100	56	15	0	0	0	0
1516	53383	26	17	0	0	0	1
4742	23000	31	17	1	1	2	1
1720	23200	28	12	0	1	0	1
0	20800	57	15	0	0	0	0
2000	10440	25	12	0	1	0	1
1709	12090	61	7	0	0	0	1
2540	28000	32	12	1	0	0	1

y*12	inc*1000	age	edu	hus	pkid	skid	mort
1250	50000	31	16	1	1	0	1
0	16567	62	12	0	0	0	0
1760	10750	44	13	0	0	0	0
1778	30200	44	13	0	0	0	1
2338	37800	26	17	1	0	0	0
973	88000	48	16	1	0	0	1
1764	22000	42	15	1	0	0	1
0	38400	64	17	0	0	0	0
353	54600	36	16	1	1	1	1
36	234500	35	17	1	2	1	1
741	50000	37	16	1	0	2	1
0	26400	35	10	0	0	2	1
1656	19000	21	13	0	1	0	0
1628	68450	35	15	1	0	0	1
0	14447	40	12	0	1	1	0
1682	74102	55	13	0	0	0	0
2250	36000	27	16	0	1	0	1
0	43169	33	14	0	0	4	0
2167	42000	34	12	1	0	2	0
2510	43160	28	14	1	0	4	1
1843	62808	58	12	1	0	0	1
0	47000	35	13	1	1	0	1
2080	48108	24	12	0	0	0	1
2100	28000	36	14	0	0	1	1
1777	24042	31	12	0	0	2	1
1568	10000	33	17	0	1	1	1
1800	56250	51	14	0	0	0	1
1837	32200	27	17	0	0	0	1
492	59100	40	15	1	0	1	1
214	85500	31	16	1	2	0	1
1636	19000	28	15	1	0	0	1
2023	28000	26	16	1	0	0	1
1616	36300	35	12	1	1	1	1
1992	19400	34	16	0	0	0	0
1200	12320	57	12	0	0	0	1
759	18798	29	12	0	0	3	1
1000	17460	62	12	0	0	0	1
0	68500	40	12	0	1	1	1

y*12	inc*1000	age	edu	hus	pkid	skid	mort
0	75400	52	12	0	0	0	0
132	52984	37	17	1	2	0	1
657	22030	36	17	1	2	0	1
2000	48200	40	12	0	0	0	1
431	28150	28	12	0	0	0	0
2226	9332	27	12	0	2	0	0
1832	26900	29	16	0	0	0	0
2684	52798	43	12	1	0	0	1
1008	28805	31	14	1	1	0	1
396	19692	58	12	0	0	0	1
593	24000	28	13	0	1	0	1
2064	21200	55	14	0	0	0	0
1791	43000	35	16	1	0	1	1
1912	26400	26	12	1	0	0	1
0	18550	32	10	1	1	1	1
0	7617	60	7	0	0	0	1
2196	30700	31	13	0	0	2	1
0	33205	35	14	1	2	1	1
0	5000	60	7	0	0	0	0
780	38000	40	12	0	1	0	0
0	186000	61	12	1	0	0	1
1482	18444	25	12	0	1	0	0
436	17000	23	12	0	2	0	0
700	51976	53	12	0	0	0	0
1365	268000	48	16	1	0	0	1
960	26463	29	16	0	2	0	1
258	35000	31	12	0	0	1	1
1800	16000	24	14	0	0	0	0
1687	37380	35	12	0	0	0	1
58	39460	44	10	1	0	0	0
2096	56434	42	12	0	0	0	1
648	4600	23	17	0	0	0	0
1750	50500	60	12	0	0	0	1
1470	32200	29	13	0	0	0	0
1224	23720	36	14	1	0	2	1
2000	0	29	12	0	1	0	1
1306	48500	32	16	1	1	0	1
892	35450	27	12	0	1	0	1

y*12	inc*1000	age	edu	hus	pkid	skid	mort
2064	46100	43	12	0	0	0	1
1920	31894	45	15	1	0	1	1
1164	30000	38	17	1	0	1	0
1800	13697	38	12	1	0	1	1
2034	45000	43	12	0	0	0	1
2072	4000	22	13	0	0	0	0
1920	56400	35	12	1	0	1	1
1040	10584	54	11	0	0	0	1
768	15830	28	14	0	2	0	1
1896	13000	26	11	1	0	0	0
1043	37900	56	16	0	0	0	0
0	24000	42	12	0	1	1	0
0	29400	46	7	0	1	1	1
1985	13288	35	11	0	0	0	1
1752	25000	28	14	1	1	0	1
1775	10000	24	12	0	1	0	0
942	12000	24	14	1	0	0	0
1992	11150	26	12	1	0	0	1
394	27820	30	12	0	0	2	1
0	47600	27	12	0	2	1	1
0	56178	51	12	1	0	0	1
1976	2400	22	12	0	0	0	0
2205	9000	48	11	0	0	0	1
1920	20000	25	11	1	1	0	1
1974	27600	41	12	0	0	0	1
2064	18150	24	14	0	1	0	1
0	7007	22	12	0	1	0	0
1800	9706	41	13	0	0	0	0
784	17994	39	10	0	0	0	1
2470	42400	39	17	0	0	1	1
1940	19800	27	17	1	0	0	0
0	17000	56	14	0	0	0	0
0	8038	24	13	1	1	0	1
0	62400	50	10	0	0	0	0
0	28600	30	10	0	0	4	0
793	7940	24	10	0	2	0	1
864	13555	26	14	0	1	0	0
1750	34224	58	12	1	0	0	0
1976	8000	30	12	0	0	0	0
1887	32899	53	12	0	0	0	0
1970	36600	29	12	1	0	2	1
1896	8272	63	12	0	0	0	0
2055	17405	24	12	0	0	2	0
2026	19350	26	14	0	1	0	0

y*12	inc*1000	age	edu	hus	pkid	skid	mort
1452	23720	26	11	0	0	0	0
1656	18000	38	17	0	1	1	1
2000	65000	33	16	1	0	0	1
2168	35600	39	17	1	0	1	1
2000	11622	48	12	1	0	0	1
0	29220	63	13	0	0	0	1
0	7704	60	12	0	0	0	1
2592	28000	35	10	1	0	0	1
1215	68000	60	17	1	0	0	0
0	98452	28	13	1	2	1	1
972	13820	18	11	0	1	0	1
0	26080	34	12	0	3	1	1
1936	26690	45	11	0	0	0	1
1242	18000	29	12	0	1	0	1
2295	43984	41	16	1	0	1	1
1644	35400	57	12	0	0	0	1
2350	15580	29	12	0	0	0	1
2160	73000	48	14	1	0	0	1
1190	31515	23	14	1	0	0	1
1000	20200	35	12	0	0	1	1
436	36000	31	17	1	1	0	1
0	28300	44	12	0	0	0	1
1331	39710	51	12	1	0	0	1
2184	67520	32	16	1	2	0	1
1263	11450	39	12	0	0	2	0
1568	27930	26	13	0	1	0	1
0	47312	49	11	0	0	0	1
1606	31200	29	15	0	0	2	1
1372	29082	24	14	0	2	0	1
2205	81000	33	14	0	0	1	1
490	41670	36	16	1	1	1	1
0	47110	47	9	0	0	0	0
1035	22000	21	14	1	0	0	0
294	49700	42	13	1	0	2	0
1904	19760	22	13	0	1	0	0
1901	44547	40	12	0	0	0	1
1386	18416	23	10	0	4	1	0
2006	26462	45	10	0	0	0	0
696	34720	63	17	0	0	0	0
0	9500	25	12	0	1	0	0
0	23310	63	12	0	0	0	0
1908	34800	29	14	1	0	0	1
691	40998	36	12	0	1	1	1
0	38299	57	14	0	0	0	1

References

AS: *Annals of Statistics*
ECA: *Econometrica*
ER: *Econometric Reviews*
ET: *Econometric Theory*
IER: *International Economic Review*
JAE: *Journal of Applied Econometrics*
JASA: *Journal of The American Statistical Association*
JBES: *Journal of Business and Economic Statistics*
JOE: *Journal of Econometrics*
JRSS: *Journal of The Royal Statistical Society, Series B*
RES: *Review of Economic Studies*

Ahn, H.T., and C.F. Manski, 1993, Distribution theory for the analysis of binary choice under uncertainty with nonparametric estimation of expectations, *JOE* **56**, 291–321.

Amemiya, T., 1974, The nonlinear two-stage least squares estimator *JOE* **2**, 105–110.

Amemiya, T., 1985, *Advanced Econometrics*, Harvard University Press.

Anderson, S.P., A. de Palma, and J.F. Thisse, 1992, *Discrete Choice Theory of Product Differentiation*, MIT Press.

Andrews, D.W.K., 1987a, Consistency in nonlinear econometric models: A generic uniform LLN, *ECA* **55**(6), 1465–1471.

Andrews, D.W.K., 1987b, Asymptotic results for generalized Wald tests, *ET* **3**, 348–358.

Andrews, D.W.K., 1988a, Chi-square diagnostic tests for econometric models: Introduction and applications, *JOE* **37**, 135–156.

Andrews, D.W.K., 1988b, Chi-square diagnostic tests for econometric models: Theory, *ECA* **56**, 1419–1453.

Andrews, D.W.K., 1991a, Asymptotic optimality of generalized C_L, cross-validation and generalized cross-validation in regression with heteroskedastic errors, *JOE* **47**, 359–377.

Andrews, D.W.K., 1991b, Heteroskedasticity and autocorrelation consistent covariance matrix estimation, *ECA* **59**, 817–858.

Andrews, D.W.K., 1994, Asymptotics for semiparametric econometric models via stochastic equicontinuity, *ECA* **62**, 43–72.

Bickel, P.J. and K.A. Doksum, 1981, An analysis of transformations revisited, *JASA* **76**, 296–311.

Bickel, P.J., C.A.J. Klaassen, Y. Ritov, J.A. Wellner, 1993, *Efficient and Adaptive Estimation for Semiparametric Models*, Johns Hopkins University Press.

Bierens, H.J., 1987, Kernel estimators of regression function, in *Advances in Econometrics*, Cambridge University Press.

Bierens, H.J., 1990, A consistent conditional moment test of functional form, *ECA* **58**, 1443–1458.

Bierens, H.J., 1994, Comments on "Artificial Neural Networks: An Econometric Perspective," *ER* **13**, 93–97.

Bierens, H.J. and H.A. Pott-Buter, 1990, Specification of household Engel curves by nonparametric regression, *ER* **9**, 123–184.

Billingsley, P., 1986, *Probability and Measure*, 2nd ed., John Wiley & Sons, NY.

Bloomfield, P. and W.L. Steiger, 1983, *Least Absolute Deviations*, Birkhauser, Boston, MA.

Börsch-Supan, A. and V.A. Hajivassiliou, 1993, Smooth unbiased multivariate probability simulators for maximum likelihood estimation of limited dependent variable models, *JOE* **58**, 347–368.

Bowden, R.J., 1973, The theory of parametric identification, *ECA* **41**, 1068–1074.

Bowden, R.J. and D.A. Turkington, 1984, *Instrumental Variables*, Cambridge University Press.

Box, G.E.P. and D.R. Cox, 1964, An analysis of transformation, *JRSS* **26**, 211–246.

Breiman, L. and J.H. Friedman, 1985, Estimating optimal transformations for multiple regression and correlation, *JASA* **80**, 580–597.

Buchinsky, M., 1994, Changes in the U.S. wage structure 1963–1987: Application of quantile regression, *ECA* **62**, 405–458.

Cameron, A.C. and P.K. Trivedi, 1986, Econometric models based on count data: Comparisons and applications of some estimators and tests, *JAE* **1**, 29–53.

Caroll, R.J. and D. Ruppert, 1988, *Transformation and Weighting in Regression*, Chapman and Hall.

Chamberlain, G., 1980, Analysis of covariance with qualitative data, *RES* **47**, 225–238.

Chamberlain, G., 1982, Multivariate regression models for panel data, *JOE* **18**, 5–46.

Chamberlain, G., 1984, Panel data, in *Handbook of Econometrics*, edited by Z. Griliches and M. Intrilligator, North Holland.

Chamberlain, G., 1986, Asymptotic efficiency in semiparametric models with censoring, *JOE* **32**, 189–218.

Chamberlain, G., 1987, Asymptotic efficiency in estimation with conditional moment restrictions, *JOE* **34**, 305–334.

Chamberlain, G., 1992, Efficiency bounds for semiparametric regression, *ECA* **60**, 567–596.

Charlier, E., B. Melenberg, and A.H.O. van Soest, 1995, A smoothed maximum scope estimator for the binary choice panel data model with an application to labor force participation, *Statistica Neerlandica* **49**, 324–342.

Chen, H., 1988, Convergence rates for parametric components in a partly linear model, *AS* **16**, 136–146.

Chow, Y.S., and H. Teicher, 1988, *Probability Theory*, 2nd. Ed., Springer-Verlag.

Cleveland, W.S., S.J. Devlin, and E. Grosse, 1988, Regression by local fitting, *JOE* **37**, 87–114.

Collomb, G. and W. Härdle, 1986, Strong uniform convergence rates in robust nonparametric time series analysis and prediction, *Stochastic Processes and Their Applications* **23**, 77–89.

Cosslett, S.R., 1987, Efficiency bounds for distribution-free estimators of the binary choice and the censored regression models, *ECA* **55**, 559–585.

Cumby, R.E., J. Huizinga, and M. Obstfeld, 1983, Two-steps two-stage least squares estimation in models in rational expectations, *JOE* **21**, 333–355.

Dagenais, D.G. and J.M. Dufour, 1991, Invariance, nonlinear models and asymptotic tests, *ECA* **59**, 1601–1615.

Das, S., 1991, A semiparametric structural analysis of the idling of cement kilns, *JOE* **50**, 235–256.

Davidson, R. and J.G. MacKinnon, 1981, Several model specification tests in the presence of alternative hypotheses, *ECA* **49**, 781–793.

De Jong, P., 1987, A central limit theorem for generalized quadratic forms, *Probability Theory and Related Fields* **75**, 261–277.

De Jong, R.M. and H.J. Bierens, 1994, On the limit behavior of a chi-squared type test if the number of conditional moments tested approaches infinity, *ET* **9**, 70–90.

Delgado, M.A. and P.M. Robinson, 1992, Nonparametric and semiparametric methods for economic research, *Journal of Economic Surveys* **6**, 201–249.

Delgado, M.A. and T. Stengos, 1994, Semiparametric specification testing of non-nested econometric models, *RES* **61**, 291–303.

Dharmadhikari, S. and K. Joag-dev, 1988, *Unimodality, Convexity, and Applications*, Academic Press.

Dudley, R.M., 1989, *Real Analysis and Probability*, Chapman and Hall.

Eckstein, Z. and K.I. Wolpin, 1989, The specification and estimation of dynamic stochastic discrete choice models, *Journal of Human Resources* **24**, 562–598.

Eicker, F., 1963, Asymptotic normality and consistency of the least squares for families of linear regressions, *Annals of Mathematical Statistics* **34**, 447–456.

Elbers, C. and G. Ridder, 1982, True and spurious duration dependence: The identifiability of the proportional hazard model, *RES* **49**, 402–411.

Engle, R.F., 1984, Wald, likelihood ratio and Lagrange multiplier tests in econometrics, in *Handbook of Econometrics II*, North-Holland.

Eubank, R., 1989, *Spline Smoothing and Nonparametric Regression*, Marcel Dekker, Inc.

Fan, J., 1992, Design-adaptive nonparametric regression, *JASA* **87**, 998–1004.

Fan, J., 1993, Local linear regression smoothers and their minimax efficiencies, *AS* **21**, 196–216.

Freedman, D.A., 1985, The mean versus the median: A case study in 4-R Act litigation, *JBES* **3**, 1–13.

Friedman, J. and W. Stuetzle, 1981, Projection pursuit regression, *JASA* **76**, 817–823.

Gallant, A.R. and D. Nychka, 1987, Semi-nonparametric maximum likelihood estimation, *ECA* **55**(2), 363–390.

Gallant, A.R. and G. Tauchen, 1989, Seminonparametric estimation of conditionally heterogeneous processes: Asset pricing applications, *ECA*, 1091–1120.

Gourieroux, C. and A. Monfort, 1993, Simulation-based inference: A survey with special reference to panel data models, *JOE* **59**, 5–33.

Gozalo, P.L., 1993, A consistent model specification test for nonparametric estimation of regression function models, *ET* **9**, 451–477.

Gregory, A.W. and M.R. Veall, 1985, Formulating Wald tests of nonlinear restrictions, *ECA* **53**, 1465–1468.

Györfi, L., W. Härdle, P. Sarda, and P. Vieu, 1989, Nonparametric curve estimation from time series, *Springer Lecture Notes in Statistics* **60**, Springer-Verlag.

Hajivassilious, V.A. and P.A. Ruud, 1994, Classical estimation methods for LDV models using simulation, *Handbook of Econometrics IV*, North-Holland.

Hall, P., 1983, Large sample optimality of least squares cross validation in density estimation, *AS* **11**, 1156–1174.

Hall, P., 1984, Central limit theorem for the integrated square error of multivariate nonparametric density estimators, *Journal of Multivariate Analysis* **14**, 1–16.

Hall, P. and J.L. Horowitz, 1990, Bandwidth selection in semiparametric estimation of censored linear regression models, *ET* **6**, 123–150.

Han, A., 1987, Nonparametric analysis of a generalized regression model, *JOE* **35**, 303–316.

Hansen, L., 1982, Large sample properties of generalized method of moments estimators, *ECA* **50**, 1029–1054.

Härdle, W., 1990, *Applied Nonparametric Regression*, Cambridge University Press.

Härdle, W., P. Hall, and J.S. Marron, 1988, How far are automatically chosen regression smoothing parameters from their optimum? *JASA* **83**, 86–101.

Härdle, W., W. Hildenbrand, and M. Jerison, 1991, Empirical evidence on the law of demand, *ECA* **59**, 1525–1549.

Härdle, W., P. Janssen, and R. Serfling, 1988, Strong uniform consistency rates for estimators of conditional functionals, *AS* **16**, 1428–1449.

Härdle, W. and O. Linton, 1994, Applied nonparametric methods, *Handbook of Econometrics IV*, edited by R.F. Engle and D.L. McFadden, Elsevier Science.

Härdle, W. and E. Mammen, 1993, Comparing nonparametric versus parametric regression fits, *AS* **21**, 1926–1947.

Härdle, W. and J.S. Marron, 1985, Optimal bandwidth selection in nonparametric regression function estimation, *AS* **13**, 1465–1481.

Härdle, W. and T.M. Stoker, 1989, Investigating smooth multiple regression by the method of average derivatives, *JASA* **84**, 986–995.

Härdle, W. and A.B. Tsybakov, 1993, How sensitive are average derivatives? *JOE* **58**, 31–48.

Hastie, T.J. and C. Loader, 1993, Local regression: Automatic kernel carpentry, *Statistical Science* **8**, 120–143.

Hausman, J.A., 1978, Specification tests in econometrics, *ECA* **46**, 1251–1272.

Hausman, J.A. and D. McFadden, 1984, Specification tests for the multinomial logit model, *ECA* **52**, 1219–1240.

Hausman, J.A., W.K. Newey, and J.L. Powell, 1995, Nonlinear errors in variables: estimation of some Engel curves, *JOE* **65**, 205–233.

Heckman, J.J., 1979, Sample selection bias as a specification error, *ECA* **47**, 153–161.

Heckman, J.J., 1984, The χ^2 goodness of fit statistic for models with parameters estimated from microdata, *ECA* **52**, 1543–1547.

Heckman, J.J. and V. Hotz, 1989, Choosing among alternative nonexperimental methods for estimating the impact of social program: The case of manpower training, *JASA* **84**, 862–874.

Heckman, J.J. and J.A. Smith, 1995, Assessing the case for social experiments, *Journal of Economic Perspective* **9**, 85–110.

Honoré, B.E., 1992, Trimmed LAD and LSE of truncated and censored regression models with fixed effects, *ECA* **60**, 533–565.

Horowitz, J.L., 1992, A smoothed maximum score estimator for the binary response model, *ECA* **60**, 505–531.

Horowitz, J.L., 1993a, Semiparametric and nonparametric estimation of quantal response models, in *Handbook of Statistics* **11**, edited by G.S. Maddala, C.R. Rao, and H.D. Vinod, Elsevier Science Publishers.

Horowitz, J.L., 1993b, Semiparametric estimation of a work-trip mode choice model, *JOE* **58**, 49–70.

Horowitz, J.L. and G.R. Neumann, 1987, Semiparametric estimation of employment duration models, *ER* **6**, 5–40.

Horowitz, J.L. and G.R. Neumann, 1989, Specification testing in censored models: Parametric and semiparametric methods, *JAE* **4**, S61–S86.

Hsiao, C., 1986, *Analysis of Panel Data*, Cambridge University Press.

Huber, P.J., 1981, *Robust Statistics*, Wiley.

Ichimura, H., 1993, Semiparametric least squares (SLS) and weighted SLS estimation of single index models, *JOE* **58**, 71–120.

Ichimura, H. and L.F. Lee, 1991, Semiparametric least squares estimation of multiple index models, in W.A. Barnett, J. Powell, and G. Tauchen, eds., *Nonparametric and Semiparametric Methods in Econometrics and Statistics*, Cambridge University Press.

Izenman, A.J., 1991, Recent developments in nonparametric density estimation, *JASA* **86**, 205–224.

Jennen-Steinmetz, C. and T. Gasser, 1988, A unifying approach to non-parametric regression estimation, *JASA* **83**, 1084–1089.

Keane, M.P., 1992, A note on identification in the multinomial probit model, *JBES* **10**, 193–200.

Kiefer, N., 1988, Economic duration data and hazard functions, *Journal of Economic Literature* **26**, 649–679.

Kim, J.K. and D. Pollard, 1990, Cube-root asymptotics, *AS* **18**, 191–219.

Klein, R.W. and R.H. Spady, 1993, An efficient semiparametric estimator for binary response models, *ECA* **61**, 387–421.

Koenker, R. and G. Bassett, 1978, Regression quantiles, *ECA* **46**, 33–50.

Kuan, C.M. and H. White, 1994, Artificial neural networks: An econometric perspective, *ER* **13**, 1–91.

Lafontaine, F. and K.J. White, 1986, Obtaining any Wald statistic you want, *Economics Letters* **21**, 35–40.

Lancaster, T., 1992, *The Econometric Analysis of Transition Data*, Cambridge University Press.

Lee, B.J., 1992, A heteroskedasticity test robust to conditional mean specification, *ECA* **60**, 159–171.

Lee, L.F., 1992, Amemiya's generalized least squares and tests of overidentification in simultaneous equation models with qualitative or limited dependent variables, *ER* **11**, 319–328.

Lee, L.F., 1995, Semiparametric maximum likelihood estimation of polychotomous and sequential choice models, *JOE* **65**, 381–428.

Lee, M.J., 1989, Mode regression, *JOE* **42**, 337–349.

Lee, M.J., 1992a, Median regression for ordered discrete response, *JOE* **51**, 59–77.

Lee, M.J., 1992b, Winsorized mean estimator for censored regression model, *ET* **8**, 368–382.

Lee, M.J., 1993a, Quadratic mode regression, *JOE* **57**, 1–19.

Lee, M.J., 1993b, Specification tests for ordered probit, manuscript.

Lee, M.J., 1994, Heteroskedasticity and symmetry tests for Tobit models, manuscript.

Lee, M.J., 1995a, A semiparametric estimation of simultaneous equations with limited dependent variables: A case study of female labor supply, *JAE* **10**, 187–200.

Lee, M.J., 1995b, Nonparametric two stage estimation of simultaneous equations with limited endogenous regressors, *ET*, forthcoming.

Lee, M.J., 1995c, Asymptotic variance of two-sample two-stage moment-based estimators with parametric and nonparametric first stage, manuscript.

Lehman, E.L., 1975, *Nonparametrics: Statistical Methods Based on Ranks*, Holden-Day, Inc., San Francisco, CA.

Luenberger, D.G., 1969, *Optimization by Vector Space Methods*, Wiley.

Mack, Y.P. and H. Müller, 1989, Derivative estimation in nonparametric regression with random predictor variable, *Sankya* **A51**, 59–72.

Mack, Y.P. and B.W. Silverman, 1982, Weak and strong uniform consistency of kernel regression estimates, *Zeitschrift für Wahrscheinlichkeitstheorie und verwandte Gebiete* **61**, 405–415.

MacKinnon, J.G., 1992, Model specification tests and artificial regressions, *Journal of Economic Literature* **30**, 102–146.

MacKinnon, J.G. and L. Magee, 1990, Transforming the dependent variable in regression models, *IER* **31**, 315–339.

Maddala, G.S., 1983, *Limited-Dependent and Qualitative Variables in Econometrics*, Cambridge University Press.

Maddala, G.S., 1986, Disequilibrium, self-selection, and switching models, in *Handbook of Econometrics III*, eds. Z. Grilliches and M.D. Intriligator, North-Holland.

Maddala, G.S., 1987, Limited dependent variable models using panel data, *Journal of Human Resources* **22**, 307–338.

Malinvaud, E., 1970, *Statistical Methods of Econometrics*, North Holland.

Manski, C.F., 1975, Maximum score estimation of the stochastic utility model of choice, *JOE* **3**, 205–228.

Manski, C.F., 1985, Semiparametric analysis of discrete response, *JOE* **27**, 313–333.

Manski, C.F., 1987, Semiparametric analysis of random effects linear models from binary panel data, *ECA* **55**, 357–362.

Manski, C.F., 1988, *Analog Estimation Methods in Econometrics*, Chapman and Hall.

Manski, C.F., 1991, Regression, *Journal of Economic Literature* **29**, 34–50.

Marron, J.S., 1988, Automatic smoothing parameter selection, *Empirical Economics* **13**, 187–208.

Matzkin, R.A., 1991, Semiparametric estimation of monotone and concave utility functions for polychotomous choice models, *ECA* **59**, 1314–1327.

Matzkin, R.A., 1992, Nonparametric and distribution-free estimation of the binary threshold crossing and the binary choice models, *ECA* **60**, 239–270.

Matzkin, R.A., 1993, Nonparametric identification and estimation of polychotomous choice models, *JOE* **58**, 137–168.

McFadden, D., 1984, Econometric analysis of qualitative response models, in *Handbook of Econometrics II*, eds., Z. Griliches and M.D. Intriligator, North-Holland.

McFadden, D., 1989, A method of simulated moments for estimation of discrete response models without numerical integration, *ECA* **57**, 995–1026.

Melenberg, B. and A. van Soest, 1993, Semiparametric estimation of the sample selection model, DP #9334, Center for Economic Research, Tilburg University, The Netherlands.

Melenberg, B. and A. van Soest, 1995a, Parametric and semiparametric modeling of vacation expenditures, *JAE*, forthcoming.

Melenberg, B. and A. van Soest, 1995b, Semiparametric estimation of equivalence scales using subjective information, *Statistica Neerlandica*, forthcoming.

Miller, R.G., 1981, *Survival Analysis*, Wiley.

Müller, H.G., 1988, Nonparametric regression analysis of longitudinal data, *Lecture Notes in Statistics* **46**, Springer-Verlag.

Müller, H.G., 1992, Goodness-of-fit diagnostics for regression models, *Scandinavian Journal of Statistics* **19**, 157–172.

Nadaraya, E.A., 1964, On estimating regression, *Theory of Probability and Its Applications* **9**, 141–142.

Newey, W., 1985, Maximum likelihood specification testing and conditional moment tests, *ECA* **53**, 1047–1070.

Newey, W.K., 1988, Adaptive estimation of regression models via moment restrictions, *JOE* **38**, 301–339.

Newey, W.K., 1990a, Semiparametric efficiency bounds, *JAE* **5**, 99–135.

Newey, W., 1990b, Efficient instrumental variables estimation of nonlinear models, *ECA* **58**, 809–837.

Newey, W., 1991, Efficient estimation of Tobit models under conditional symmetry, in W.A. Barnett, J. Powell, and G. Tauchen, eds., *Nonparametric and Semiparametric Methods in Econometrics and Statistics*, Cambridge University Press.

Newey, W., 1993, Efficient estimation of models with conditional moment restrictions, in *Handbook of Statistics* **11**, edited by G.S. Maddala, C.R. Rao, and H.D. Vinod, Elsevier Science Publishers.

Newey, W., 1994, The asymptotic variance of semiparametric estimators, *ECA* **62**, 1349–1382.

Newey, W. and D. McFadden, 1994, Large sample estimation and hypothesis testing, *Handbook of Econometrics IV*, North-Holland.

Newey, W. and J.L. Powell, 1990, Efficient estimation of linear and type I censored regression models under conditional quantile restrictions, *ET* **6**, 295–317.

Newey, W. and T.M. Stoker, 1993, Efficiency of weighted average derivative estimators and index models, *ECA* **61**, 1199–1223.

Newey, W. and K. West, 1987a, A simple positive semidefinite, heteroskedasticity and autocorrelation consistent covariance matrix, *ECA* **55**, 703–708.

Newey, W. and K. West, 1987b, Hypothesis testing with efficient method of moment estimation, *IER* **28**, 777–787.

Nijman, T., 1990, Estimation of models containing unobserved rational expectations, *Advanced Lectures in Quantitative Economics*, Academic Press.

Nolan, D. and D. Pollard, 1987, U-Processes: Rate of convergence, *AS* **15**, 780–799.

Nolan, D. and D. Pollard, 1988, Functional limit theorems for U-processes, *Annals of Probability* **16**, 1291–1298.

Pagan, A.R. and A. Ullah, 1988, The econometric analysis of models with risk terms, *JAE* **3**, 87–105.

Pagan, A.R. and F. Vella, 1989, Diagnostic tests for models based on individual data, *JAE* **4**, S29–S59.

Pakes, A. and D. Pollard, 1989, Simulation and the asymptotics of optimization estimators, *ECA* **57**(5), 1027–1057.

Park, B.U. and J.S. Marron, 1990, Comparison of data-driven bandwidth selectors, *JASA* **85**, 66–72.

Phillips, P.C.B. and J.Y. Park, 1988, On the formulation of Wald tests of nonlinear restrictions, *ECA* **56**, 1065–1083.

Pinske, C.A.P., 1993, On the computation of semiparametric estimates in limited dependent variable models, *JOE* **58**, 185–205.

Pollard, D., 1984, *Convergence of Stochastic Processes*, Springer-Verlag.

Pötcher, B.M. and I.R. Prucha, 1989, A uniform LLN for dependent and heterogeneous data processes, *ECA* **57**(3), 675–683.

Powell, J.L., 1984, Least absolute deviations estimation for the censored regression model, **JOE 25**, 303–325.

Powell, J.L., 1986a, Symmetrically trimmed least squares estimation for Tobit models, *ECA* **54**, 1435–1460.

Powell, J.L., 1986b, Censored regression quantiles, *JOE* **32**, 143–155.

Powell, J.L., 1994, Estimation of semiparametric models, in *Handbook of Econometrics IV*, edited by R.F. Engle and D.L. McFadden, Elsevier Science.

Powell, J.L., J.H. Stock, and T.S. Stoker, 1989, Semiparametric estimation of index coefficients, *ECA* **57**, 1403–1430.

Prakasa Rao, B.L.S., 1983, *Nonparametric Functional Estimation*, Academic Press.

Prakasa Rao, B.L.S., 1987, *Asymptotic Theory of Statistical Inference*, Wiley.

Press, W.H., B.P. Flannery, and S.A. Teukolsky, 1986, *Numerical Recipes: The Art of Scientific Computing*, Cambridge University Press.

Ramsey, J.B., 1969, Tests for specification errors in classical linear least squares regression analysis, *JRSS* **31**, 350–371.

Rilstone, P., 1991, Nonparametric hypothesis testing with parametric rates of convergence, *IER* **32**, 209–227.

Robinson, P.M., 1987, Asymptotically efficient estimation in the presence of heteroskedasticity of uknown form, *ECA* **55**, 875–891.

Robinson, P.M., 1988a, Root-N consistent semiparametric regression, *ECA* **56**, 931–954.

Robinson, P.M., 1988b, Semiparametric econometrics, *JAE* **3**, 35–51.

Robinson, P.M., 1991, Best nonlinear three-stage least squares estimation of certain econometric models, *ECA* **59**, 755–786.

Rosenblatt, M., 1956, Remarks on some nonparametric estimates of a density function, *Annals of Mathematical Statistics* **27**, 832–837.

Rosenblatt, M., 1991, Stochastic curve estimation, NSF-CBMS Regional Conference Series in Probability and Statistics, Vol. 3, Institute of Mathematical Statistics, Hayward, CA.

Rothenberg, T.J., 1971, Identification in parametric models, *ECA* **39**, 577–591.

Rothenberg, T.J., 1973, *Efficient Estimation with a priori Information*, Yale University Press.

Ruppert, D. and M.P. Wand, 1994, Multivariate locally weighted least squares regression, *AS* **22**, 1346–1370.

Rust, J., 1987, Optimal replacement of GMC bus engines, *ECA* **55**, 999–1033.

Rust, J., 1994, Structural estimation of Markov decision processes, *Handbook of Econometrics IV*, North-Holland.

Ruud, P.A., 1984, Test of specification in econometrics, *ER* **3**, 211–242.

Scott, D.W., 1992, *Multivariate Density Estimation*, Wiley.

Serfling, R., 1980, *Approximation Theorems of Mathematical Statistics*, Wiley.

Sherman, R., 1993, The limiting distribution of the maximum rank correlation estimator, *ECA* **61**, 123–137.

Sherman, R., 1994, Maximal inequalities for degenerate U-processes with applications to optimization estimators, *AS* **22**, 439–459.

Silverman, B.W., 1978, Weak and strong uniform consistency of the kernel estimate of a density function and its derivates, *AS* **6**, 177–184 (Appendum 1980, *AS* **8**, 1175–1176).

Silverman, B.W., 1984, Spline smoothing: The equivalent variable kernel method, *AS* **12**, 898–916.

Silverman, B.W., 1986, *Density Estimation for Statistics and Data Analysis*, Chapman and Hall.

Speckman, P.E., 1988, Regression analysis for partially linear models, *JRSS* **B50**, 413–436.

Staniswalis, J.G., 1989, The kernel estimate of a regression function in likelihood-based models, *JASA* **84**, 276–283.

Staniswalis, J.G. and T.A. Severini, 1991, Diagnostics for assessing regression models, *JASA* **86**, 684–692.

Stern, S., 1992, A method for smoothing simulated moments of discrete probabilities in multinomial probit models, *ECA* **60**, 943–952.

Stoker, T.M., 1986, Consistent estimation of scaled coefficients, *ECA* **54**, 1461–1481.

Stoker, T.M., 1991, Equivalence of direct, indirect and slope estimators of average derivatives, in W.A. Barnett, J. Powell, and G. Tauchen, eds., *Nonparametric and Semiparametric Methods in Econometrics and Statistics*, Cambridge University Press.

Stone, C.J., 1984, An asymptotically optimal window selection rule for kernel density estimates, *AS* **12**, 1285–1297.

Tauchen, G., 1985, Diagnostic testing and evaluation of maximum likelihood models, *JOE* **30**, 415–443.

Tibshirani, R., 1988, Estimating transformations for regression via additivity and variance stabilization, *JASA* **83**, 394–405.

Vinod, H.D. and A. Ullah, 1988, Flexible production function estimation by nonparametric kernel estimators, *Advances in Econometrics* **7**, 139–160, JAI Press.

Watson, G.S., 1964, Smooth regression analysis, *Sankhya A* **26**, 359–372.

Whaba, G., 1990, Spline models for observational data, CBMS-NSF #59, SIAM Society, Philadelphia, PA.

Whang, Y.J. and D.W.K. Andrews, 1993, Tests of specification for parametric and semiparametric models, *JOE* **57**, 277–318.

White, H., 1980, A heteroskedasticity-consistent covariance matrix estimator and a direct test for heteroskedasticity, *ECA* **48**, 817–838.

White, H., 1982, Maximum likelihood estimation of misspecified models, *ECA* **50**, 1–25.

White, H., 1984, *Asymptotic Theory for Econometricians*, Academic Press.

White, H. and I. Domowitz, 1984, Nonlinear regression with dependent observations, *ECA* **52**, 143–161.

Winkelmann, R. and K.F. Zimmermann, 1995, Recent developments in count data modelling, *Journal of Economic Surveys* **9**, 1–24.

Wooldridge, J.M., 1992, A test for functional form against nonparametric alternatives, *ET* **8**, 452–475.

Index